AF430881

Comprehensive Approach to
Practical Organic Chemistry
(Qualitative Analysis, Synthesis and UV, IR, NMR & MS Spectral Identification)

Comprehensive Approach to
Practical Organic Chemistry

(Qualitative Analysis, Synthesis and UV, IR, NMR & MS Spectral Identification)

Dr. Venkateswarlu Peesapati PhD CChem FRSC, FTAS

UGC Professor (Retd.) Osmania and JNT University,
Visiting Professor NIPER, Hyderabad.

PharmaMed Press
An imprint of BSP Books Pvt. Ltd.
4-4-309/316, Giriraj Lane,
Sultan Bazar, Hyderabad - 500 095.

Comperhensive Approach to Practical Organic Chemistry

by Dr. **Venkateswarlu Peesapati**

© 2023 *by Publisher,* All rights reserved.

No part of this book or parts thereof may be reproduced, stored in a retrieval system or transmitted in any language or by any means, electronic, mechanical, photocopying, recording or otherwise without the prior written permission of the publishers.

Published by

PharmaMed Press

An imprint of BSP Books Pvt. Ltd.

4-4-309/316, Giriraj Lane, Sultan Bazar, Hyderabad - 500 095.

Phone: 040-23445688; Fax: 91+40-23445611

E-mail: info@pharmamedpress.net

www.bspbooks.net/www.pharmamedpress.net

ISBN: 978-93-95039-74-1 (Hardback)

Dedicated to
the everlasting memory of
my **Departed Parents and**
my Teacher
Padmabhushan Prof. T. R. Seshadri (FRS)

FOREWORD

My association with Professor Venkateswarlu Peesapati, CChem, FRSC, is almost a decade old. Both of us are assiduously and jointly involved in Royal Society of Chemistry (London)- Deccan Chapter activities in India since 2012 and have successfully conducted seminars, workshops and practical training programs, covering students from high schools to universities. During our intellectual interactions, he broached his ambition to bring out a convenient and concise practical manual, keeping in view the Indian educational system and institutions.

After getting an M.Sc. degree from Andhra University (A.P) in the year 1968, Peesapati joined the prestigious Prof. T R Seshadri's (FRS) research group in Delhi University for his PhD. program in the topic "Synthesis of Furanocoumarins". Having decorated with a doctorate in organic chemistry, he went to the USA and UK and worked as a Post-doctoral & Senior Research Fellow in different universities for 14 years. After returning from the UK, he served as a professor in BITS-Pilani, Osmania, and JNT Universities as a UGC Professor and the head of the chemistry department at NIPER, Hyderabad. I have enjoyed associating myself as a course coordinator and Dean in NIPER, Hyderabad, with Dr Peesapati. He has a highly student-friendly personality and is sincerely desirous of teaching the techniques and mechanisms in organic chemistry to the student community in general and post graduates in particular.

With such an active and knowledgeable background, it is not surprising that his ever-devotional attitude towards teaching has made possible the birth of this laboratory manual. During the writing of this book, Peesapati has referred to several laboratory manuals and other related books and has presented the topics combining his vast and valuable experience and abroad spectrum of knowledge. As such, I feel this book suits well for all educational institutions.

The book is categorized into two sections. Different experimental techniques, analysis, organic synthesis, isolation of natural products and qualitative assays are grouped into the first section, which contains nine chapters. The second section having two chapters deals with a brief description of spectroscopy and spectrometric identifications of organic compounds. Almost 60 significant preparations of organic compounds are included in chapter 6. In this book, he brought out most of his laboratory experiences and covered actual examples of compounds, which were synthesized during his teaching tenure. They are grouped systematically into aliphatic and aromatic compounds, as well as name reactions and re-arrangements. Some of the experiments in this book include multi-step synthesis. The unique feature of this section and the manual is that he has explained the step-by-step mechanisms involved in each synthesis.

As spectroscopy can potentially identify organic compounds in the academic syllabus and the industrial sector, spectral identification is especially emphasized in this book. Twenty-one samples are given to students for them to analyse and practice. Moreover, an exciting aspect of the manual is the problems and solutions chapter, featured as the last chapter, which includes student's exercises on spectral data of compounds, from which the students may think and elucidate the structures from the provided data. In general, this enlightens and satisfies the student with immense knowledge and the practice of day-to-day laboratory topics and skills.

We should appreciate Dr. Peesapati for his devotional involvement and zeal. He tried to bring novelty and utility at each point of the presentation of this laboratory manual, which will be undoubtedly a book of choice for both the academic and industrial sectors.

I wish Prof. Peesapati good luck and hope he may bring out more such valuable books.

Dr. Y.V.D Nageswar, PhD FRSC
Retired chief scientist (IICT, Hyderabad)
Retired Dean (NIPER, Hyderabad)

FOREWORD

The practical organic chemistry is an essential part of course syllabus in UG and PG MSc Chemistry in many universities and colleges.

There have been few comprehensive practical organic chemistry books in the market including well known books such as Vogel's, Mann and Saunders and many others, despite the increasing interest of organic chemists, bio-chemists, pharmaceutical specialists and medicinal chemists. Most of the books that appear have been short ones except few exceptions, either specialized or aimed at the advanced level for under graduates.

Dr. Peesapati is filling this gap in the present book. The book is comprehensive and includes mechanistic approach for synthetic compounds and identification of organic molecules through spectroscopy, which added advantage for teachers and students. Author is well qualified for the task, having 40 years of teaching and research experience and I have had the pleasure of collaborating with him in the area of synthesis of prostaglandins at the University of Edinburgh in the mid nineteen eighties.

I have gone through the manuscript of this book. His understanding of the subject is thorough and extensive which reflects itself in his writing.

This book covers the entire area of practical organic chemistry including synthesis, mechanism, isolation of natural products, separation techniques, especially identification of spectral analysis of organic compounds and their properties.

I believe graduate students and others interested in learning practicals beyond what they find in other text books will greet this book with applause.

Dr. Norrie H Wilson

Retd. Professor, Organic Division
Department of Pharmacology,
Edinburgh University, Scotland UK
4th August 2022

Preface

This practical organic chemistry book consists of analysis, a series of experiments and identification of unknown compounds through spectral data to illustrate and teach the students some of the techniques utilized by organic chemists. The book is designed to build on the expertise gained in previous levels, and to emphasise points of mechanism on most of synthetic compounds. The experiments comprise a series of transformations carried out on known starting materials. Using the knowledge of organic chemistry and spectroscopy, the products obtained from these transformations can be identified. Thus, when a product is obtained, the melting or boiling point should be determined, the IR spectrum recorded and purity can be assessed by thin layer chromatography (t.l.c.). Most of the experiments will give practice in the assignment of spectra (IR, ^{1}H, & ^{13}C NMR and MS) and a full structural assignment of each compound on the basis of spectral data.

A laboratory course is very important and essential to all levels for chemistry students. **"For me chemistry without the practical work is only half chemistry and half chemistry is no chemistry"** The purpose of this laboratory book is to teach students the techniques involved in organic practices. The lab manual is an attempt to provide a meaningful course of practical skills to the students which meets the requisite standards of various universities and institutions. The present book attempts to give to the student, hands on experience, learning organic analysis, synthesis of important organic compounds, isolation of some natural products and sets of spectra confirming structure and properties of the compounds. The experiments have been carefully selected to give the student a wide range of practical experience in organic chemistry. A large majority of these experiments are class tested and followed in class syllabi. The subject matter of the book is sufficiently comprehensive to permit the teacher to cover any reasonable course of instruction. The essence of the book, then, is simplicity, easy to follow and concise. It is earnestly hoped that the edition of practical organic chemistry will be highly useful to aspiring chemists.

I have been fortunate to have had the opportunity to teach at both the graduate and undergraduate level students not only at BITS, Pilani, Osmania, J N T. Universities, NIPER-Hyderabad and Strathclyde University in UK as well. I came in close contact with the both postgraduate and MS students and my fellow teachers during my teaching and research career. Out of this experience and my ongoing desire to teach the subject matter, including laboratory skills, and in a lively and understandable manner, this new book on practical organic chemistry was born.

Although the book is primarily designed for organic/medicinal chemistry students, it is anticipated that students of other disciplines, like pharmacy students will profitably employ the text as a reference book or for review.

The study of organic chemistry coupled with hands on experience by doing practical work to ensure that the student has an intellectual grasp of the subject to prepare him/her not only his qualifying examination but for various competitive examinations, including job interviews as well. However, a student may wish to pursue individual topics further, and thus relevant books or references are noted at the end.

The laboratory text book is consists of 9 chapters which are outlined in the "Contents" section.

-Author

Acknowledgement

The author is extremely grateful to Dr Y.V.D. Nageswar for meticulously going through the entire draft of the manuscript and enriching it with his valuable comments and editions. I thank him for readily accepted in writing the foreword for the book. The author takes this opportunity to express his gratitude to Prof. P.S.N. Reddy for going through part of the first draft giving his valuable suggestions. Thanks to Dr. N. Shankaraiah and Dr. R. Srinivas for their valuable feedback on spectrometric identification of organic compounds and Dr. Narendra Talluri for his suggestions in HPLC techniques.

Thanks are also due to my former students Dr B. Rupavani and Dr N. Lingaiah for their help in drawing few structures. I would like to thank Prof. Y. Anjaneyulu for his valuable suggestions. Thanks to Mr B. Veeraiah and Mrs T. Neelima for their tireless and superlative efforts in typing the manuscript. I would like to thank all my colleagues and friends who encouraged me throughout this task of producing the book.

My special thanks to my former colleague Dr. Norrie Wilson, Edinburgh University, Scotland for his valuable comments and suggestions in particular spectral analysis data and contents of the book.

I want to express my special appreciation to Professor P.N. Sarma, whose detailed and constructive suggestions extended far beyond the call of duty of a reviewer, helped with the checking of the entire page proof.

Finally, sincere thanks to my wife Venkata Lakshmi, not only for her understanding during the writing process but also helping in typing much of the text and supporting me in every respect in finishing the book. Without her help this work would not have been possible.

I would like to thank K. Venkateshwar Rao proprietor of KV Xerox and G. Balram for their help.

The author records his appreciation to his sons Mr Sreedhar Peesapati and Mr Vidyadhar Peesapati for their encouragement during the period of writing. My sincere thanks to Mr Anil Shah, Director PharmaMed Press, and MR D. Naresh, Production Manager & other staff for their support and cooperation in bringing this book.

Any error in facts and figures is, however, the sole responsibility of the author.

Dr. P. Venkateswarlu

Contents

CHAPTER 4

CHROMATOGRAPHY

CHAPTER 5

ESTIMATION OF ASSAY

CHAPTER 7

ISOLATION OF NATURAL PRODUCTS

CHAPTER 8

IDENTIFICATION OF
ORGANIC MOLECULES BY SPECTROSCOPY

CHAPTER 9

PROBLEMS AND SOLUTIONS (UV, IR, NMR, MS)

PROBLEMS

SOLUTIONS TO PROBLEMS

Appendices

1

EXPERIMENTAL TECHNIQUES

1.1 Safety Precautions

The handling of many chemicals in a laboratory is attended with considerable hazards unless proper precautions are observed. It is the duty of all members of laboratory staff to co-operate in the prevention of accidents. In addition to the welfare of the staff of the laboratory there is concern for preservation of the building, equipment furnishing and apparatus. It is the duty of the employing authority to provide safe working conditions in the form of adequately ventilated laboratories with suitable equipment such as benching with non-absorbent creak-free surfaces, fume cabinets, protective clothing and facilities for hand washings. It is recommended that a senior member of staff is appointed as a safety officer and that he should have responsibility for the accident book in which all accidents and their after effects are recorded shortly after they occur. Most of the dangers encountered in laboratories start from inferior material, prone chemicals fire, faulty electric wiring, faulty apparatus and last but not the least, careless working.

One of the most important aspects of concern in the organic chemistry laboratory is the handling of inflammable materials. The precautions to be observed cannot be over emphasized.

It is essential that each chemistry laboratory is to be provided with a first aid box, fire alarm, fire extinguisheres, waste disposable bins, showers and eye washers etc.

The laboratory work will be concerned with the principles involved as well as the acquisition of the techniques. Thus it is mandatory that each student study each experiment prior to undertaking any laboratory procedure.

Chemical Hazards

Most compounds are highly toxic when ingested orally. Many chemicals also poisonous, corrosive, carcinogenic or explosive. Corrosive chemicals such as acids and alkalis are stored in low shelves and opened with care. Dangerous chemicals obtained from commercial sources usually carry a warning printed on the bottle. These warnings should be followed. Certain substances, can produce serious burns upon contact with the skin. One should never taste any compound and odour of substances should be detected with extreme care.

Mouth pippetting is always potentially dangerous and some form of safety pippette must be used instead. Procedures involving boiling solvents, toxic gases and vapours must be carried out in an efficient fume cupboard.

Sensitive tissues, for example, the eyes should not be needlessly exposed to vapours. One should never place his/her face directly over a reaction mixture. In case of any chemical get his/her into the eyes, they should be flooded immediately with copious amounts of water. In order to decrease any possibility of damage to the eyes, all students will be required to wear safety glasses while working in the laboratory.

Fire

Fire is one of the most serious and most likely hazards to occur in a laboratory. All the staff should know where the fire extinguishers are and how to use them. The most generally useful fire extinguisher in the laboratory is the carbondioxide cylinder which can be safely used with most chemicals and electric equipment, and is clean. Dry powder extinguishers, like sand, are also useful but are messy in use. Asbestos blankets are useful for smothering small fires and burning clothing.

Most organic compounds are combustible. Those with low boiling points and high vapor pressure at room temperatures may present a serious fire hazard. Ether, which has a boiling point of 35°C may be ignited by a flame removed by sixteen feet. Hence, it is never permissible to heat over an open flame any substance in an open vessel containing such volatile liquids. Steam bath is ideal for this purpose.

Summary of safety precautions: The following simple rules have been drawn up for your own and others protection. Read them before starting on practical work:

1. No smoking is allowed in chemical laboratories.
2. Every student must wear protective eye shields at all times in the laboratory. This is to protect you from your neighbour's mistakes as well as your own.
3. Report any accident immediately to a demonstrator (even if it does not involve personal injury, as in spillage of chemicals or breaking of glasses).
4. Carryout experiments which produce toxic chemicals or vapours, and / or are likely to be violet, in a fume cupboard.
5. Fire is a serious hazard in the laboratory and is usually caused by the careless handling of organic solvents. These must not be heated using a Bunsen burner.
6. Be familiar with the placing of fire extinguishers in the laboratory.
7. Do not point a test tube which is being heated, or in which a reaction is occurring, at any person in the neighbourhood.

8. Do not peer into the mouth of a test tube which is being heated or in which a reaction may be occurring.

9. If the clothing is splashed by a corrosive liquid, strip the clothing and treat the skin immediately. As a first treatment washing with water is generally appropriate, call a demonstrator to assist you.

10. Wear a laboratory coat at all times in the practical laboratory to protect your cloths.

11. Always carry a small towel to the laboratory to assist you in handling hot objects in addition to tongs.

12. Bunsen burners may only be used in the fume cupboard or keep it away from the inflammable solvents.

1.2 Risk Assessment of Some General Reagents

No.	Compound	Hazard	Spillage	Disposal
1.	Acids (All)	Corrosive, toxic	Wash off with Plenty of water	Pour down sink with a lot of water
2.	Bases	Corrosive, toxic	-do-	-do-
3.	Benzil	Irritant	-do-	Return to teacher
3a.	Benzene	Toxic, avoid inhalation carcinogenic	-do-	-do-
4.	Bromine	Serious burns upon contact with the skin	-do-	Pour down sink with a lot of water
5.	Benzoyl chloride	Corrosivelachrymatory, avoid inhalation	-do-	Return to teacher
6.	Bromine in acetic acid	Corrosive wear gloves	-do-	use volume specified in manual
7.	Chloroform	Toxic, avoid inhalation	-do-	Place in waste solvent bottle for halogenated solvents.
8.	Dichloro-methane	Toxic, avoid inhalation	-do-	Solvents
9.	Concentrated ammonia soln	Corrosive irritant wear gloves	-do-	Pour down sink with lot of water

Table *contd...*

No.	Compound	Hazard	Spillage	Disposal
10.	Conc. Sulphuric acid	Corrosive, toxic, wear gloves	-do-	-do-
11.	Cyclohexane	Flammable	-do-	Use waste bottle provided
12	2-Chloro Phenol	Toxic, corrosive, causes burns	-do-	Pour NaOH soln., down sink with a lot of water
13.	H_2O_2 soln.,	Oxidiser, irritant	-do-	Pour down sink with water
14.	Iodine	Irritating to eyes, skin	-do-	Neutralise with sodium thiosulphate
15.	Naphthalene	Flammable, irritant	-do-	Return to demonstrator
16.	Benzaldehyde	Toxic, irritant	-do-	Wash test soln., down sink with water
17.	Benzyl amine	Corrosive, lachrymatory	-do-	-do-
18.	Brady's reagent	Flammable, corrosive	-do-	-do-
19.	2-Naphthol solution	Irritant	-do-	-do-
20.	t-Butanol	Flammable, corrosive	-do-	-do-
21.	Ethanol	Flammable	-do-	-do-
22.	Petroleum ether	Flammable	-do-	Place residue in waste bottle
23.	Nitrobenzoic acid	Irritant	-do-	Return used sample to the teacher
24.	Sodium dichromate soln.,	Cancer suspect agent	-do-	Pour down sink with water
25.	Sodium nitrite	Toxic	-do-	-do-
26.	Phenolphthalein in ethanol	Toxic, flammable	-do-	-do-
27.	Starch soln., (under toluene)	Flammable	-do-	-do-
28.	2-Naphthol solution	Irritant	-do-	-do-
29.	Silver nitrate solution	Toxic	-do-	-do-
30.	Unknown amines	Irritant, toxic	-do-	Return unused sample to teacher

Table *contd…*

No.	Compound	Hazard	Spillage	Disposal
31.	Ethanoic anhydride	Corrosive, lachrymatory	-do-	-do-
32.	2,4-Dichloro Phenol	Corrosive, wear gloves	-do-	Return the sample to the teacher
33.	Phenols amines, acetanilide and ethanoic acid	Corrosive toxic wear gloves	-do-	-do-

1.3 Glass and Plastic Laboratory Ware

Borosilicate glass is now almost exclusively used for the manufacture of apparatus they will be used for conducting a chemical reaction or be used for measuring volumes. This glass consists of about 80% silica and 13% boric oxide with the oxides of sodium, aluminium and other metals. It is more of sodium, aluminium and other metals. It is more resistant to thermal shock than ordinary soda glass but should not be abused to the extent, for example, heating beakers containing liquid in flame without the protection of a gauze or plunging hot glass into cold water.

Generally, volumetric glassware is calibrated for use at 20°C (27°C for use in tropical countries).

1.4 Apparatus

Pipettes: The undesirability of pipetting by mouth has been increasingly recognised in recent years. This has led to the introduction of various devices to fill and empty conventional pipettes and of disposable-pipettes. The pipette fillers include rubber bulbs and various valve-operated designs. Only the most commonly used pipettes are described below, there are, of course, many other designs of pipettes available.

1. **(a) One-mark pipettes:** Also referred to as volumetric or transfer pipettes. These pipettes are governed by BS1583. The capacity is defined as the volume of water at 20°C delivered by the pipette when used in the prescribed manner.

   (b) Graduated pipettes: They consist of pipettes calibrated for delivery from zero down to any graduation line.

   (c) Pasteur pipettes: These are uncalibrated pipettes that can easily be made in the laboratory or obtained commercially.

   (d) Disposable-tip pipettes: A number of pipetting systems have been made using disposable plastic tips. They eliminate the necessity of mouth pipetting and are quicker to use than conventional pipettes. Fixed - or variable-volume patterns are available.

(e) Dispensers: For the repeated dispensing of fixed volume of solution, the advantage of using them include speed of operation, reproducibility and safety.

2. **Burettes:** Burettes with PTFE (polytetrafluoroethylene) stopcocks are recommended particularly when alkalis are to be used. PTFE stopcocks require no lubrication and do not seize up.

3. **Graudated measuring cylinders:** Measuring cylinders can have spouts or be stoppered. Cylinders are not occurate, they should not be used in making solutions where a moderate degree of accuracy is required.

4. **Volumetric flasks:** Volumetric flasks to hold 5 ml, 10 ml, 50 ml, 100 ml, 250 ml, 500 ml and 1 litre of solutions are commonly used.

5. **Conical (Erlenmeyer) flasks:** These are useful when making solutions, particularly when boiling is required. They are commonly used in titration.

6. **Buchner (Filter) flasks:** These are similar to conical flasks but have a side arm which can be attached to a vaccum pump. A Buchner funnel has a sintered glass platform or a wire-mesh platform used for supporting filter paper.

7. **Filter funnels:** The ordinary conical filter funnel can have plain or fluted sides.

8. **Separating funnels:** Separating funnels may be spherical, conical or cylindrical. Separating funnels are used in extraction procedures and to separate immiscible liquids. When mixing the contents during an extraction procedure, the pressure that builds up should be released occasionally by inverting the funnel and slowly opening the stop cock. After extraction is complete, clamp the funnel in a vertical position and allow the fluids to separate. Remove the stopper and drain out the lower liquid.

   Care should be taken to avoid the formation of emulsions when using a separating funnel; gentle inversions will be found preferable to shaking.

9. **Thermometers:** There are greater variety of thermometers designed for particular purposes over various temperature ranges.

10. **Filter pumps:** Filter (Venturi) pumps are available in glass, metal and plastic. Depending, of course, upon the water pressure available, an ultimate vacuum of about 15 torr is possible with this type of pump.

11. **Desiccators:** Glass and plastic desiccators are available, may be of vacuum or non-vacuum type. The non-vacuum type can be used to maintain a dry atmosphere in which chemicals that have been previosly dried can be kept. The vacuum type can be used to dry solids.

   Commonly used desiccants are a) phosphorus pentoxide, very powerful but use only if strictly necessary b) Silicagel, reusable, safe, incorporates an indicator; normally blue changing to pink when wet and reusable after heating in a oven. c) Anhydrous calcium chloride, satisfactory for storage of dehydrated chemical.

12. **Plastics:**Most of the apparatus described here, as well as many other items used in the laboratory, can now be manufactured in plastic.

13. **Condensers:**The condensers are used for refluxing and ordinary distillation. The air condenser is employed if the liquid distilling has a very high boiling point.

14. **Flasks:**These are common type of flasks for a variety of purposes. Round bottomed flasks are employed for refluxing and distillation purpose. The Erlenmeyer (conical) flask is useful when making solutions and titrations.

1.5 Cleaning Glass Apparatus

It is very important that clean glassware must be used for carrying out any type of reaction. Presence of impurities may have an undesirable effect on the particular reaction as well as the purity of the final end compound.

(A) **Chromic acid cleaning method**: Glass is soaked in a solution of sodium dichromate ($Na_2Cr_2O_7.2H_2O$) 7 g; concentrated sulphuric acid 100ml. This solution should be prepared and used **with care** and stored in a glass bottle. The glass-ware is soaked overnight in the cleaning fluid, then rinsed **several times** in distilled water. The apparatus is then dried, open-end down or in a hot-air oven. Acid washing is particularly good for removing silicone grease and certain organic compounds.

(B) **Washing laboratory glass-apparatus with detergents**: A number of detergents are available commercially which for most purposes are (at least) as efficient as acid, safer to use and often cheaper. The glassware should be soaked in an aqueous solution of the detergent. After soaking, the glassware is brushed (if necessary), washed with tap water and then rinsed with distilled water and dried.

1.6 Purification of Solvents

Commercially available grades of organic solvents are of adequate purity for use in many reactions provided that the presence of small quantities of water is not harmful to the course of the reaction (unless the reaction needs dry conditions), and also the presence of other impurities (e.g., ethanol in diethyl ether and thiophene in benzene) is unlikely to cause undesirable side reactions. The commercially available solvents for general use are often accompanied by specifications indicating the amount and nature of the impurities present in it. When the levels of impurities, including moisture or water content, are not acceptable for a particular reaction, it is more economic to purify the commercial grade than to purchase the more expensive analar grade solvent.

Common drying agents

(i) For Alcohols:Anhydrous potassium carbonate, calcium sulphate or magnesium sulphate, calcium oxide.

(ii) For Saturated and aromatic hydrocarbons and ethers:Anhydrous calcium chloride, calcium sulphate phosphoric acid.

(iii) For Aromatic Aldehydes, Anhydrous calcium, magnesium and sodiumsulphates

Ketones

(iv) For Amines: Solid potassium or sodium hydroxide, calcium and barium oxide

(v) For Organic acids:Anhydrous calcium, sodium & magnesium sulphates

The preliminary treatment is infact essential for the vast majority of organic solvents, unless it is certain that the water content is very low, before using the more powerful drying agents (such as a reactive metal, e.g., sodium or a metal hydride, e.g., calcium hydride, lithium aluminum hydride. Attention should be drawn to the considerable fire or explosion hazards of these highly reactive drying agents, particularly at the end of a solvent distillation when residual material has to be disposed of.

It is often convenient to remove final traces of water with the aid of a molecular sieve and store the dried solvent in the presence of the sieve. The term, molecular sieve, applies to a group of dehydrated synthetic sodium and calcium aluminosilicate adsorbent (Zeolites). Currently four principal types are available in the market, namely types 3A, 4A, 5A and 13A, representing an effective pore diameter of approximately 0.3, 0.4, 0.5 and 1.0 mm respectively.

Note: Almost all organic solvents are flammable. Apart from taking the obvious precautions of avoiding all flames in the vicinity of a solvent distillation, it must be remembered that faulty electrical connections or even contact with hot metal surface may ignite the vapour of volatile solvents. It is advisable to use double surface condensers in many cases for solvent distillation.

Purification of commercial ether ($C_2H_5OC_2H_5$): The commercial ether is usually contaminated with water and ethanol. Furthermore, when ether is allowed to stand for sometime in contact with air and exposed to light slight oxidation occurs with the formation of the highly explosive diethyl peroxide (Et_2O_3). If present, the peroxide may be removed by shaking 1 litre of ether with 10-20 ml of concentrated solution of an iron (III) sulphate, prepared by dissolving 60 g of iron (II) sulpahte in a mixture of 6 ml conc. sulphuric acid and 110 ml of water. Remove the aqueous solution and pour the ether portion in a clean winchester bottle and add 50-100 g of anhydrous calcium chloride. Allow the mixture to stand for 24 hrs with occational shaking; the water and ethanol are largly removed during this period.

Finally the ether should be redistilled. Filter it through a large fluted filter paper into a 500 ml round bottomed flask, add two or three anti-bumping granules, and arrange the flask for distillation by passing cold water through condenser. Note that the end of the condenser must lead into a flask surrounded by ice. The ether is most suitably distilled by placing the distilling flask on (but not in) an electrically heated constant-head water-bath, collecting the fraction boiling between 34° and 38°C.

Benzene (C_6H_6): Benzene has ben identified as a carcinogen (CAUTION). Commercial grade benzene may contain thiophene (b.p. 84°C), which cannot be separated by distillation. The commercial benzene is shaken two or three times with about 15 percent of its volume of conc. sulphuric acid (i.e., 15 ml per 100 ml of benzene) in a stoppered separating funnel (Alternatively, the mixture may be stirred mechanically for 20-25 minutes) until the acid layer is colourless or very pale yellow on standing. After each shaking, the mixture is allowed to settle and the lower layer is drawn off. Next, the benzene is shaken twice with water and once with 10% sodium carbonate solution in order to remove most of the acid and finally dried with anhydrous calcium chloride. After filtration, the benzene is distilled and the fraction, b.p. 80-81°C, collected. The distilled benzene may either be stored over sodium wire or left in the presence of a type 5A molecular sieve. Pure benzene has b.p. 81°C / 760 mmHg. and m.p. 5.5°C.

Chloroform ($CHCl_3$): Chloroform is a suspect carcinogen, whereever possible it should be replaced by dichloromethane as an alternative solvent. The commercial grade may contain upto 1% of ethanol which is added as a stabiliser. The ethanol may be removed by either one of the following procedures.

(a) The chloroform is shaken with about half its volume of water, then dried over anhydrous calcium chloride for at least 24 hours, and distilled.

(b) The chloroform is passed through a column of basic alumina (10 g per 14 ml of solvent). This procedure removes not only traces of water but also acid and eluate may be used directly.

Dichloromethane (CH_2Cl_2): Commercial grade is purified using 5% sodium carbonate solution and water. It is dried over anhydrous calcium chloride and then distilled. The fraction b.p. 40-41°C is collected. Dichloromethane is a useful substitute for diethyl ether in extraction process.

***t*-Butyl alcohol (C_4H_9OH):** This and other higher alcohols may be purified by drying with anhydrous potassium carbonate or with anhydrous calcium sulphate and fractionated after filtration. The fraction with the boiling point 81-83°C is to be collected.

Acetone (CH_3COCH_3): The commercial grade usually contains appreciable quantities of methanol, acetic acid and water. Commercial acetone is heated and refluxed with solid potassium permanganate until the voilet colour remains. It is further dried by keeping the solvent overnight, with anhydrous potassium carbonate, filtered and redistilled. Pure acetone distilles at 56.5°C / 760 mmHg.

Acetonitrile (CH₃CN): Some commercial grades of acetonitrile is usually contaminated with water, acetamide and ammonium acetate. Water may be removed with activated silica gel or type 4A molecular sieves. [Further, this partially dried solvent is stirred with phosphorus pentoxide until hydrogen evolution stops. The solvent is decanted from the solid and fractionally distilled at atmospheric pressure. The pure acetonitrile has b.p. 81-82°C/760 mmHg.

Dimethyl sulphoxide (DMSO, Me₂S=O): The commercial grade may be purified by standing overnight over freshly activated alumina, barium oxide or calcium sulphate. The filtered solvent is then fractionally distilled over calcium hydride under reduced pressure (12 mmHg) and stored over type 4A molecular sieve. The pure solvent has b.p. 189°C.

N,N-Dimethylformamide (DMF) (Me₂NCHO): Dimethyl formamide can be purified first by drying over anhydrous calcium sulphate or type 3A molecular sieve for 72 hours, followed by distillation under reduced pressure. The pure solvent has b.p. 153°C.

Ethyl acetate (CH₃COOC₂H₅): Commercial grade usually contains water, ethanol and acetic acid as impurities. A mixture of ethyl acetate (1000 ml), acetic anhydride (100 ml) and concentrated sulphuric acid (10 drops) is heated under reflux for 4 hours and then distilled. It is further dried by shaking with 20-30 g of anhydrous potassium carbonate; filtered and redistilled. Pure ethyl acetate has b.p. 77°C/760 mmHg.

Dioxane (1,4-dioxane, diethylene dioxide, (CH₂CH₂O)₂: The commercial grade dioxane contains small quantities of acetaldehyde and glycol acetate. A mixture of 1 litre technical grade dioxane, 14 ml of conc. hydrochloric acid and 100 ml of water is refluxed for 6-12 hours; while a slow stream of nitrogen is bubbled into the solution to remove acetaldehyde formed. After cooling to the room temperature, it is treated with excess potassium hydroxide pellets until some remain undissolved and the strongly alkaline aqueous layer is separated from the dioxane. Dioxane layer is decanted and refluxed with excess sodium metal for 8-12 hours. Finally, the dioxane is distilled over the fractionating column and receiving flask being covered with black paper. The pure compound has b.p. 101.5°C/760 mmHg.

Removal of any peroxide present as impurity on storage, can be eliminated by passing the above distilled dioxane through a column of basic activated alumina (100 g for 300 ml of dioxane).

Ethyl alcohol (Et OH): Ethanol is used as a solvent/reagent in a wide variety of reactions. Alcohol is a mixture consisting of 95.6% by weight of ethanol and the remainder being water. This is known as rectified spirit. Ethyl alcohol of a high degree purity is frequently required in many organic preparations. Absolute alcohol of 99.5 percent grade may be purchased or it may be conveniently prepared by the dehydration of rectified spirit with calcium oxide. In the laboratory, the alcohol is heated to reflux with calcium oxide (quicklime) for 6 hours and distilled. It is very hygroscopic and precautions have to be taken to guard against the absorption of water during the distillation and subsequent storage. The distillate, which still contains about 0.5% water, is known as absolute alcohol. The last traces of water can be removed by refluxing with magnesium. Pure absolute alcohol b.p. 78°C.

Methanol (MeOH): The methanol now available in the market is suitable for most purposes without purification. Synthetic methanol may contain a small amount of acetone and may be removed by the following procedure.

A mixture of 250 ml methanol, 12.5 ml furfural and 30 ml of 10% sodium hydroxide solution is refluxed in a 1 litre round bottomed flask, fitted with a double surface condenser, for 6 to 10 hours. A resin is formed during heating which removes all the acetone present. The alcohol is then distilled. Pure methyl alcohol has b.p. 65°C.

Petroleum ether: The fractions of petroleum, which are commonly used, have b.p. 40-60, 60-80, 80-100 and 100-120°C. They contain some unsaturated hydrocarbons (chiefly aromatic) and may be removed by shaking with 10% conc. sulphuric acid. The solvent is then shaken with a mixture of a concentrated solution of $KMnO_4$ in 10% sulphuric acid to remove oxidizable impurities. The solvent is washed with sodium carbonate and thoroughly with water, dried over anhydrous calcium chloride and distilled.

Pyridine (C_6H_5N): Pyridine of analytical reagent grade is satisfactory for most purposes. If it is desired to get the dry product, it is heated under reflux over calcium hydride, potassium or sodium hydroxide pellets or over barium oxide and then distilled with careful exclusion of moisture. Pure pyridine has b.p. 115°C/760 mmHg. It is highly hygroscopic and should be stored over calcium hydride or type 4A molecular sieve.

Tetrahydrofuran (C_4H_8O): The commercial grade solvent contains water and peroxide as impurities. Peroxide, if present, may be removed by running through column of alumina. It is purified by distillation over calcium hydride or lithium aluminium hydride. Pure tetrahydrofuran has b.p. 65-66°C. It should not be stored for more than a few days unless an anti-oxidant is added.

1.7 Recrystallisation

Recrystallisation is a process of purification and always involves a separation of the solid that is wanted from the liquid that contains unwanted material of course the liquid may contain some wanted material too, but loss of this is the price that is paid for the purity of the remainder.

Learning to choose a suitable solvent for recrystallisation is part of the course. It is convenient to class solvents by polarity, and by capacity for hydrogenbonding, as shown in the diagram below. Generally, solvents near to foot of the diagram are useful for non-polar materials, while those higher up will dissolve a wider range of substances. Solids whose molecules can participate in hydrogen bonds are generally more soluble in solvents that also can form hydrogen bonds than in solvents that lack this capacity.

Size of apparatus

Generally, chemicals are easiest to handle if the container is about one-third full, or less for narrow items like test tubes. thus, for 10 ml of liquid, a 25 ml flask is more appropriate than either a 10 ml flask (risk of spillage) or a 100 ml flask (loss due to wetting the surface of the container). The same principle applies when collecting a solid on a filter; choose a funnel of suitable size for the solid to be collected.

If the quantity of liquid is small-say less than 5 ml, it is often easier to transfer the liquid using a pipette, rather than by pouring it. By extration, filtration on this scale can be avoided by removing the liquid with a pipette, leaving behind the solid which should then be washed with fresh solvent.

Selected solvents, classified by polarity (vertical scale) and by capacity for hydrogen-bonding (horizontal scale).

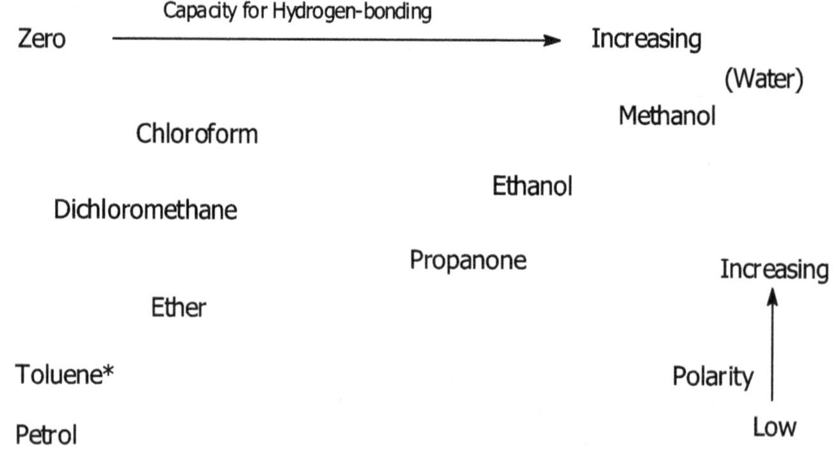

* *Boiling point 110°C; cannot be removed on a steam bath at atmospheric pressure. Propanone and the solvents to the right of it are miscible with water.*

1.8 Determination of Boiling Point on a Semi-Micro Scale

Using a small flame, draw out the centre part of a melting-point capillary so that the diameter is reduced to about 1/3 of the original. Cut the tube into two at the narrowest point and seal the larger end of each tube. You should have two tubes looking like (1) below.

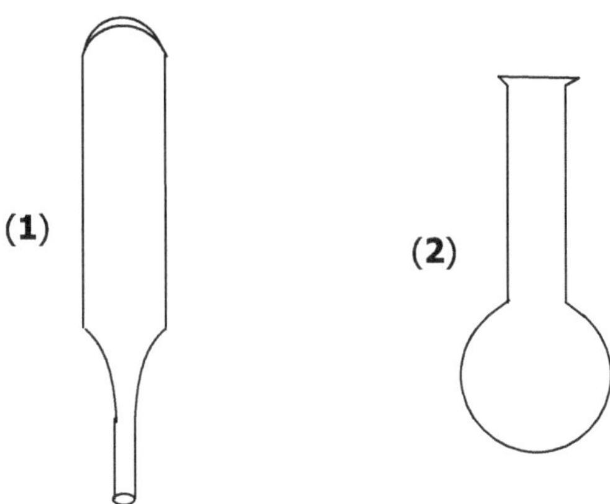

Fig. 1.1

Prepare a heating bath by taking a long-necked flask like (2) which must be dry and adding liquid paraffin till the level is about 5 mm (1/4") below the shoulder. Fit a cork to a thermometer and, using a file, cut a notch in the side of the cork to allow air to flow past it.

Put 1 or 2 drops, no more, of liquid in a small (5 mm x 25 mm) test tube and insert the prepared capillary, open end down. Join the test tube to the thermometer with a nylon clip so that the sample of liquid is beside the bulb. Using a tripod and gauze, and a Bunsen with a small flame, heat the bath. As the temperature rises bubbles of air escape from the capillary. When the bubbling becomes rapid, remove the flame and allow the temperature to fall till liquid re-enters the capillary. Note the temperature, t_1. Heat the bath again gently till a rapid stream of bubbles issues from the capillary, at temperature t_2. The boiling point can be taken as half-way between t_1 and t_2. If the values lie more than 10° apart, repeat the cycle of heating and cooling.

1. Choose for practice one of the following:

Toluene,	b.p.	110°C
Methanol	b.p.	68°C
Ethanol b.p.	78°C	

2. Obtain an unknown from a demonstrator.

3. No liquid unknown you will do in this excercize has a b.p. higher than 260°C.

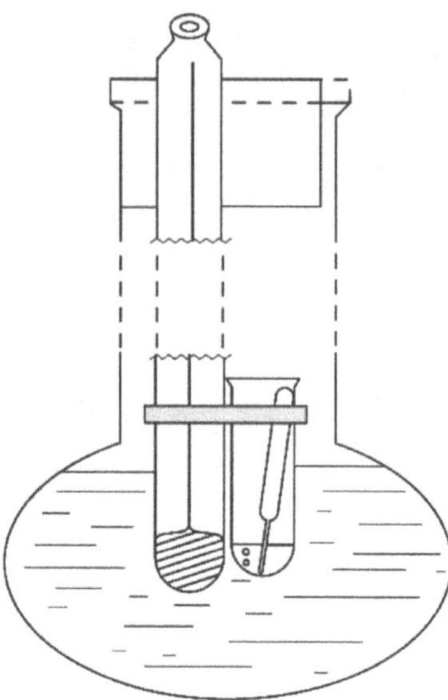

Fig. 1.2 Boiling point apparatus

1.9 Melting Point Determination

The melting point is the property of an organic solid which is most frequently used as a criterion of purity. A pure compound has a sharp m.p. (i.e., melts over a narrow temperature range).

(a) **Determination:** Crush a small amount of the sample on a filter paper with a small spatula. Introduce a little of the powder into the open end of a capillary tube and shake it down to the closed end by gentle stroking with a file. The column of solid should be no more than 3 mm in length and should be tightly packed.

Place the sample in a melting point apparatus, heat, and note the temperature at which melting occurs. This is the **approximate** melting point of the sample. Prepare a new sample and repeat the melting point measurement, this time allowing the temperature to rise by only **approximately** 2°C per min., as the melting point is approached. Record the temperature at which melting begins, and the temperature at which the last traces of solid disappear. This is the melting point range of the sample, normally expressed in the form, e.g., sample x, m.p. = 125-128°C. You will not obtain accurate melting points if the sample is heated too rapidly.

(b) **Melting point of a pure sample:** Obtain small samples of two unknown solids **y** and **z.** using the above method. Determine their melting points, and record your values. If in doubt about your technique, check with your demonstrator that your values are acceptable.

1.10 Identification of an Unknown Sample by Mixed Melting Points

Theory: If one mixes two solids, melting points A and B, than the melting point of the mixture will be lower than either A or B. For example, 2 solids with m.p's 100^0C and 105^0C will, when mixed, give a sample of melting point well below 100^0C. This phenomenon, called melting point depression, is very useful for distinguishing between solids of a very similar point. A mixture of two compounds not only melts at a temperature below either of the pure compounds but also over a greater temperature range.

The following compounds will be available:

Benzamide	(A)
Cinnamic acid	(B)
Urea	(C)

Determine the melting point of each of these compounds, thus showing that their melting points lie too close together to enable any one compound to be identified by a simple melting point determination.

An unknown compound (**U**) will be supplied that is one of the members of this series. It should be identified by mixed melting point determination.

Suppose the compounds each have m.p. 132-135° and the unknown compound is found to have m.p. 132°C. Mix a small amount of **U** with a small amount of **A** and determine the m.p. of the mixture. Suppose the mixture has m.p. 121-127°, then **U** has lowered the m.p. of **A** and so **U** is not identical with **A**. Now mix **U** with **B** and find the m.p. of the mixture. If this is 132-134°, then **U** has not lowered the melting point of **B**.SoU must be identical with **B**. Confirm by determining the m.p. of a mixture of **U** and **C** and if you are correct should be lower than 132°.

Conclusion: **U** is identical with **B**

1.11 Purification of Organic Compounds by Recrystallisation

Crystallisation is a very efficient method for the purification of solid organic substances. It is first necessary to find a suitable solvent, i.e., one which will readily dissolve the material when hot, but only to a small extent when cold.

The technique of recrystallisation thus involves dissolving the pure organic material in the minimum volume of hot solvent. Insoluble impurities, which will not dissolve in the hot solvent, are filtered off. The filtrate is than allowed to cool. The organic material crystallises out off solution leaving any soluble impurities in the solution. The crystals are then filtered off using a Buchner funnel. It is important to free the crystals from then mother liquors by washing with a little cold solvent and by sucking dry.

The solvents normally used for recrystallisation are:

	b.p. (°C)	
Distilled water	100	Non-inflammable
Acetone	56	Inflammable
Ethyl alcohol (industrial spirit)	78	Inflammable
Benzene	81	Inflammable
Light petroleum	60-80	Inflammable
Acetic acid	118	Not readily Inflammable
Chloroform	61	Non-inflammable

(a) **Solubility and Crystallisation:** Test the solubility of 0.1 g samples of the following compounds in water, ethanol, and benzene noting the solubility in the cold and hot solvent and also how well the compound crystallises when a hot solution is cooled.

   Salicylic acid

   Benzanilide

(b) **Purification Exercise:** Apparatus required: Conical flask (25, 50 and 100ml), Filter funnel, Buchner funnel and flask

Obtain from the demonstrator the sample of impure material and carry out small scale solubility tests in a small test tube to determine the best recrystallising solvent as above. Weigh out about 2.0 g of the material into a suitable flask, add a small quantity of the selected solvent and a boiling stone and heat to boiling point.

N.B. If an inflammable solvent is being used the flask should be fitted with a condenser and the heating carrried out on the steam bath, solvent being added down the inside of the condenser. If only a small quantity of solvent has been added initially the solid will not be completely soluble even at b.p. Therefore add more solvent a little at a time (using a dropper) until the main bulk of the solid just dissolves completely in the minimum volume of boiling solvent. It is now necessary to filter the hot solution to remove insoluble impurities.

The filtration is being carried out rapidly either through a 'fluted' filter paper or through a Buchner funnel using the water pump. The material in solution is always liable to crystallise out during the filtration thus blocking the funnel. To avoid this, it is best to preheat the funnel in the oven immediately before use. The filtrate is set aside to cool. Slow cooling promotes the growth of large crystals; rapid cooling (under tap) gives small crystals. When crystallisation is complete, collect the crystals by suction filtration using a Hirsch or Buchner funnel, wash with a little cold solvent, allow to dry, and determine m.p. and weight of dry crystals.

Ideally the material should be recrystallized until the m.p. remains constant and is not altered by further crystallization.

Write a brief account of the process in your lab notebook and submit this report together with the dry recrystallized material (in a labelled sample tube) to the demonstrator.

(c) **Solubility test:** Test the solubility of the impure material in the following solvents - chloroform, light petroleum (b.p. 68-80°), methanol, and water. Place about 0.02 g (a small spatula tip full) of the compound in a small test tube and add about 0.5 ml of the solvent to be tested. Shake and observe the solubility of the compound in the cold solvent. If the compound is not appreciably soluble in the cold solvent, warm the test tube on a steam bath and observe its solubility (use a low Bunsen flame in a fume cupboard when water is used as the solvent). (N.B. A good solvent is one in which the compound is fairly soluble in hot solvent but only sparingly soluble in cold solvent). Record your observation and deduce which of the solvent(s) you have used are suitable for recrystallization.

1.12 Purification of a Liquid by Fractional Distillation

Fractional Distillation

Distillation and recrystallisation are the two chief methods of purifying organic compounds. Several kinds of distillations are possible, including simple distillation, fractional distillation, and steam distillation. This experiment is concerned with the separation, by fractional distillation, of two miscible liquids, methylene chloride and carbon tetrachloride.

It will be seen from the diagram below (curve A) that it is not possible to separate these liquids completely by simple distillation. Use of fractionating column, however, results in an efficient separation of the two components (curve B). In the following experiment the efficiency of the separation is checked by gas-liquid chromatography (G.L.C.).

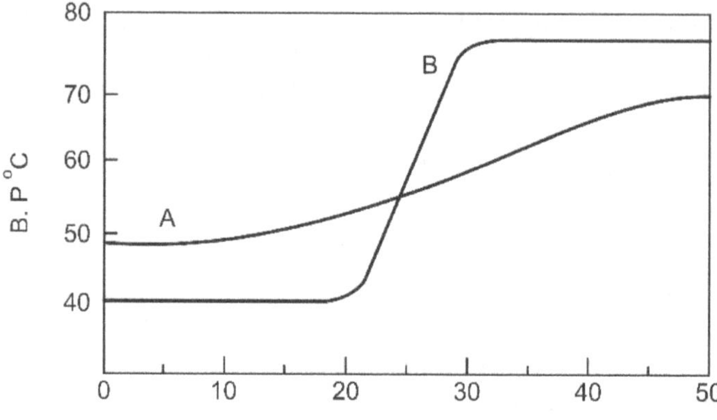

Fig. 1.3 Volume of distillate (ml.)

Distillation curve for 50 ml. of a 1:1 mixture of methylene chloride and carbon tetrachloride: (A) – simple distillation   (B) fractional distillation.

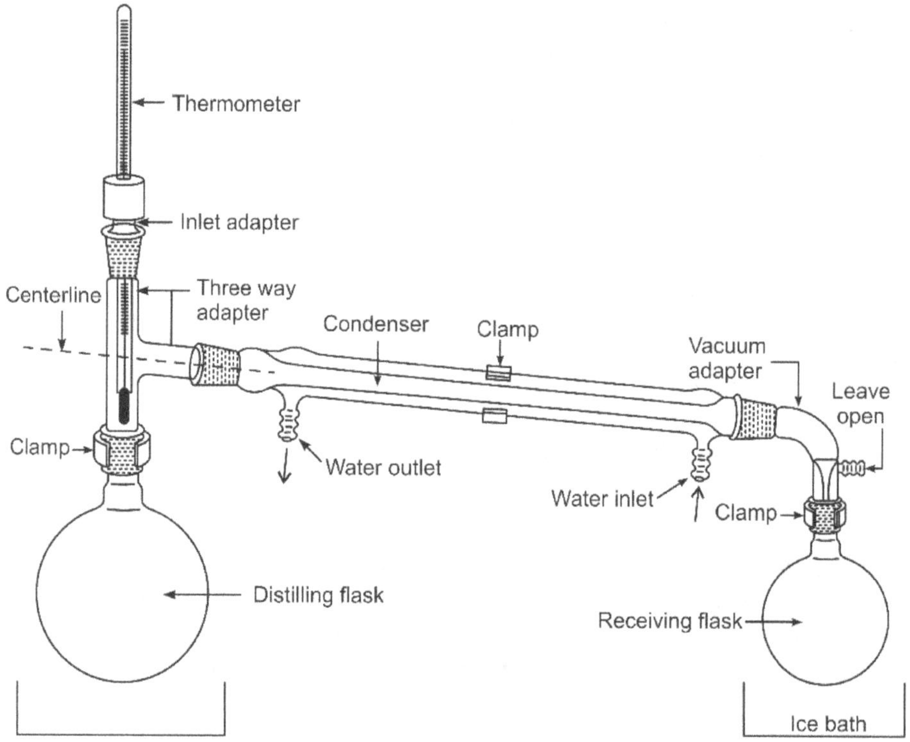

Fig. 1.4 Simple distillation apparatus

Fractional Distillation

Place 50 ml of the given mixture in the distilling flask and add a small boiling stone (to prevent super heating and to facilitate even boiling).Connect the flask to the apparatus and heat it gently with a smallnon-luminous Bunsen flame.

The slower the distillation, the better will be the separation. Collect the 5 ml fraction of the distillate in a small measuring cylinder and note the temperature at which each fraction distils. Transfer each fraction to a clean, dry, numbered test-tubes; stopper the test-tubes and set them aside for subsequent g. l. c. examination.

Plot a graph of the volume of distillate against the boiling point.

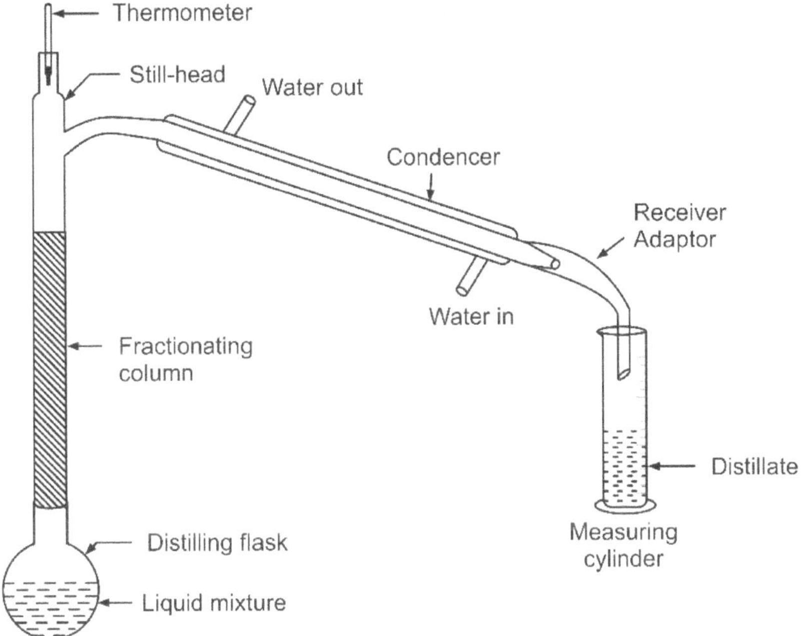

Fig. 1.5 Fractional distillation apparatus

1.13 Vacuum Distillation
(Distillation at Reduced Pressure)

High boiling liquids, or those which decompose at or normal boiling point, are generally distilled at lower temperature under reduced pressure.

Use a round bottom flask as the distillation pot. The size of the flask depends on the volume to be distilled, but ideally it should be between one quarter and one half full – and no more. In order to prevent the *extremely bad bumping* that occurs with *every* vacuum distillation, the liquid should be stirred with a magnetic stirrer bar. The distillation pot is then fitted with a still head, quickfit thermometer (see demonstrator), condenser and a pig (3-way receiver adapter) with 3 round bottom flasks (see diagram below).

Place the liquid in the distillation pot and assemble the apparatus, all the ground glass joints should be lightly greases, with vacuum grease <u>not</u> Vaseline. Remember to support the collection flasks, until the system is under vacuum, gravity rules! Start stirring vigorously but do n not heat. Turn the tap for the water pump on full before connecting the tubing to the system. Once connected, use the stopcock on the trap to carefully apply the vacuum. Ensure that the contents of the pot do not froth over and contaminate the collection flasks. If this happens, dismantle the apparatus and start again. Only when the apparatus is fully evacuated can the heater be switched on.

Therefore, if you purify a sample by distillation at reduced pressure, it is acceptable to quote the boiling range at which you collected the sample as long as you quote the pressure. With a water pump a pressure of 10-20 mmHg can be obtained, under these conditions, boiling points are reduced by approximately 100^0C.

Vigorous stirring is maintained throughout the distillation and fractions are collected by rotation of the pig at the appropriate point. When distillation is complete, turn of the heating and allow the apparatus to cool. Then, holding the collecting flasks, release the vacuum by opening the stopcock on the trap. Donot swish the water pump off before releasing the vacuum.

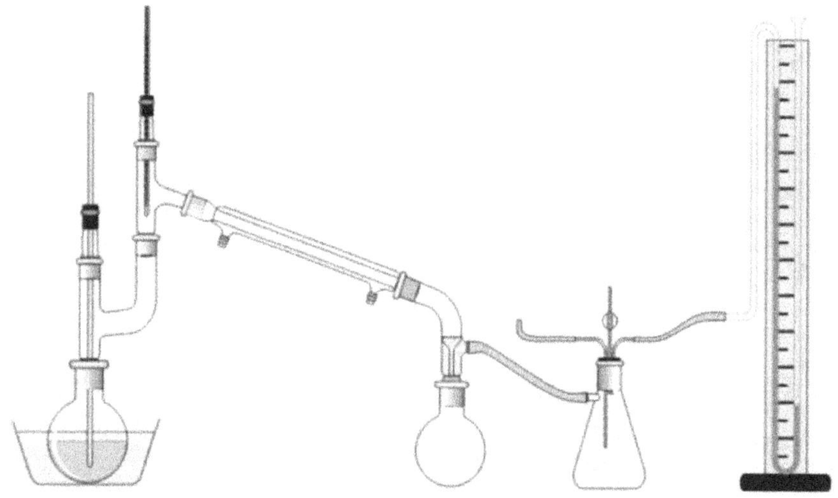

Fig. 1.6 Vacuum distillation set-up

LABORATORY REAGENTS

2.1 Preparation of Laboratory Reagents

1. **Ammonium acetate(1N):** Dissolve 73 g of ammonium acetate (CH_3COONH_4) in water and dilute to 1 liter.

2. **Ammonium carbonate (1N):** Dissolve 96 g of the analar grade salt $(NH_4)_2 CO_3.H_2O)$ (1N) in 170 ml of 1N NH_4OH and dilute to 1 liter.

3. **Ammonium chloride (1N):** Dissolve 53.5 g of NH_4Cl in 1 liter of water.

4. **Ammonium nitrate (1N):** Dissolve 80 g of NH_4NO_3 in 1 liter of water.

5. **Ammonium oxalate (0.5 N):** Dissolve 35.5 g of ammonium oxalate $((NH_4)_2C_2O_4.H_2O)$ in water. Dilute to 1 liter.

6. **Ammonium sulphate (0.5 N):** Dissolve 33 g $(NH_4)_2SO_4$ in 1 liter of water.

7. **Ammonium sulphide, colourless (1N):** Treat 68 ml of conc. (NH_4OH) ammonium hydroxide solution with H_2S until saturated, keeping the solution cold. Add 68 ml of conc. NH_4OH and dilute to 1 liter.

8. **Ammonium molybdate (0.5 N):** Dissolve 45 g of pure ammonium molybdate in a mixture of 40 ml of aqueous ammonia (d 0.88) and 60 ml of water and then adding 120 g of ammonium nitrate and dilute the solution to 1 liter with water.

9. **Ammonium thiocyanate (1N):** Dissolve 76 g of ammonium thiocyanate (NH_4CNS) in water and dilute to 1 liter.

10. **Ammonium vanadate:** Dissolve 60 mg of ammonium vanadate in 200 ml of water.

11. **Ammoniacal silver nitrate solution (Tollen's reagent):** Tollen's reagent is unstable and must be prepared just prior to use. Dissolve 5 g of silver nitrate in 100 ml of water (**solution 1**) and 5 g of sodium hydroxide in 100 ml of water (**solution 2**). Keep these two solutions separately until required. When the reagent required, mix equal volumes (say, 1 ml each) of **solution 1** and **solution 2** in a clean test tube, and then add dilute ammonia solution (concentrated ammonium hydroxide diluted with 5 volumes of water) drop by drop with shaking until the silver oxide is just dissolved. Only small volume should be prepared just before use and any residue washed down the sink with a large amount of water (Note : An excess ammonia will decrease the sensitivity of the reagent). Great

precautions must be taken in the preparation and use of this reagent, **which must not be heated**.

12. **Barium chloride (0.25M, 0.5N)**: Dissolve 61 g of barium chloride (BaCl$_2$.2H$_2$O) in water. Dilute to 1 liter.

13. **Benedict's reagent (Test for reducing substances)**: Dissolve 43 g crystalline sodium citrate and 25 g of anhydrous sodium carbonate in 175 ml of water. Filter if necessary and add a solution of 4.32 g of crystalline copper sulphate in 25 ml of water with stirring. Make up to 250 ml water.

    (a) **Barfoed's reagent for reducing mono saccharides**. Dissolve 6.6 g of cupric acetate and 1 ml of glacial acetic acid in water and dilute to 100 ml.

14. **Calcium chloride (0.25M, 0.5N)**: Dissolve 55 g of CaCl$_2$.6H$_2$O in water. Dilute to 1 liter.

15. **Chlorine-water**: Pass chlorine gas (KMnO$_4$ + Conc. HCl) into cold water till a saturated solution is obtained and keep it in a dark brown bottle.

16. **Ceric ammonium nitrate**: Dissolve 20 g of ceric ammonium nitrate in 500 ml warm dilute (2 N) nitric acid.

17. **Dichromate solution**: Dissolve 20 g of potassium or sodium dichromate in a mixture of 150 ml of water and 50 ml conc. sulphuric acid.

18. **2,4-Dinitrophenylhydrazine (2,4-DNP)**: Dissolve 2 g of 2,4-DNP in 15 ml conc. sulphuric acid added cautiously to 500 ml of 30 percent alcohol.

19. **Disodium hydrogen phospahte (1N)**: Dissolve 60 g disodium hydrogen phosphate (Na$_2$HPO$_4$.12H$_2$O) in water. Dilute to 500 ml.

20. **Ferric chloride (0.5M, 1.5N)**: Dissolve 135 g of (FeCl$_3$) ferric chloride in one liter of water containing 20 ml of conc. HCl.

21. **Ferrous sulfate (0.5M, 1N)**: Dissolve 13.9 g of ferrous sulphate (FeSO$_4$.7H$_2$O) in water containing 1.0 ml of conc. sulphuric acid. Dilute to 100 ml; solution doesn't keep well.

22. (a) **Fehling's solution, Solution No.1**: Dissolve 34.6 g of copper(II) sulphate crystals in water containing a few drops of dilute sulphuric acid, and dilute the solution to 500 ml.

    (b) **Fehling's solution, Solution No.2**: Dissolve in 200ml of water, while cooling, 60 g of pure sodium hydroxide and 173 g of pure sodium potassium tartrate (Rochelle salt) in water, filter if necessary through a sintered glass funnel and make up the filtrate and washings to 500 ml.

    Keep the two solutions separately in tightly stoppered bottles and mix exactly equal volumes (say 1 ml each) immediately before use.

23. **Ferrous ammonium sulpahte (Mohr's salt):** Dissolve 19.6 g of $(FeNH_4SO_4)_2.6H_2O$ in water containing 1 ml of conc. H_2SO_4. Dilute to 100 ml. prepare fresh solutions for best results.

24. **Formaldehyde-sulfuric acid (Marquis' reagent for alkaloids):** Add carefully 10 ml of formaldehyde solution to 50 ml of sulphuric acid.

25. **Iodine solution:** Add 2 g of powdered iodine to a solution of 4 g of potassium iodide in a minimum amount of water and make upto 300 ml with water.

26. **Lead acetate (0.5M, 1N):** Dissolve 19.0 g of lead acetate in water and dilute to 100 ml.

27. **Lead nitrate (0.25M, 0.5N):** Dissolve 8.3 g of $Pb(NO_3)_2$ in water. Dilute to 100 ml.

28. **Lime water (Calcium hydroxide),** (0.02M, 0.04N): Dissolve 1.5 g of calcium hydroxide or calcium oxide in water and make upto 1 liter.

29. **Lucas reagent:** To concentrated hydrochloric acid (5 ml) add anhydrous zinc chloride (2 g) and boil for 1 minute, cool to room temperature.

30. **Mercuric chloride (0.1N):** Dissolve 1.35 g mercuric chloride $(HgCl_2)$ in water and dilute to 100 ml water.

31. **Molish's reagent (General test for carbohydrate):** Dissolve 10 g of 2-naphthol in 100 ml ethanol (95%).

32. **Nessler's reagent (for ammonia):** Dissolve 5 g of KI in the smallest quantity of cold water (5 ml). Add a saturated solution of mercuric chloride (about 2.2 g in 35 ml of water) until an excess is indicated by the formation of a precipitate. Then add 20 ml of 5 N NaOH and dilute to 100 ml. Let settle, and draw off the clear liquid.

33. **Neutral ferric chloride:** Prepare the neutral ferric chloride solution (i.e., free from HCl) by adding dilute sodium hydroxide solution to the bench ferric chloride reagent until a slight precipitate of iron (III) oxide is formed. Filter off the precipitate and use the clear solution.

34. **Ninhydrin solution:** Dissolve 0.2 g ninhydrin in 100 ml distilled water.

35. **Potassium dichromate (0.125M):** Dissolve 37 g potassium dichromate $(K_2Cr_2O_7)$ in 1 liter of water.

36. **Potassium ferricyanide (0.2N):** Dissolve 2.1 g of solid $K_2(Fe(CN)_6.3H_2O)$ in water and dilute to 1 liter.

37. **Potassium iodide-iodine solution (0.5M, 0.5N):** It is prepared by dissolving 20 g of potassium iodide and 10 g of iodine in 100 ml of water.

38. **Potassium hydroxide (alcoholic) solution:** A mixture of powdered potassium hydroxide (10 g) and rectified spirit (100 ml) is refluxed for 30 minutes, cool and decant through a filter, if solid material remains.

39. **Potassium permanganate solution (0.02M)**: Dissolve 3.16 g of $KMnO_4$ in 1 litre water.

40. **Picric acid**: Prepare a saturated solution of picric acid in ethanol or benzene and filter the clear solution from undissolved portion of the solid.

41. **Potassium iodide mercuric iodide (reagent for proteins)**: Dissolve 5 g of KI in 50 ml of water, and saturated with mercuric iodide (about 12 g). Dilute to 100 ml.

42. **Potassium ferrocyanide (0.5M, 2N)**: Dissolve 21.1 g of potassium ferrocyanide in water. Dilute to 1 litre.

43. **Potassium carbonate (0.75M, 1.5N)**: Dissolve 20.7 g of K_2CO_3 in 100 ml of water.

44. **Potassium chloride (0.5M, 0.5N)**: Dissolve 37 g of KCl in 1 litre of water.

45. **Sodium carbonate (0.75M, 1.5N)**: Dissolve 79.5 g of Na_2CO_3 in water. Dilute to 1 litre.

46. **Sodium chloride (0.5M, 0.5N)**: Dissolve 29 g of NaCl in 1 litre of water.

47. **Sodium hydrogen phosphate (0.167M, 0.5N)**: Dissolve 60 g of sodium hydrogen phosphate ($Na_2HPO_4.12H_2O$) in 1 litre of water.

48. **Sodium nitrate (0.05M, 0.05N)**: Dissolve 4.3 g of $NaNO_3$ in 100 ml of water.

49. **Sodium sulfide (0.05M, 0.5N)**: Dissolve 60 g of solid $Na_2S.9H_2O$ in water and dilute to 500 ml of water.

50. **Stannous chloride (0.05M, 0.1N)**: Dissolve 11.3 g of $SnCl_2.2H_2O$ in 17 ml of conc. HCl, using heat if necessary. Dilute with water to 100 ml.

51. **Sodium hypobromite**: Dissolve 100 g of NaOH in water by external cooling in ice-water and make upto 1 litre, add slowly 25 ml of bromine with stirring the mixture.

52. **Sodium hydrogen sulphite solution**: Dissolve 300 g of $NaHSO_3$ in water, make up to 500 ml and pass sulphur dioxide for a few minutes to remove sodium sulphite.

53. **Starch solution**: A solution of starch which can be kept indefinitely is made as follows : mix 500 ml of saturated NaCl solution (filtered), 80 ml of glacial acetic acid, 20 ml of water and 3 g of starch. Bring slowly to boil and boil for two more minutes and filter if necessary; store in a glass stoppered bottle.

OR

Make a paste of 4 g of soluble starch in 5 ml of cold water. In a separating beaker boil 150 ml water and add cold paste till a clear solution is obtained.

54. **Sodium bicarbonate (10%) solution**: Dissolve 10 g of sodium bicarbonate in 100 ml of water.

55. **Sodium nitroprusside**: Dissolve 1 g of sodium nitroprusside in 100 ml of distilled water.

56. **Silver nitrate (0.05N)**: Dissolve 8.5 g of $AgNO_3$ in water dilute to 100 ml.

57. **Sodium acetate (1N) solution**: Dissolve 136 g of sodium acetate in water and dilute to 1 liter.

58. **Schiff's reagent**: Add one gram of sodium metabisulphite to a solution of 0.1 g of p-rosaline hydrochloride and 2 ml of conc. HCl in 100 ml of water.

59. **Tollen's reagent**: See ammoniacal silver nitrate solution.

2.2 Solutions

Types of Solutions

1. **Molar solution (M):** A molar solution is one which contain one mole or one gram-molecular weight of the solute in 1 litre of solution. The term molar is used for a solution containing 1 mole in 1000 grams of solvent.

2. **Normal solution (N)**: A normal solution contains the gram-equivalent weight of solute in 1 litre of solution.

    N = Weight of solute per litre of solution/Equivalent mass

    Relation between N and M

    N = M x n

    Where n = number of e- change

    N = Normality ; M = Molarity

3. **Isotonic solutions**: Isotonic solutions are those that exert equal osmatic pressures. Isotonic saline is one contain 8.5 g of NaCl in 1 litre of aqueous solution.

4. **Aqueous**: Aqueous solutions are, of course, those using water as solvent. Where no solvent has been mentioned it can be assumed to be water.

5. **Percentage solution**: Percentage solution of solids refer to grams per 100 ml of solution, percentage mixtures of liquids refer to milli liters per 100 ml of mixture (example 60% of ethanol is 60 parts of ethanol plus 40 parts of water).

6. **Weight/volume and volume/volume**: The abbreviations w/v and v/v are sometimes added after the strength of a solution. Thus trichloroacetic acid 15% w/v is a solution containing 15 g of trichloroacetic acid per 100 ml of solution, and 75% ethanol v/v is a solution containing 75 ml of ethanol per 100 ml of solution.

7. **Buffer solution**: Buffer solution that tend to resist change in pH on the addition of small quantities of acid or alkali. Many buffer solutions are

mixtures of either a weak acid and one of its salts or a weak base and one of its salts. Commonly used buffer systems are:

(a) Phosphate: Potassium hydrogen orthophosphate (KH2PO4) + disodium hydrogen orthophosphate (Na2HPO4).

(b) Acetate: Acetic acid + Sodium acetate

(c) Citrate: Citric acid + Sodium citrate

(d) Barbitone (Veronal): Sodium diethylbarbiturate + HCl

8. **Dilution:** Diluting from one strength to another. A simple rule to remember when diluting from one strength to another is as follows: If **S** is the strength of the given solution and **t** is the required strength, take **t** parts of the given solution and make up to **S** parts with solvent. For example, a 10% aqueous solution may be obtained from a 25% solution by taking 10 ml of the 25% solution and making upto 25 ml with water.

If a definite volume of dilution is needed use the following formula.

$$tV/S = Vs$$

Where **V** = Volume required, **Vs** = Volume of solution of concentration **S** to be diluted.

Example: Given a 36% solution, prepare 500 ml of a 9% solution.

Substitution in the formula above

$$9 \times 500/36 = Vs = 125$$

Therefore take 125 ml of the 36% solution and dilute to 500 ml.

2.3 Acids and Bases

Acids and Bases		**ml. per litre to make approx:-**			
		Approx Normality	**1 N (N/1)**	**0.1N (N/10)**	**0.01N (N/100)**
Acids :	S.G	N	ml	ml	ml
Glacial acetic acid	1.06 (99%)	16	57	5.7	0.6
Hydrochloric acid	1.16 (36%)	12	86	10	1.0
Nitric acid	1.42 (69%)	16	63	6.3	0.7
Sulphuric acid	1.84 (95%)	36	28	2.8	0.3
Bases :					
Ammonium hydroxide	0.88	20	50	5.0	0.5
Potassium hydroxide	--	--	58 g	5.8 g	0.6 g
Sodium hydroxide	--	--	42 g	4.2 g	0.42 g

Approximate Normality

Dilute acids: Use the amount of concentrated acids or bases mentioned and dilute to one litre to make appropriate normality.

(i)   **Dilute acids for general use**

   (a) **Hydrochloric acid (3N):** Add 258 cc of conc. hydrochloric acid to water and dilute to 1 litre.

   (b) **Nitric acid (3N):** Add carefully 189 cc of conc. nitric acid, by external cooling, to water and make up the volume to 1 litre.

   (c) **Sulphuric acid (3N):** Add slowly 84 cc of con. sulphuric acid, by external cooling, to cold water and make upto the volume to 1 litre.

   (d) **Acetic acid (3N):** Add 172 cc of glacial acetic acid to water and dilute to 1 litre.

   (e) **Dilute hydrochloric acid (5% v/v, 1.4N):** Dilute 5 cc of conc. HCl with water until total volume is 100 cc.

   (f) **Dilute acetic acid (5% v/v, 0.8N):** Dilute 5 ml of glacial acetic acid with water until total volume is 100 cc.

   (g) **Dilute sulphuric acid (5% v/v, 1N):** Add 5 cc of conc. H_2SO_4 to cold water and make upto 100 cc with water.

   (h) **Dilute nitric acid (5% v/v, 0.8N):** Add 5 cc of conc. HNO_3 carefully to water by external cooling, and make upto 100 cc with water.

(ii) **Dilute bases for general use**

   (a) **Sodium hydroxide (10% aqueous, 2.5N):** Dissolve 10 g of NaOH in water and dilute to 100 ml.

   (b) **Sodium hydroxide (30% aqueous, 7.5N):** As above, using 30 g of NaOH.

   (c) **Ammonium hydroxide (dilute):** Dilute 1 volume of conc. ammonia (d, 0.88) to 3 vols. of water.

   (d) **Potassium hydroxide (3N):** Dissolve 176 g of the pellets in water and dilute to 1 litre.

2.4 Indicators

(a) **Litmus (Indicator) solution:** Extract litmus powder three times with boiling alcohol, each treatment consuming an hour. Discard or reject the alcohol extract. Treat residue with an equal amount of cold water and filter; then extract with five times its weight of boiling water, cool and filter. Combine the aqueous extracts.

(b) **Methyl orange:** Dissolve 0.1 g of methyl orange in 100 ml of water filter, if necessary.

(c) Methyl red:

   (i) Dissolve 0.1 g of methyl red in 60 ml of alcohol and dilute with 40 ml of water.

   (ii) Methyl red modified : Dissolve 120 mg of methyl red and 80 mg of methylene blue in 100 cc of 95% alcohol.

(d) Phenolphthalein: Dissolve 1 g of phenolphthalein in 50 ml of alcohol and add 50 ml of water with constant stirring.

(e) Thymol-blue: Dissolve 0.5 g of thymol-blue in 100 ml of ethanol and add 200 ml of water.

CHEMICAL EXAMINATION, IDENTIFICATION AND CHARACTERISATION OF ORGANIC COMPOUNDS

3.1 Organic Qualitative Analysis

The elements usually found in organic compounds, besides carbon and hydrogen, are oxygen, nitrogen, sulphur, the halogens, a few metals such as sodium, magnesium and calcium. In more advanced work other elements become important, but these lie outside the scope of this book. Instructions for the detection of the principal elements will now be given.

The organic chemist is always faced with the problem of characterising and finally elucidating the structure of unknown organic compounds. The problem of the elucidation of structure has been revolutionised by the use of a wide range of spectroscopic techniques which are now available; and have been applied extensively to confirm the unknown structures.

In qualitative organic analysis the chief concern is with the detection and identification of an unknown organic compound. The unknown organic compound(s) may be a solid or liquid or both or a mixture. The first part contains various preliminary investigations, particularly the study of physical state, colour, odour, acidic, basic or neutral, extra elements greatly assist the further search to identify the characteristic functional group or groups present. The second part is devoted to different classes of organic compounds, and a description is given of the general reactions of members of particular classes and the compounds may then be converted into an appropriate solid derivative, which may confirm the identity of the compound. Some important members of each class are given below to ensure that students are not assigned unreasonably unknown compounds. Compounds marked with asterisk are generally more difficult than the others in each group and should only be used in the latter stages of the course.

ALCOHOLS	PHENOLS
C_1-C_4 (all isomers)	phenol
	o,m and p-cresol
cyclohexanol	p-chlorophenol
benzyl alcohol	o and p-bromophenol
cinnamyl alcohol	2,4-dichlorophenol
benzhydrol alcohol	o and p-nitrophenol
	a and b-naphthol
	p-t-butylphenol
	2,6-dimethylphenol (2,6-xylenol)
	2,4-dinitrophenol
	3,5-dimethylphenol (3,5-xylenol)
ACIDS	**ALDEHYDES**
Acetic acid	n-butyraldehyde and isobutyraldehyde
Propionic acid	crotonaldehyde
Crotonic acid	phenylacetaldehyde
Benzoic acid	benzaldehyde
o and p-toluic acid	o and p-chlorobenzaldehyde
phenylacetic acid	anisaldehyde
o and p-chlorobenzoic acid	p- tolualdehyde
p-bromobenzoic acid	cinnamaldehyde
o, m and p-nitrobenzoic acid	
cinnamic acid	
3,5-dinitrobenzoic acid	
Anisic acid	
Phthalic acid	
KETONES	**AMINES**
acetone	aniline
diethyl ketone	o,m and p-toluidine
ethyl methyl ketone	2,5-dimethylaniline
methyl-n-propyl ketone	3,4-dimethylaniline
methyl isopropyl ketone	o,m and p-chloroaniline
methyl isobutyl kentone	2,5-dichloroaniline
cyclohexanone	p-bromoaniline
acetophenone	o,m and p-anisidine
p-methylacetophenone	ethyl-p-aminobenzoate
p-methoxy acetophenone	m-nitroaniline
propiophenone	ortho and p-nitroaniline
benzophenone	N-methylaniline
benzalacetophenone*	N,N-dimethylaniline
benzil*	N-ethylaniline
	N,N-diethylaniline
	benzylamine
	dibenzylamine

Contd...

ESTERS methylethyl benzoate ethyl p-nitrobenzoate methyl and ethyl cinnamate methyl and ethyl phenyl acetate dibutyl phthalate methyl and ethyl crotonate phenylbenzoate diethyl phthalate benzyl acetate benzyl phenyl acetate methyl phenyl acetate dimethyl oxalate dimethyl succinate	**AMIDES** benzamide acetamide urea
ANILIDES acetanilide benzanilide p-chloroacetanilide p-bromoacetanilide p-methylacetanilide o and p-nitroacetanilide	**NITRILES** phenyl acetonitrile benzonitrile
CARBOHYDRATES glucose fructose sucrose maltose starch	**ETHERS** diethyl ether di-n-propylether di-isopropyl ether anisole guiacol methyl 2-naphthyl ether
QUINONES benzoquinone anthraquinone	**NITRO-COMPOUNDS** nitrobenzene p-nitrotoluene m-dinitrobenzene

HYDROCARBONS
benzene
toluene
naphthalene
anthracene
chlorobenzene
bromobenzene
	AMINO ACIDS
	I amino-aliphatic carboxylic acid
	   glycine, cysteine
	II amino-aromatic carboxylic acid
	   anthranilic acid,
	   sulphanilic acid.

3.2 The Identification of an Unknown Compound

Methods of qualitative analysis in organic chemistry cannot have the precision of those available in inorganic chemistry, nevertheless, the method, which involves a series of tests on the given compound, is systematic and when all the results are considered, a complete identification of the compound should be possible.

3.3 Preliminary Tests

3.3.1 Physical State and Colour of the Compound

Solid, liquid, crystalline form, colour, odour etc. The smell of a compound will very often give some indication of the group to which it belongs. For example, phenols have a characteristic antiseptic smell. The lower aliphatic aldehydes have a very pungent irritating smell. The lower aliphatic ketones have a smell similar to acetone, the first member of the series. The lower alipahtic carboxylic acids have a pungent smell like acetic acid but more unpleasant, Aliphatic amines have a characteristic fishy odour. Aromatic amines have a characteristic smell (cf.aniline). Benzaldehyde and nitrobenzene smell like bitter almonds. Esters often have a pleasant, fruity smell.

Impurities frequently discolour compounds which would otherwise be colourless. Observe whether the colour presists after purification.

Yellow -   Aromatic nitro compounds e.g., nitrobenzene (liquid), o-nitrophenol (solid), n-nitrosoamines.

Orange red -   Compounds containing the -N=N- chromophore e.g. azobenzene (solid).

Green -   Some aromatic nitroso compounds, e.g., p-nitrosodimethyl aniline (solid)

Brown / Purple - Aerial oxidation of amines and phenols after prolonged storage produces coloured impurities.

3.3.2 Melting Point or Boiling Point

If the compound is solid, purify by recrystallization from a suitable solvent and determine its melting point (m.p).

If the compound is liquid, purify by distillation and hence determine its boiling point (b.p). Observe and record carefully whether:

(a) Compound melts (or boils) over a narrow range, around 2°; the compound is likely to be pure.

(b) Solid does not melt below 350°C, it may be a salt.

(c)   Solid melts first, then decomposes e.g., it becomes dark, or water is given off (useful information in a number of cases).

(d)   Solid decomposes before it melts (e.g. carbohydrates), record this as, decom. 188°C.

3.3.3 Ignition Test

Place about 0.1 g of the compound in a porcelain crucible. Heat gently at first, and then to dull redness. Observe:

(i)   Whether the substance melts, sublimes, explodes or is inflammable (nature of flame?)

(ii)   Whether gases or vapours are evolved and their odour (Caution!).

(iii)   Whether residue is left. Dissolve residue in water and test solution with litmus

Aliphtic substances will burn readily with a clear blue or yellow flame. If they contain a high proportion of halogen, they will ignite only on strong heating. CCl_4 will not burn.

Aromatic compounds (benzene, aniline etc.,) will burn with a very smoky flame depositing soot. If a residue remains after strong heating, this is probably the carbonate or oxide of a metal. The original compound must therefore have been the metallic salt of a carboxylic acid, sulphonic acid or phenol, or the bisulphite compound of an aldehyde or ketone.

3.3.4 Fusion with Soda-Lime

Fusion with soda-lime produces the following reactions:

Ammonium salts, acid amides and urea : gives ammonia (identified by smell and by testing with litmus).

Glycine (H_2N-CH_2COOH)   :   gives methylamine

Carboxylic acids and their salts give the corresponding hydrocarbons.

Examples:

$CH_3.COOH$ (acetic acid)   :   methane + CH_3COOCH_3

$HOC_6H_4.COOH$ (salicylic acid)   : C_6H_5OH (phenol)

HCO_2H (formic acid)   : $H_2O + CO$

$(COOH)_2$ (oxalic acid)   : H_2

$CCl_3.CH(OH)_2$ (chloral hydrate)   : $CHCl_3$

Test: Thoroughly mix about 0.1 g of compound with 5 times its bulk of soda-lime, place in a 3" pyrex test-tube and heat cautiously on a Bunsen flame. Observe whether gases or vapours are produced; odour (caution!) and flammability of vapours).

3.3.5 Lassaigne Test

Detection of elements - The sodium fusion test: Some polynitro compounds can react violently with molten sodium. Always wear safety spectacles and carry out the fusion with a safety screen or in the fume cupboard.

When provided with a substance for identification, it is first necessary to determine what elements it contains. It is not usually necessary to test for CARBON and HYDROGEN in a substance which is known to be organic. The presence of nitrogen, sulphur, the halogens (Cl, Br, I) and phosphorus can be detected by means of the Lassaigne test.

Procedure: A small cube (3 mm) of freshly cut sodium metal is placed in a small test tube or ignition tube (3 x 3/8") and a small quantity of the unknown substance is added. The tube is held firmly by means of tongs and warmed gently on a bunsen flame. Vigorous reaction may take place at first and as this subsides the tube is heated more and more strongly until red hot is attained. The tube is heated for a few seconds more to make sure that all the excess sodium combines with the glass. The red hot tube is then plunged into water(10 cc) contained in a small beaker (**extreme care should be taken during this part of the operation as any unreacted sodium will react violently with the water and may explode!**). The ignition tube shatters on contact with the water and the contents are dispersed. The resulting solution is boiled and filtered through a small Hirsch or Buchner funnel into a filter tube (use suction pump). The residue in the test tube is boiled with 2 cc of water and filtered once more. The resulting combined filtrate is strongly alkaline and will contain the N, S, halogens or phosphorous (if they are present originally) in ionic form - during the sodium fusion, if the compound contains nitrogen,

Equation : $Na + (N + C) - \text{from compound} \rightarrow NaCN$

Similarly if S or Halogen (X) or P is present

Equation : $2Na + S \rightarrow Na_2S$

$Na + Ha\log en[X] \rightarrow NaX$

The solution may therefore contain cyanide, sulphide, and/or halide or phosphide ions and these are tested as follows:

(i) **Test for Nitrogen**: A portion of the solution (1-2 c.c.) is added to a test tube containing solid Ferrous sulphate (about 100 mg) (**if sulphur is also present a black precipitate of FeS will be obtained at this stage, otherwise a green precipitate of Fe(OH)$_2$ will be observed**). The solution is heated gently to boiling point, then, without cooling, acidified with dilute sulphuric acid. If nitrogen is present a precipitate of prussian blue (Ferric ferrocyanide) is obtained. If nitrogen is absent, a pale yellow solution results.

Equation : $6NaCN + FeSO_4 \rightarrow Na_4Fe(CN)_6 + Na_2SO_4$

also, during boiling $FeSO_4$, i.e., Fe^{2+}. is oxidised (air) to $Fe(SO_4)_3$, i.e., Fe^{3+}

then : $3Na_4Fe(CN)_6 + 2Fe_2(SO_4)_3 \rightarrow Fe_4[Fe(CN)_6]_3$

and on acidification (H_2SO_4) the prussian blue $(Fe[Fe(CN)_6]_3)$ colour appears in the solution.

(ii) **Test for sulphur**: This element may be tested by diluting a portion of the fusion solution (3-4 drops) with about 3 cc of water and adding a few drops of fresh solution of sodium nitroprusside, $Na_2[Fe(CN)_5NO]$. An immediate purple or red-violet colouration is obtained if sulphur is present, the coloration slowly fades on standing.

(iii) **Test for halogens**: A portion of the solution (2 cc) is treated with a slight excess of silver nitrate solution (about 3 cc). A black precipitate of AgOH $(+Ag_2S + AgCN)$ forms if sulphur and nitrogen have been detected above + AgX, if halogen is present) is obtained. An excess of conc. nitric acid is then added and the mixture is heated to boiling point. H_2S and HCN (both are gases) will be expelled. (Boil for 1-2 minutes). If a white or yellow precipitate remains after this a halogen is probably present. If no precipitate is obtained on boiling, halogen is absent. If a precipitate remains, it is allowed to settle and the supernatant liquor is decanted. More conc. HNO_3 is added to the precipitate and boiling is repeated (1-2 min.). If the precipitate again remains it is silver halide. To determine which halogen is present proceed as follows:

The liquor is decanted from the halide precipitate which is then washed with water (twice) by decantation. The washed precipitate is treated with conc. ammonia solution (3-5 cc). If the precipitate is **white** and dissolves easily in the ammonia-chlorine is present. If the precipitate is **pale yellow** and dissolves **with difficulty** in the ammonia - bromine is present. A yellow precipitate which **is insoluble** in ammonia indicates the present of iodine.

Note: This test can be used **only** when **one** halogen is present; it is unsatisfactory if the original compound contains more than one halogen.

(iv) Test for phosphorous: The presence of phosphorus may be indicated by a smell of phosphine during the sodium fusion and the immediate production of a jet-black colour when a piece of filter paper moistened with silver nitrate solution is placed over the mouth of the ignition tube after the sample has been dropped on the hot sodium. Boil sodium fusion solution (2 cc) with conc. nitric acid (1 cc) for 1 minute; cool and add an equal volume of ammonium molybdate reagent and allow the tube to stand in boiling water for several minutes. If phosphorus is present, a yellow precipitate of ammonium 12-molybdophosphate will separate.

$$P \rightarrow PO_4^{3-} \rightarrow (NH_4)_3[PMo_{12}O_{40}]$$

3.3.6 Procedure for Solubility Tests

Solubility of compound: All solubility tests are carried out at the room temperature in small test tube but of sufficient size to permit free shaking of the solvent and the solute.

A compound is most soluble in that solvent to which it is most closely related in structure, i.e., n-hexane is sparingly solubly in water, dissolves in 3 volumes of methanol, and is completely miscible with n-butyl and the higher alcohols.

The monohydroxy- and monocarboxy- derivatives of the hydrocarbons are soluble in ether; the lower homologs (upto Ca. C_5) are soluble in water with two OH groups the compound becomes more water-like; consequently the solubility in ether decreases.

Most salts are soluble in water and insoluble in ether.

Amount of material required for solubility: It is always advisable to employ an arbitrary ratio of 0.05 g of solid or 0.10 ml of liquid for 3.0 ml of solvent.

(i) Solubility in water (cold and hot): Treat a 0.10 g portion of the solid/liquid with 1.0 ml portions of water, shaking vigorously after each addition, until 3.0 ml have been added.

If substance is soluble or partly soluble in cold water, test the solution with litmus:

(a) *Alkaline reaction*: Some amines and alkali-metal salts of weak acid e.g., sodium acetate.

(b) *Acid reaction*: Carboxylic and sulphonic acids; acid halides and anhydrides, some esters; amine salts, e.g., aniline hydrochloride.

If soluble in hot water, cool the solution and examine any crystals which separate and test filtrate with litmus.

(ii) Solubility in ether: Use 0.10 g of solid or 0.10 ml of a unknown liquid in a dry test tube and add 3.0 ml of solvent ether.

(iii) Solubility in dilute sodium hydroxide solution (5%): Carboxylic acid $(R.CO_2H)$, sulphonic acids $(R.SO_3H)$, phenols $(Ar.OH)$ and aryl sulphonamides $(Ar.SO_2NH_2)$ are examples of a few very common classes of compounds which contain an acidic group of sufficient strength to react with alkalis.

(a) *Cold*: Acids and phenols dissolve readily. Ammonium salts liberate NH_3, Amine salts give the free amine which may separate as a discrete layer - if not, they can be extracted with ether.

(b) *Hot*: Ammonia is liberated from ammonium salts amides, imides and nitrites. Esters will be hydrolysed to acid + alcohol, cool and acidify with mineral acid - crystal of organic acid may separate.

Aliphatic aldehydes polymerise to give resins.

Aromatic aldehydes give benzyl alcohol + sodium benzoate and on acidification (HCl) benzoic acid will precipitate.

If test (iii) is positive, observe reaction with dilute sodium bicarbonate solution.

(iv) Solubility in dilute sodium hydrogen carbonate solution (5%) : Add a small amount of the unknown compound to about 3 cc of dilute sodium bicarbonate solution in a test-tube. Carboxylic and sulphonic acids dissolve with evolution of CO_2. Phenols may dissolve but no effervescence is observed - hence distinction from acids [Note : Nitro-phenols may be sufficiently acidic to react with Na_2CO_3 and give CO_2].

(v) Test with Mineral Acid

Solubility in dilute hydrochloric acid (5%)

(a) Most amines are more soluble but amine salts sometimes precipitate. This test may indicate the presence of a basic nitrogen atom in the molecule. However, in purely aromatic secondary and tertiary amines (e.g., diphenylamine and triphenylamine) the basic character of the nitrogen atom is diminished to such an extent that do not form salts with dilute hydrochloric acid.

If this test is positive, then determine reaction of compound with nitrous acid. Since nitrous acid is unstable, it is best generated *in situ*.

Test : Cool the solution of the compound (preferably in dilute hydrochloric acid) in ice. Add a small crystal of ice to the solution, together with an ice-cold solution of sodium nitrite in water. The mixture will become pale blue. Observe the reaction, primary aliphatic amines liberate gas (N_2). Secondary aliphatic amines give oily yellow nitrosoamines. Tertiary aliphatic amines give no visible reaction. Primary aromatic amines produce diazonium salts $RN^{2+}X$. The presence of these may be detected by adding a cold solution of b-naphthol in dilute sodium hydroxide, when a red azodye is produced. Secondary aromatic amines yield nitrosamines. Tertiary aromatic amines will yield a dark orange-red solution on treatment with nitrous acid.

(b) Bisulphite compounds give SO_2

(c) Salts of acids (e.g, sodium benzoate), give free acid and if a solid it precipitates. If it forms a liquid acid it will be probably soluble in water, and it may be extracted with ether or chloroform.

(d) Amphoteric behaviour of certain classes of compounds. e.g., amino acids should be considered.

$$R-CH-COOH \qquad 2\ H_2N-C_6H_4-COOH$$
$$\qquad |$$
$$\qquad NH_2$$

But note that sulphanilic acid $P - H_2N - C_6H_4 - SO_3H$ only exhibits the solubility reactions of the $-SO_3H$ group.

Solubility in conc. sulphuric acid: Olefins, ethers and alcohols dissolve; saturated hydrocarbons and aromatic hydrocarbons are insoluble. Sugars and other carbohydrates give charring on standing. Urea gives $\{CO_2 + (NH_4)_2SO_4\}$.

3.3.7 Miscellaneous Class Reactions

Determine the effect of the following class reagents upon small samples of the unknown compound.

Tests for unsaturated compounds

(a) ***With bromine solution***: The test may be carried out using a solution of bromine in water, acetic acid, or carbon tetrachloride for the compound. Decolourisation indicates an addition of bromine to a double or triple bond.

$$e.g.,\quad R_1R_2C=CR_3R_4 \quad + \quad Br_2 \quad \longrightarrow \quad R_1R_2\overset{|}{C}-\overset{|}{C}R_3R_4$$
$$\qquad\qquad\qquad\qquad\qquad\qquad\qquad\qquad\qquad Br\ Br$$

or substitution of bromine in an aromatic ring. The latter will, however, be accompanied by evolution of HBr and this can be observed if test is carried out in CCl_4. Phenols and aromatic amines react in this way e.g., **phenol** gives tribromophenol, **Aniline** gives tribromoaniline. In aqueous solution these tribromoderivatives precipitate.

Test: Dissolve 0.2 g compound in 2 ml carbon tetrachloride and add a solution of bromine in carbontetrachloride, dropwise, until the colour persists for a minute. Blow across the mouth of the test tube to detect any HBr.

Repeat using aqueous bromine and the neat substance, or aqueous solution of the compound.

(b) ***With potassium permanganate solution*** : Alkenes are oxidised to glycols, and the permanganate solution becomes decolourised.

e.g., $R^1R^2C=CR^3R^4C \xrightarrow{[O]}$ $R^1R^2\text{-}C\overset{O}{-}C\text{-}R^3R^4 \xrightarrow{H_2O}$ $R^1R^2\text{-}\underset{OH}{C}\text{-}\underset{OH}{C}\text{-}R^3R^4$

Easily oxidisable substances e.g., aldehydes, alcohols and amines will also decolourise $KMnO_4$ solution more slowly, unless the decolorisation is instantaneous.

Test : A small amount of the substance (0.1 g) is dissolved in ethanol or acetone (2 cc) and a 2% aqueous solution of $KMnO_4$ is added dropwise. Compound containing a double bond will be decolorised instantaneously due to the oxidation of double bond by the MnO_4^-.

(c) ***Test with ferric chloride solution***: Use a neutral or very, weak acid solution of $FeCl_3$. Colour reactions are given by solutions of phenols in water (violet colour) or in alcohol (green colour). (**NB.** Enols also give these reactions).

$$6\ C_6H_5OH + FeCl_3 \longrightarrow [Fe(OC_6H_5)_6]_3 + 3HCl$$
Violet complex

The above tests will not give any direct evidence for the unknown compound being a satuatated hydrocarbon, a simple benzene derivative, an ether, or an ester. The latter class of compounds can be recognised by alkaline hydrolysis to an acid. However a much more rapid determination can be made with the following.

(d) *Test for ester linkage*: The common esters are usually fragrant-smelling liquids. One or two drops of the substance is treated with three or four drops of a saturated solution of hydroxylamine hydrochloride in ethanol. To this mixture is added three drops of 20% methanolic potassium hydroxide. Heat gently to boiling point, cool, acidify with dilute HCl and then add ferric chloride solution dropwise. A deep red or purple colour indicates the presence of an ester.

The reactivity of acid chlorides and anhydrides under certain of the above test conditions should allow their easy identification. Thus in addition to their hydrolysis to acids (under ester or milder condtions), their ready reaction with primary amines would be distinguishing.

(e) *Acidic and neutral nitrogen compounds* fall into a wide variety of classes of organic compounds. Simple primary amides when warmed with 10% sodium hydroxide solution give ammonia readily together with the salt of the corresponding acid.

$$R \cdot CO \cdot NH_2 \ + \ NaOH \ \longrightarrow \ RCOONa \ + \ NH_3 \uparrow$$

Substituted amides $RCONR^1R^2$ suffer hydrolysis with more difficulty and may require boiling for 2-3 hr. with (10 to 70%) sulphuric acid solution.

Nitriles may be hydrolysed by boiling with either 30-40% sodium hydroxide solution or 50-70% sulphuric acid for several hours, but the reaction takes place less readily than for primary amides.

$$RCN \ + \ HOH \ \xrightarrow{\ NaOH\ } \ R \cdot CO \cdot ONa \ + \ NH_3 \uparrow$$

$$RCONH_2 \ + \ NaOH \ \longrightarrow \ R \cdot CO \cdot ONa \ + \ NH_3 \uparrow$$

(f) *Test for nitro group* **(-NO₂)**: A simple colour test will indicate the presence of a nitro group.

A small quantity of the substance (0.1 g) is heated with 5 cc of an approximately 5% solution of titanous chloride in dil HCl (a bluish-purple solution). The heating should be done on the steam bath or over a small flame. Decolourisation of the reagent with in 1-2 minutes indicates the presence of a nitro group.

$$R{-}NO_2 \ + \ Ti^{++} \ \longrightarrow \ R{-}NH_2 \ + \ Ti^{++++}$$

(g) *Test with metallic sodium (CAUTION*)* : Metallic sodium will detect the presence of a free -OH group in alcohols, carboxylic acids, sulphonic acids and phenols.

$$\text{e.g., } 2ROH \ + \ 2Na \ \longrightarrow \ 2\,RONa \ + \ H_2 \uparrow$$

Test: Treat 1 ml of the compound (or 1 ml of a solution of the compound in dry ether or dry benzene) with a SMALL thin slice of freshly cut sodium in a small, dry test tube observe whether hydrogen is evolved and the sodium reacts.

*NB. Metallic sodium must be handled with extreme *caution*and must not be allowed to come into contact with water. Thus, whether the above test is positive or negative the residual sodium should be destroyed by adding the contents of the test tube to a beaker containing ethanol or methylated spirit in the fume cupboard.

Acetyl chloride will also detect the presence of free -OH groups, and also primary and secondary amines. The reactions are usually vigorous. Hydroxyl compounds (except usually tertiary alcohols) react with evolution of HCl

$$\text{e.g.,} \quad ROH \ + \ CH_3COCl \ \longrightarrow \ RO{\cdot}CO{\cdot}CH_3 \ + \ HCl$$

With primary and secondary amines the HCl is absorbed by excess amine, causing salt precipitation.

$$\text{e.g.,} \quad 2RNH_2 + CH_3COCl \rightarrow RNH.CO.CH_3 + RNH_2$$

3.4 Functional Group Identification

The above tests should give an indication of the class to which the compound belongs, i.e., any functional groups (-OH, -COOH, -NH$_2$ etc.) present in the compound will be detected. For complete identification of the compound, it is now necessary to consult a series of tables which list the physical constants (m.p., b.p etc.,) of possible compounds under the appropriate headings (alcohols, acids, amines etc.). This should narrow the search to two or three possibilities and any further distinguishing tests suggested in the text should be carried out. For final identification it is necessary to prepare at least two solid derivatives of the compound and determine if the m.p. are in agreement with those given in the tables.

The groups which are most likely to encounter are listed below together with short summaries of their reactions.

3.4.1 Acidic Compounds

The principle classes of acidic compounds are listed above. The difference between strong acids and weakly acidic pseudo acids (e.g., phenols, enols, nitroalkanes) should be made by observing the nature of the reaction with sodium hydrogen carbonate solution.

(A) **Carboxyl group - carboxylic acids (R-CO.OH):** The possibilities are formic, acetic, propionic, chloroacetic, oxalic, succinic, lactic, tartaric, citric, benzoic, salicylic, phthalic, cinnamic, toluic, phenyl acetic, chloro-, bromo- and nitro- benzoic acids and acetyl salicylic acid.

Reactions: (i) Soluble in cold dilute NaOH (ii) soluble in dilute sodium hydrogen carbonate with evolution of CO_2 (distinction from phenols) (iii) may react with conc. sulphuric acid (iv) loss of CO_2 when heated with soda-lime and give corresponding decarboxylated compound.

Confirmatory test for carboxylic acid-ester formation: Warm a small amount of the unknown acid with 2 parts of ethanol and few drops of concentrated sulphuric acid for two minutes, cool and pour the contents into aqueous sodium carbonate solution contained in an evaporating dish, and smell immediately. An acid usually yields a sweet, fruity smell of an ester.

Derivatives :Amide, anilide, esters (if solid), S-benzyl-isothiouronium salt and p-toluidides.

3.4.2 The Hydroxyl Group (-OH)

(a) *Alcohols*: R-OH where R is an alkyl group. Most of these aliphatic alcohols are liquids at room temperature and you may expect to encounter any in the series from C_1 to C_6 (methanol to hexanol). Other possibilities are ethylene glycol, benzyl alcohol ($C_6H_5.CH_2OH$), and cyclohexanol ($C_6H_{11}.OH$).

   Reactions of alcohols: (i) The lower alcohols react vigorously with sodium in the cold with evolution of H_2. The higher alcohols also react but may require heating (Note: The alcohol must be free from water). This reaction with sodium should have been noted during the Lassaigne test. (ii) Alcohols react with acetyl chloride with evolution of heat (iii) methyl carbinols (CH_3-CH(OH)-R) give iodoform test- thus isopropanol, sec-butanol etc. react.

   Derivatives - 3,5-dinitrobenzoate, acetates.

(b) *Phenols*: R-OH where R is an aromatic (aryl) radical. The following may be encountered : Phenol, o-, m-, and p-cresol, - and -naphthol, chloro-, bromo- and nitrophenols, catechol, resorcinol and hydroquinone, salicylic acid, salicylate esters, vanillin.

Reactions: (i) Soluble in dilute NaOH but no effervescence with dilute sodium bicarbonate (ii) Often give a white precipitate of bromo derivative with bromine water (iii) Colour reaction with ferric chloride (iv) Form azodye when coupled with benzene diazonium chloride solution (v) some phenols will form phenolphthalein when fused with phthalic anhydride and conc. H_2SO_4.

Confirmatory tests for phenols - Ferric chloride solution test: Dissolve or suspend 0.05 g of the unknown compound in water or aqueous methanol and add 2 drops of neutral ferric chloride solution and observe the colour.

Phenols and enols give colours ranging from green through red to purple.

Bromine water: Many phenols yield solid bromo compounds on the addition of bromine water. Dissolve or suspend 0.20 g of the unknown compound in 5 ml of dilute hydrochloric acid or of water, and add bromine water dropwise until decolourisation is slow, a white precipitate of the bromophenol may form.

Phthalein test: Many phenols yield phthaleins, which give characteristic colourations in alkaline solution, when fused with phthalic anhydride and a small amount of concentrated sulphuric acid. Place in a dry test-tube 0.2 g of the compound and 0.1 g of phthalic anhydride, mix well together and add 1 drop of conc. sulphuric acid warm the test-tube gently over a low Bunsen-burner flame until a clear, deepred liquid is obtained. Allow the test tube cool and add 5 ml of 5 percent sodium hydroxide solution. After stirring to dissolve all the solid matter pour the contents of the tube into 200 cc of water in a large beaker and examine the colour. Fluorescein gives a yellow-brown solution in alkali which has a strong green fluorescence. If possible examine the solution against a black background or in a dark room by the light of an ultra-violet lamp.

Liebermann's test: A mixture of sodium nitrite (2-3 crystals) and the given compound 0.5 g is heated gently for 20-30 seconds in a dry test tube; cool, and add 1 ml of concentrated sulphuric acid. An intense green (or greenish blue) colouration will be developed. The colour changes to red on addition of cold water (30 ml); Further, the colour becomes deep blue or green upon adding excess of sodium hydroxide solution.

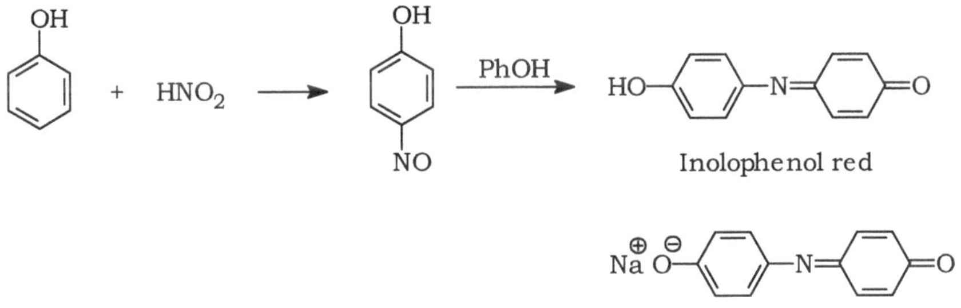

Inolophenol red

Blue indophenol

Derivatives: Benzoyl, acetyl, p-toluene sulphonyl, 2,4-dinitrophenyl ethers, aryloxy acetic acids and bromo derivetive.

3.4.3 Basic Compounds

Organic compounds that dissolve in dilute hydrochloric acid usually contain nitrogen. The most important basic nitrogen compounds are the primary, secondary and tertiaryamines.

Amines : Primary ($R-NH_2$), secondary ($R-NH-R^1$), tertiary (R^3-N).

(a) *Primary aliphatic amines*: React with nitrous acid ($NaNO_2 + HCl$) to give corresponding alcohol and / or olefin $+ N_2$.

(b) *Primary aromatic amines*: Anilines, chloro-, bromo- and nitroanilines, toluidines, anisidines.

Reactions : (i) Often give precipitate of bromo-derivative with bromine water (ii) Best recognised by azo-dye formation for primary aromatic amines.

Confirmatory tests for primary aromatic amines - carbylamine test : Place in a test tube about 0.5 ml of chloroform and a few drops of the amine. Add about 2-3 ml of alcoholic potassium hydroxide solution and warm gently. They often give smell of phenyl isocyanide that is produced.

$$R\,NH_2 + CHCl_3 + 3\,KOH \longrightarrow RNC + 3KCl + 3H_2O$$

Note: Never pour the mixture into a sink; first add an excess of concentrated hydrochloric acid which will hydrolyse the isocyanide to the odourless amine. This may then be safely discarded.

Derivatives : Acetyl, benzoyl, p-toluene sulphonyl, picrate, and azo-dye.

(c) ***Secondary amines*** : Diethyl amine, piperidine, methyl aniline, ethyl aniline, diphenylamine.

   Reations: Do not give the isocyanide reaction but give acyl derivatives. With cold HNO_2 solution ($NaNO_2$ + dil. HCl) they give insoluble, sometimes oily, yellow nitrosamines ($R_2N\text{-}N{=}O$). Do not give azo-dye.

   Derivatives: Acetyl, benzoyl and toluene p-sulphonyl and benzene sulphonyl derivatives.

(d) ***Aromatic tertiary amines*** : Dimethyl aniline, diethyl aniline.

   Reaction: With HNO_2 give nuclear substituted p-nitroso compounds which turn green on addition of NaOH.

   Derivatives: p-Nitroso compound, Picrate.

3.4.4 Neutral Compounds

The carbonyl group $\left(\!>\!C{=}O\right)$

(A) ***Aldehydes R-CHO***: The following may be encountered : The lower aliphatic aldehydes C, to C_4 (acetaldehyde to butyraldehyde), chloral, benzaldehyde; chloro- and nitro- benzaldehydes, salicylaldehyde cinnamaldehyde, anisaldehyde and vanillin.

Reactions: (i) Precipitate with Brady's reagent; (ii) Restore colour to Schiff's reagent. (iii) Form bisulphite addition compounds. (iv) with hot concentrated NaOH solution, aliphatic aldehydes (except H.CHO) polymerise, aromatic aldehydes give the Cannizzaro reaction; (v) Reduce Fehlings solution (vi) Reduce Tollen's reagent; (vii) with alkaline $KMnO_4$, they are oxidised to corresponding acid (viii) acetaldehyde gives iodoform.

Derivatives: 2,4-Dinitrophenyl hydrazone, semicarbazone, oxime.

(B) ***Ketones R-CO-R¹***: The possibilities are acetone, methyl ethyl ketone, diethyl ketone, acetophenone ($C_6H_5\text{-}CO\text{-}CH_3$), benzophenone

($C_6H_5.CO.C_6H_5$) cyclohexanone, cyclopentanone, benzoin ($C_6H_5.CH(OH).CO.C_6H_5$), and benzil ($C_6H_5.(CO)_2.C_6H_5$).

Reactions: (i) Precipitate with Brady's reagent; (ii) Methyl ketones (CH_3-CO-R) give iodoform; (iii) some methyl ketones form bisulphite compounds; (iv) Do not give Schiff's test (some methyl ketones are said to do so slowly); (v) Do not reduce Tollen's reagent or Fehling's solution.

Silver mirror test or **Tollens test:** The ammoniacal solution of silver hydroxide required for the test is not stable and is made up fresh as required. The best results are obtained when a large excess of ammonia is avoided and when the test tube used is thoroughly clean.

Test: To 3 cc of dilute silver nitrate solution add drop by drop, a very dilute solution of ammonia (1 cc of conc. ammonium hydroxide solution in 10 cc of water) until the first precipitate just dissolves. The resulting solution is known as Tollen's reagent. Add a single drop of the pure liquid to be tested, or 3,4-drops of a solution. Quickly mix by shaking, and let the tube stand undisturbed. If the separation of silver is slow the tube may be warmed gently in a beaker of hot water. This is **Tollen's test**.

Schiff's test: Aldehydes produce a pink colour, while ketones show no effect. Some aromatic aldehydes (e.g., Vanillin) give a negative result.

Schiff's reagent is a solution of p-rosaniline hydrochloride, which is normally megenta in colour, but by dissolving freshly prepared, saturated aqueous solution of sulphur dioxide, resulting in a removal of the colour or pale yellow. Aldehydes oxidize the colourless compound back to the coloured dye and is itself reduced to an alcohol. Many other substances also do this, but ketones do not, except for acetone.

Test: Dilute about 1 cc of the colourless Schiff's reagent with 5 cc water and add the drop of the substance to be tested (or a few drops of a solution) and observe the result.

Confirmatory test for ketones - Iodoform test: Methyl ketones give a positive iodoform reaction. Some aldehydes, especially compounds containing the $CH_3.CO$ grouping (e.g., acetaldehyde) give positive iodoform test. Alc. having the structure CH_3-CH(OH). R, which undergo oxidation to the corresponding methyl ketone, also slowly gives a positive test. Dissovle 0.2 g or 9 to 10 drops of the unknown compound in 4 ml of water, if it is insoluble in water, add sufficient dioxane to produce a clear solution. Add 4 ml of 5 percent sodium hydroxide solution and then add a potassium iodide-iodine reagent dropwise with shaking until a definite dark colour of iodine persists. If no iodoform separates after 2-3 minutes, warm the test tube in a beaker of water at 60°C. Add more drops of iodine reagent until the dark colour is not discharged after 2 minutes heating at 60°C. Allow to stand for 10-15 minutes. The test is positive if a yellow precipitate of iodoform is deposited.

Confirmatory test for the carbonyl group

Most aldehydes and ketones react with 2,4-dinitrophenyl hydrazine in acid (Brady's reagent) to form yellow, orange or red precipitate.

The equation for the preparation of this derivative is

$$R_2C{=}O + \text{(2,4-dinitrophenyl hydrazine)} \longrightarrow \text{(hydrazone)} + H_2O$$

Test: Add 0.5 cc (10 drops) of the aldehyde or ketone to 10 to 15 cc of Brady's reagent (2,4-dinitrophenyl hydrazine in methanol). If precipitation of the product does not take place immediately or on scratching, the mixture should be heated for 5 minutes on the steam bath. Cool thoroughly and collect the crystalline precipitate on the Buchner funnel. Wash with a little methylated spirit, dry, and determine melting point.

NB: Occationally the precipitate is oily at first, but this becomes a crystalline solid upon standing.

Note: The above reagent is very dilute and is intended for qualitative test. It is hardly suitable for the preparation of crystalline derivatives except in very small quantities.

If the test is positive, it will be necessary to test the original substance with Fehling's solution, Schiff's reagent, ammonical silver nitrate (Tollen's test) etc., in order to determine whether the substanace is an aldehyde or a ketone.

NB: Aromatic amines sometimes give a precipitate with Brady's reagent. This precipitate is the amine sulphate formed by reaction of the amine with sulphuric acid present in the Brady's reagent.

Fehling's solution: This solution is prepared from copper sulphate, sodium hydroxide,and sodium potassium tartrate, and contains a complex cupric salt, which is probably produced from the interaction of the cupric hydroxide first formed with the tartrate. The copper is reduced by the aldehyde from the cupric (Cu^{++}) to the cuprous (Cu^{+}) condition and precipitates as red cuprous oxide (Cu_2O). The solution does not keep well and hence is prepared whenever required by mixing a solution of copper sulphate ('**solution 1**') with a solution containing the alkali and the tartrate ('**solution 2**').

Test: To a mixture of 5 cc each of solutions **1** and **2** in a test tube add three drops of the unknown substance to be tested (or a correspondingly greater quantity of a solution), heat the tube gently in a beaker of boiling water and observe the result.

Silver mirror test: The ammoniacal solution of silver hydroxide required for the test is not stable and is made up fresh as required. The best results are obtained when a large excess of ammonia is avoided and when the test tube used is thorougly clean (boil it out with nitric acid, wash, let it soak for a minute or two with 10 percent sodium hydroxide solution, and wash thoroughly).

Iodoform Test: Methyl ketones, i.e. compounds containing the CH_3CO grouping give a positive iodoform reaction. Dissolve 0.1 g or 4-5 drops of the given compound in 2 ml of water, if it is insoluble in water, add sufficient dioxane to produce a homogeneous solution. Add 2 ml of 5 percent sodium hydroxide solution and then introduce a potassium iodide – iodine reagent dropwise with shaking until a definite dark color of iodine persists. Allow to stand for 2-3 minutes if no iodoform separates at room temperature, warm the test tube in a beaker of water at 60°C for a few minutes. Iodoform CHI_3, will precipitate as yellow crystals or a yellow turbidity if CH_3CO is present. Filter off the yellow precipitate, dry upon pads of filter paper and determine the melting point of iodoform melts at 120°C.

(C) **Quinones:** The possibilities are benzoquinone, anthraquinone, 1,2-naphthaquinone, 1,4-naphthaquinone etc.

The following tests will be found useful for the detection of simple quinones. Quinones are coloured crystalline solids; they are usually insoluble in water but soluble in ether. The carbonyl groups of quinones often do not react in a normal way with carbonyl group reagents, because of their oxidising properties. Thus p-benzoquinone reacts with 2,4-dinitrophenyl hydrazine hydrochloride in hot alcoholic solution to give 2, 4′ -dinitrophenyl-4-azophenol, m.p. 185-186°C.

$$(NO_2)_2\ C_6H_3\text{-}NH\text{-}NH_2 + C_6H_4O_2 \rightarrow (NO_2)_2\ C_6H_3\text{-}N{=}N\text{-}C_6H_4OH$$

Test for quinones - Hydroiodic acid: Substances of the p-benzoquinone type liberates iodine from hydroiodic acid. Dissolve 0.1 g of the unknown quinone in a little methylated spirit and add 10 ml of 10 percent aqueous potassium iodide solution to a mixture of 5 cc of ethanol and 5 cc of concentrated hydrochloric acid and observe. Iodine is liberated. Anthraquinone does not liberate iodine from potassium iodide solution.

Reduction with zinc powder and sodium hydroxide: Dissolve or suspend 0.1 g of the unkown quinone in 10 percent sodium hydroxide solution and add zinc powder upon boiling the contents of the mixture a red colour is produced; upon shaking the colour of the solution disappears owing to aerial oxidation and the quinone reappears as almost colourless.

Oxidation - Tollen's solution: To 3 cc of ammoniacal silver nitrate solution add, a few drops of cold aqueous quinone solution and observe. A silver mirror or usually a dark precipitate of metallic silver is formed in the cold.

Derivatives of quinones: Semicarbazone, oximes and reductive acetylation.

Ester group - R-CO-OR[1]: Methyl, ethyl, propyl, butyl, phenyl and benzyl esters of the acids listed above may be encountered. Most are liquids. The presence of an ester functional group may be established chemically by applying the hydroxamic acid test. (Note : the hydroxamic acid test is also given by acid chlorides and some primary aliphatic amides which are readily converted by hydroxyl amine hydrochloride with hydroxamic acids).

Reactions: (i) Esters are generally insoluble in water, soluble in ether and fairly inert. (ii) Recognised by the hydroxamic acid test (see the test for ester linkage mentioned above) (iii) To identify, hydrolysis to the corresponding acid and alcohol and identify these separately.

Derivatives of esters: Acid and alcohol, amide, anilide (i) Esters are hydrolysed by refluxing with aqueous or alcoholic solution of potassium or sodium hydroxide (4-5 M). The reaction may be completed in 30 minutes.

(D)   Amide Group - R-CO-NH$_2$: Aliphatic amides are generally colourless crystalline solids, readily soluble in water e.g., Acetamide, urea etc.

Acetamide (CH$_3$CONH$_2$)

(i)   To 0.1 g of freshly prepared mercuric oxide, add 5 cc of an aqueous strong solution of acetamide. The yellow solid dissolves forming (CH$_3$CONH$_2$)$_2$Hg.

(ii) On boiling with bromine water and sodium hydroxide solution, methylamine, having ammonical smell is produced.

Urea (NH$_2$CONH$_2$)

(i)   *Biuret test* : Heat gently 0.5 g of solid in a dry test tube until the molten mass once again solidifies (ammonia gas will be evolved). Cool the mass, add 2 ml of water and warm. Add one drop of very dilute copper sulphate solution followed by excess of aqueous sodium hydroxide solution, a violet colour results.

Confirmatory test for amides: Amide group can be detected easily by heating with caustic soda, it yields ammonia. Suspend 0.2 g of the solid in 2 ml of water. To this add 2 ml of 10 percent sodium hydroxide solution and warm the mixture. Ammonia given off, if the original substance contains an amide group.

$$RCONH_2 + H_2O \longrightarrow R\text{-}COOH + NH_3 \uparrow$$

Derivatives of amides: Nitrates, oxalates, picrates.

Anilides: Acetanilide-(CH$_3$CONH-C$_6$H$_5$). White rhombic plates, sparingly soluble in water.

Test (i): Isocyanide reaction: To a strong aqueous solution of given compound add slightly excess sodium hydroxide solution and shake well. Emulsion is formed. Add 5 drops of chloroform and heat the mixture gently. Intolerable offensive smell of isocyanide is given out.

***Test* (ii)**: Place a mixture of original substanace and add about double the weight of sodium nitrite, in a basin. Moisten with 3 drops of conc. hydrochloric acid, when yellow colour changing over to green on heating.

Derivatives of Anilides(i) p-Bromo-acetanilide, (ii) p-nitro-acetanilide.

These are not essential if the acid and amine obtained on hydrolysis are crystalline compounds : but crystalline derivatives of the acid and amine can be prepared by the standard procedure for acids and amines respectively.

(E) Alcohols and Ethers (R-OH and R-O-R^1): If the unknown, neutral, oxygen containing substance does not give the reactions for aldehydes, ketones, esters and anhydrides, it is more likely either an alcohol or an ether. Chemically, alcohols and ethers may be simply differentiated by the use of two reagents - metallic sodium and acetyl chloride.

Reaction with metallic sodium: Sodium reacts with alcohols with the evolution of hydrogen.

$$2\,\text{R-OH} + 2\,\text{Na} \rightarrow 2\,\text{RNO}^-\,\text{Na}^+ + \text{H}_2\uparrow$$

Whereas dry ethers do not react with sodium.

Test: Treat 1.0 cc of the dried unknown compound with a thin slice of freshly cut metallic sodium (handle with care) in a small dry test tube. Observe whether hydrogen is evolved when the sodium reacts.

The acetyl chloride test: Acetyl chloride reacts readily with primary and secondary alcohols with the evolution of hydrogen chloride. Whereas ethers are not reactive by acetyl chloride. In a dry test tube place 0.5 cc of the dried unknown compound with 0.3-0.4 cc of pure acetyl chloride and note whether reaction has taken place. Add 3 cc of water and neutralise the aqueous layer with solid sodium hydrogen carbonate and observe whether the product is different from that of the original alcohol.

Differentiation between primary, secondary and tertiary alcohols: The three classes of alcohols differ in their behaviour on oxidation with hot acidic dichromate solution. Primary alcohols yield aldehydes, and secondary alcohols yield ketones. Tertiary alcohols are uneffected under the conditions of the reaction.

Derivatives of alcohols - 3,5-Dinitrobenzoates, p-nitrobenzoates

Derivatives of ethers :The low reactivity of aliphatic ethers gives problem in the preparation of appropriate crystalline derivatives. Commonly encountered aromatic ethers (e.g., diphenyl ether) are limited in number.

Cleavage with a hydriodic acid: Aromatic ethers undergo fission when heated with constant boiling point hydriodic acid.

$$ArOR + HI \longrightarrow Ar.OH + RI$$

The cleavage products are a phenol and an alkyl iodide, which will serve to identify the ether.

Picrates: Ethers of many polynuclear aromatic systems are characterised as their picrates prepared similar to the method described for the corresponding derivates in aromatic hydrocarbons.

(F) The Nitro Group (R-NO₂): Only aromatic nitrocompounds such as the following will be encountered viz., nitrobenzene, nitrotoluenes, nitroanisoles,   m-dinitrobenzene, nitroaniline, nitrobenzoic acids, chloro and bromo nitrobenzenes. The nitro compounds are reduced by zinc and ammonium chloride solution to the corresponding hydroxylamines, which may be detected by their reducing action upon an ammonical solution of silver nitrate or Tollen's reagent.

$$R . NO_2 + 4[H] \xrightarrow[\text{NH}_4\text{Claq}]{\text{Zn}} R.NHOH + H_2O$$

Derivatives of nitro groups: Amines

(G) Carbohydrates {Cx(H2O)}y: Mono and di-saccharides (or related compounds) are colourless solids or syrupy liquids which are freely soluble in water, practically insoluble in ether and other organic solvents and neutral in reaction. Polysaccharides are generally insoluble in water because of their molecular weight treatment with concentrated sulphuric acid usually produces excessive charring.

Confirmatory test for polyhydric alcohols (Carbohydrates)

(i) *Molisch Test*: This is a general test for carbohydrates. Place 10 mg of the given compound in a test tube containing 1 cc of water and add 4 drops of 10% solution of 2-naphthol in alcohol or in chloroform. Pour carefully 2 ml of conc. sulphuric acid down the side of the inclined test tube so that the acid forms a layer beneath the aqueous solution without mixing it. If a carbohydrate is present, a red ring appears at the common surface of the liquids, the colour quickly changes on standing or shaking a dark solution being formed.

(ii) *General reaction with sulphuric acid* : Warm about 0.1 g of the given compound with few drops of conc. sulphuric acid using a **small flame**. Immediate blackening is observed. As the temperature is raised, CO_2, CO and SO_2 are given off.

Monosaccharides : Glucose, Fructose

Disaccharides : Maltose, Lactose, Sucrose (Cane sugar)

Poly saccharides : Starch and cellulose

Test for mono saccharides and disaccharides

(iii) *Barfoed's reagent* : Heat the test tube containing 2 ml of Barfoed's reagent and 2 ml of aqueous solution of the carbohydrate on a water bath.

If red copper(I) oxide is formed within 2 minutes a monosaccharide is present. If there is no red precipitate a disaccharide is present. Disaccharides on prolonged heating (about 10 minutes) may also cause reduction, owing to partial hydrolysis to mono saccharides.

(iv) *Fehling's solution*: Place 5 cc of Fehling's solution in test tube and heat to gently boiling. Add a solution of 0.1 g of the given compound in 2 ml of water and boil it for 2 to 3 minutes and observe the results (a) yellow or red precipitate glucose, fructose and maltose (b) No precipitate for cane sugar and starch.

(v) *Test for glucose and fructose*: Take equal quantities of selenium dioxide and the given compound in a test tube. Add 2 cc of dilute hydrochloric acid and heat it in a water bath. (a) Red precipitate indicates the presence of fructose (b) no red precipitate indicates glucose.

(vi) *Test for cane sugar and starch*: Take 1 cc of aqueous solution of the original solid in a test tube and add 1 cc of dilute solution of iodine. (a) blue colour indicates the presence of starch (b) no blue colour for cane sugar.

Derivatives of carbohydrates: 2,4-Dinitrophenyl hydrazone, osazone, acetates.

(H) **Hydrocarbons (benzene, toluene, naphthalene, anthracene etc.)**: Hydrocarbons may be differentiated by their solubility in sulphuric acid since **unsaturated hydrocarbons are soluble** in concentrated sulphuric acid. The two tests employed for the detection of unsaturation are decolourisation of a dilute solution of bromine in chloroform or dichloromethane and reaction with dilute aqueous potassium permanganate.

(i) *Bromine test*: Dissolve 0.2 g or 0.2 cc of the given unknown substance in 2 cc of dichloromethane or chloroform and add a 2% solution of bromine in CH_2Cl_2 or $CHCl_3$ dropwise until the bromine colour persists for one minute.

(ii) *Potassium permanganate test*: Dissolve 0.2 g or 0.2 cc of the given compound in 2 cc of acetone. and add 2% potassium permanganate solution dropwise. The test is negative if no more than 3 drops of the reagent are decolourised.

$$Ph.CH=CH.Ph \xrightarrow[H_2O]{KMnO_4} Ph.CHOH-CHOH.Ph$$

(iii) *Aluminium chloride test*: Take 0.5 gm in a dry test tube and heat till sublimes; cool it and add from the sides of the test tube freshly prepared solution of 0.1 gm or 2 or 3 drops of hydrocarbon in chloroform (1 ml) and observe the characteristic colour develops

Orange to red	-	Monocyclic
Blue colour	-	Bicyclic
Purple or green	-	Tricyclic

Differentiation between alkanes and aromatic hydrocarbons: The most satisfactory reagent is fuming sulphuric acid; place 2 cc of 20% fuming sulphuric acid in a dry test tube, add 0.5 cc of the unknown hydrocarbon and shake vigorously. Only the aromatic hydrocarbons dissolves completely, with the evolution of heat, but excessive charring should be absent warm the test tube gently, cool and pour carefully on to ice water; the aromatic hydrocarbons which undergo sulphonation gives a homogenous aquous solution.

No satisfactory test available for alkanes. A pure sample may be characterised by consideration of such physical properties such as the boiling point, the refractive index or the density etc.

(I) Amino Acids: glycine, cystine, anthranilic acid and sulphanilic acid etc. These are in general insoluble in organic solvents, sparingly soluble in ethanol, usually soluble in water. Soluble in sodium hydrogen carbonate with evolution of carbon dioxide. They do not exhibit sharp melting points, but decompose on heating. Readily soluble in both dilute alkali and dil. acid.

Confirmatory tests for the presence of amino acid

(a) *The Ninhydrin test*: Heat a solution of the given substance with a few drops of 0.25 percent aqueous solution of ninhydrin and observe. a-amino acids give a blueviolet colouration. This highly sensitive test is also given by some β-amino acids, particularly on warming.

(b) *Nitrous acid test*: The conditions of the test are the same as described for the classification of primary amines, but using AcOH in place of HCl.

(c) Derivatives of α-amino acids: Benzoyl derivative and toluene-p-sulphonyl derivatives.

3.5 Preparation of Derivatives for Identification

Before the development and wider application of spectroscopic methods for the determination of structure, confirmation of the class of an unknown organic substrance was completed by the preparation of two or more crystalline functional derivatives. Derivative preparation provides the organic chemist with an important area of study for the following reasons.

Firstly, the combination of spectroscopic data and reactivity tests enables the classification of the compound to be made with greater accuracy, leading to derivative selection to be made with confidence.

Secondly, the description of the general procedure given under provides an excellent opportunity for the student to explore on the small scale preparations with optimum reaction conditions. The requirements of a satisfactory derivative needs the following.

(1) Always prepare **solid** derivatives because (a) the most convenient physical property to measure is the m.p.; (b) b.p.'s are much dificult to determine, particularly on a small scale; (c) liquids are more difficult to purify than solids; (d) solids in general can be satisfactorily purified by crystallisation and recrystallisation from a suitable solvent.

(2) The most satisfactory derivatives (i) are easy to prepare; (ii) are easily purified; (iii) have m.p.'s well separated from those of other possible derivatives in the same series.

(3) Derivatives with m.p.'s below 50-60°C, although they may be easily prepared, are sometimes difficult to crystallise. It is therefore often advisable to choose derivatives with m.p.'s above 60°C.

Thus when a known substance (e.g., liquid) has been investigated and the tests and reactions indicate that it might be, for example, alcohol, then the identification is completed by preparing derivatives and determining their m.p.'s. Example for ROH, suitable derivatives would be (1) Iodoform m.p. 119°; (2) the 3,5-dinitrobenzoate, m.p. 92°C (the ethyl ester of 3,5-dinitrobenzoic acid), (3) the 2,4-dinitrophenyl ether, m.p. 87°C (2,4-dinitrophenyl-ethyl ether). Normally only **two** derivatives will be required.

3.5.1 Alcohols (R-OH)

(i) 3,5-Dinitrobenzoate

$$R{-}OH \quad + \quad \underset{O_2N}{\overset{O_2N}{\bigcirc}}{-}COCl \quad \longrightarrow \quad \underset{O_2N}{\overset{O_2N}{\bigcirc}}{-}COOR \quad + \quad HCl$$

Procedure: About 200 mg 3,5-dinitrobenzoyl chloride taken with 100 mg of given alcohol and the mixture boiled gently on a water bath for 10 minutes, 10 ml distilled water is added and the ester becomes solid on cooling in an ice bath. The precipitate is collected on a filter, washed with 5 to 10 ml of 2% sodium carbonate solution and recrystallised from a mixture of ethyl alcohol and water. After cooling the crystals have been removed by filtration and dried on a filter paper or porous plate, the melting point is determined.

Note: Other benzoates also can be prepared by taking appropriate acid chlorides.

(ii) Acetates

Procedure: About 50-100 mg of alcohol is mixed with acetic anhydride (1 ml) along with few drops of pyridine and the mixture is boiled for 3 to 5 minutes (if possible use air condenser). The mixture is cooled and poured into a beaker containing a few ml of cold water. The acetyl derivative is removed by filtration, washed with cold 2% hydrochloric acid, and then washed with water. Crystallise from a suitable solvent (water, aquous alcohol or alcohol).

3.5.2 Phenols (R-OH), where R is an Aromatic Ring

Derivatives: Benzoates, acetyl, p-toluene sulphonyl, 2,4-dintrophenyl ethers, aryloxy acetic acid, bromoderivatives, bromobenzoates and acetates can be prepared by using the procedure described under alcohols.

Benzoate: It can be prepared from phenol with benzoyl chloride in aqueous sodium hydroxide.

(i)   **Procedure for benzoyl derivative**: To a mixture of 200 mg of the given phenol with 2% sodium hydroxide solution (3 ml) is added benzoyl chloride (3 × 1 ml). After each addition of benzoyl chloride the mixture is shaken thoroughly and cooled under the tap if the mixture gets warm. Finally shake for 8 to 10 minutes when the smell of the benzoyl chloride disappears and the solution may be alkaline (check this). If the benzoyl chloride is still present, add some more of sodium hydroxide solution (2 ml), shake for a further 5 minutes. A solid separates at this stage; filter the solid on Buchner funnel, wash with water, and crystallise from alcohol or rectified spirit. The melting point is determined.

(ii) **Aryloxyacetic acids**: Place the given phenol (0.5 g) in a test tube along with 30% sodium hydroxide solution (2.5 ml) and add chloroacetic acid (0.75 g). The mixture is shaken thoroughly and if necessary water may be added in order to dissolve the sodium salt of the phenol. The test tube containing the mixture is then kept in a beaker of boiling water for 45 minutes to one hour. The mixture is cooled, diluted with water (5 to 10 ml), acidified to congo red with dilute hydrochloric acid and extracted with 25 ml of ether. The ether solution is washed with 10 ml of cold water and is then shaken with 5% sodium carbonate solution. The sodium carbonate solution is acidified with dilute hydrochloric acid; the aryloxy acetic acid is then collected on a filter under suction and recrystallised from hot water. The melting point is determined.

$$ArOH + ClCH_2COOH \xrightarrow[-Cl]{NaOH} ArOCH_2COO^- \xrightarrow{H^\oplus} ArOCH_2COOH$$

(iii) Toluene p-sulphonate: Toluene p-sulphonyl chloride reacts with phenols to give toluene p-sulphonates.

$$p - CH_3C_6H_4SO_2Cl + ArOH \rightarrow p - CH_3C_6H_4SO_2OAr + HCl$$

Procedure: Place 0.2 g of the phenol in a test tube along with 0.5 ml of pyridine and 0.4 g of toluene p-sulphonyl chloride. The mixture is heated on a water bath for 15 minutes.* Pour the contents into a beaker containing 25 ml of cold water and stirred until the oil solidifies. Filter the solid under a suction pump, wash with cold dilute hydrochloric acid followed by dilute sodium hydroxide (to remove pyridine and phenol respectively) and then with cold water. Recrystallise from methanol or ethanol.

3.5.3 Carboxylic Acids

Derivatives: Amides, anilides and p-toluidides.

(i) **Amide:** Acids are best converted into amides or anilides using the acid chloride, acid anhydrides, react directly with ammonia and amines.

The acid chloride is conveniently prepared by refluxing the acid (100-150 mg) with excess of thionyl chloride (2 ml) for 15-25 minutes*. Then add the acid chloride (1 ml) dropwise to strong ammonia solution (1-2 ml, cool it before). The precipitate amide is collected by filtration, washed with a little water and crystallised from a suitable solvent/ethanol, aqueous ethanol or water (occasionally).

$$R-CO_2H + SOCl_2 \longrightarrow R-COCl + SO_2 + HCl$$

$$R-COCl + 2NH_3 \longrightarrow R-CONH_2 + NH_4Cl$$

* It is advisable to place a plug of cotton wool.

(ii) **Anilide**: Add aniline (0.2 ml) to the acid chloride prepared above (1 ml), shake and after a few minutes add dilute hydrochloric acid to dissolve the excess of amine and stir the mixture thoroughly. Filter the crude product, washed with cold water. Recrystallise from hot water, dilute ethanol or ethanol.

$$R-COCl + H_2N-\langle\rangle \longrightarrow R-CO-NH-\langle\rangle + HCl$$

(iii) **p-Toluidides:** Proceed as under anilides, but substitute p-toluidine for aniline.

3.5.4 Primary, Secondary and Tertiary Amines

Derivatives: Acetyl, benzoyl, p-toluene sulphonyl, picrate and azo-dye.

(i) **Acetyl derivatives:** Primary and secondary amines are best acetylated with acetic anhydride.

$$NH_2 \quad + \quad (CH_3CO)_2O \quad \longrightarrow \quad NHCOCH_3 \quad + \quad CH_3COOH$$

Procedure: Dissolve in a test tube 100 mg of amine in glacial acetic acid (0.2 ml) and acetic anhydride (0.2 ml) and heat the mixture gently on a water bath for 5 minutes. After cooling, add water (2 ml) dropwise while shaking the reaction mixture. If the product does not appear at this stage, basify the solution with strong ammonia solution. Cool in ice to crystallize the derivative, then filter off and recrystallise from water or aqueous ethanol.

(ii) **Benzamides of primary and secondary amines**

$$RNH_2 + C_6H_5COCl + NaOH \longrightarrow RNHCOC_6H_5 + NaCl + H_2O$$

$$R_2NH + C_6H_5COCl + NaOH \longrightarrow R_2NCOC_6H_5 + NaCl + H_2O$$

Procedure: A mixture of about 0.5 g of the unknown primary or secondary amine, 10 ml of 2 M sodium hydroxide solution and 0.5 ml of benzoyl chloride (fuming cupboard) are placed in a stoppered test tube and shaken thoroughly. The stopper should be removed continously at regular intervals to release any pressure build-up. The tube is agitated continuously for few minutes. After the odour of benzoyl chloride has disappeared, the reaction mixture is cooled. The solid obtained is filtered and washed with water thoroughly. It is recrystallised from ethanol.

Note: Benzoyl chloride is carcinogenic. Hence any solution containing unreacted benzoyl chloride must be neutralized with ammonium hydroxide prior to its disposal down the sink.

(iii) **p-Toluene sulphonyl derivative:**

Procedure: Place the unknown amine (0.5 ml) or (0.5 g) in a test tube along with 8 ml of 10% sodium hydroxide solution and 0.6 to 0.7 ml of p-toluene sulphonyl chloride. After addition of sulphonyl chloride, stopper the test tube, shake well for 2-3 minutes (if the reaction mixture heats up considerably it should be cooled). Test the solution to make sure that it is alkaline (with litmus paper). After all the p-toluene sulphonyl chloride has reacted, cool the solution and filter or decant from any residue. (Note that the solubility of the residue in HCl indicates that the given compound was a tertiary amine). Acidify the clear filtrate with dilute hydrochloric acid. Some times it is necessary to scratch the test tube to get crystals. Filter the solid and crystallise from 95% ethanol.

(iv) Azo-dye of primary amines (secondary amines do not give azo-dye): These are coloured compounds which may be used as dyestuffs, and are therefore known as azo-dyes on reaction with phenol, for example, p-hydroxyazobenzene, a brown solid. With 2-naphthol gives a red solid called 1-phenylazo-2-naphthol is formend.

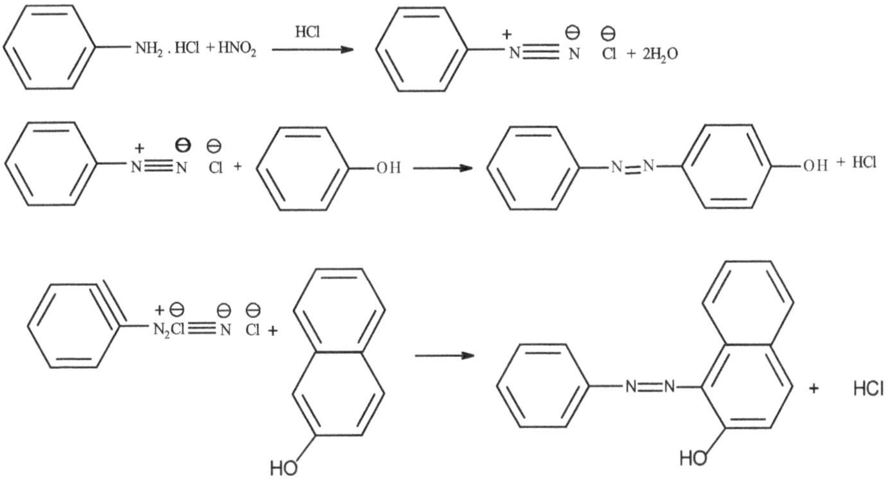

Procedure: In a 100 ml conical flask are introduced 0.5 gm of amine, 3 ml of water and 1.5 ml of concentrated hydrochloric acid. The mixture is heated on a boiling water bath for 5 minutes to form amine hydrochloride. This suspension is cooled in ice water. A cool solution of 0.25 g of sodium nitrite (not sodium nitrate) in 2 ml of water is added in one portion and stirred. A solution of 0.6 g of ß naphthol in 5 ml of 10 percent sodium hydroxide is added with stirring, and the resulting suspension is allowed to stand for 10 minutes. Then 0.5 gm of sodium acetate dissolved in 2 ml of water is added. If the dye is brownish-red in colour, a small amount or few drops of dilute hydrochloric acid should be added (but not an excess) to restore the bright red colour and neutralize excess base. After 10 minutes the dye is filtered and thoroughly washed with water. The product is thinly spread on a large watch glass by breaking any large particles.

Aromatic tertiary amines

Derivatives: p-Nitroso compound, picrate, methiodides. The picrates are suitable for all three types of aromatic amines.

(v) Picrates procedure: Add a sample of the given compound (0.3 g) to 5 ml of 95% ethanol or rectified spirit and 5 ml of cold saturated solution of picric acid in ethanol and heat the solution on a water bath for 5 minutes. Cool the solution and remove the yellow crystals of the picrate by filtration and recrystallise from ethyl alcohol, dilute alcohol or boiling water.

Caution: Some picrates explode when heated.

(vi) **p-Nitroso compound:** Dissolve 0.2 g of tertiary amine in 2 ml of dilute hydrochloric acid (1:1), cool to 0.5°, and slowly add a solution of 175 mg of sodium nitrite in 1 ml of water. After 25 minutes, filter off the precipitated yellow hydrochloride* and wash it with a little dilute HCl. Dissolve the precipitate in the minimum volume of water, add a solution of sodium carbonate or sodium hydroxide to decompose the hydrochloride (until alkaline) and extract the free base with ether. After evaporation of ether, the residual crystals are crystallised from light petroleum ether (b.p. 60-80°) or from benzene.

*Precipitation may not occur with all dialkyl amines. Add a slight excess of sodium carbonate or sodium hydroxide to the solution, extract free base with ether, etc.

$$\text{C}_6\text{H}_5{-}\text{N}{-}\text{R}_2 \ + \ \text{HONO} \ \longrightarrow \ \text{R}_2{-}\text{N}{-}\text{C}_6\text{H}_4{-}\text{NO} \ + \ \text{H}_2\text{O}$$

(vii) **Methiodides of tertiary amines**

Procedure: Mix the amine (100 mg) with methyl iodide (0.2 ml) and keep at room temperature for 5 to 8 minutes and then at 55°C for 5 minutes. Cool the solution in ice water and scratch the test tube if necessary to induce crystallisation. Recrystallise from rectified spirit.

3.5.5 The Nitro Group (R-NO$_2$)

Derivatives : Reduction to primary amines, nitration to a poly-nitro compound.

(i) **Reduction to amines:** Add 0.25 g of the given compound to 0.5 g of granulated tin in a small flask which is connected to a reflux condenser. Add 0.5 ml of 10% hydrochloric acid, in small portions, with shaking after each addition. Warm the mixture on a steam bath for 10 minutes. Decant the hot solution into another flask containing 5 ml of water and add sufficient 40% sodium hydroxide solution to dissolve the tin hydroxide. Extract the solution several times with ether. Evaporate the etherial solution after drying over anhydrous sodium sulpahte.

$$\text{R{-}NO}_2 \ \xrightarrow[\text{HCl}]{\text{Sn}} \ \text{R{-}NH}_2 \ + \ \text{SnCl}_4 \ + \ \text{H}_2\text{O}$$

The primary amine obtained above may be converted to one of the derivatives described under amines.

(ii) **Nitration to poly-nitro compound:** Aromatic mononitro compounds may be characterised by converting them into the corresponding dinitro or trinitro derivatives. The nitration of an aromatic compound must be conducted **with great care**, since many aromatic compounds react violently. It is not possible to predict the conditions necessary to nitrate an unknown compound.

Procedure: Add 0.25 g of the given compound to 1.0 ml of concentrated sulphuric acid. Introduce carefully 1.0 ml of concentrated nitric acid drop by drop, with shaking after each addition. Alternatively, slowly add concentrated sulphuric acid (1 ml) to concentrated nitric acid (1 ml) in a dry boiling tube and cool the mixture to room temperature. While shaking the tube, slowly add the unknown compound (0.25 g). Heat the contents, carefully, on a water bath at 50°C for 5 minutes. Pour the reaction mixture on to 10 g of crushed ice and collect the solid obtained by suction filteration. Recrystallise from dilute ethanol.

$$C_6H_5-NO_2 \ + \ H_2SO_4 \ \xrightarrow{HNO_3} \ O_2N-C_6H_4-NO_2$$

3.5.6 Aldehydes and Ketones

Derivatives : (i) 2,4-Dinitrophenyl hydrazone (ii) semicarbazone, (iii) oxime.

(i) **2,4-Dinitrophenylhydrazone**: The given carbonyl compound (100-200 mg) is dissolved in minimum amount of ethanol (if it is solid) and add to a an acidic solution of 2,4-dinitrophenyl hydrazine (4 to 6 ml) and shake the mixture. If there is no precipitate even after 5 minutes, heat on the water bath for a further period of 5 minutes and cool. If a precipitate still does not form, add a little dilute sulphuric acid. Crystallise from ethanol after filtration by suction. Dilute ethnol or ethyl acetate may be used for recrystallisation. Note the melting point.

$$R{\cdot}CHO \ + \ C_6H_3(NO_2)_2NHNH_2 \ \longrightarrow \ RCH{=}N{-}\overset{H}{\underset{|}{N}}{-}C_6H_3(NO_2)_2 \ + \ H_2O$$

$$R_2{\cdot}CO \ + \ C_6H_3(NO_2)_2NHNH_2 \ \longrightarrow \ R_2C{=}N{-}\overset{H}{\underset{|}{N}}{-}C_6H_3(NO_2)_2 \ + \ H_2O$$

(ii) **Semicarbazone:** Dissolve semicarbazide hydrochloride (100 mg) sodium acetate (150 mg) and aldehyde or ketone (100 mg) in a few drops of water. If the given compound is not soluble in water, add a little ethyl alcohol. If a precipitate form at this stage it may be sodium chloride. Shake the mixture thoroughly and place the test tube in a beaker of boiling water and allow it to cool. Cool it further by keeping it in a beaker of ice-water. If ethyl alcohol was added, it may be essential to dilute with water. Collect the crystals of the semicarbazone by filtration and recrystallise from an appropriate solvent (methanol, ethanol, water or aqueous ethanol).

Semicarbazone

$$RCHO \ + \ NH_2NHCONH_2 \cdot HCl \ \longrightarrow \ RCH{=}NNHCONH_2 \ + \ H_2O$$

(iii) Oximes: Aldehydes form oximes readily but ketones form oximes only after long standing or refluxing.

Procedure: The given carbonyl compound (100 mg) is dissolved in ethanol (1.0 ml) and mix it with a solution of sodium acetate (150-200 mg) and hydroxylamine hydrochloride (100 mg) in water (0.5 to 1.0 ml). Slowly heat the mixture in a boiling tube for 5-10 minutes, then cool the solution and neutralise with an acid (HCl). Filtrer the solid under suction pump and recrystallise from ethanol.

Oximes

$$R\ CHO + NH_2OH.HCl \rightarrow R\ CH = NOH + H_2O$$

$$R_2\ CO + NH_2OH.HCl \rightarrow R_2\ C = NOH + H_2O$$

3.5.7 Esters (R-CO-OR[1])

Esters are usually characterised by hydrolysis followed by identification of the alcoholic and acidic products.

Derivatives: Acid and alcohol, amide, anilide esters are hydrolysed by refluxing with aqueous or alcoholic solution or potassium or sodium hydroxide (4-5 M). The reaction may be completed in 1/2 hour to 1 hour.

Procedure: Place 1 g of the given compound ester and 15 ml of 30% sodium hydroxide solution in a round bottomed flask, and boil them together gently under reflux until the molten ester has completely disappeared (about one hour). The mixture now contains sodium phenoxide and sodium benzoate. Cool the solution in ice and liberate the phenol by addition of dilute sulphuric acid until a faint precipitate of benzoic acid appears and the solution is acidic to litmus paper. Filter the acid under the suction and recrystallise from hot water.

Alternatively, after the solution is acidic to litmus paper, add sodium carbonate solution, stirring vigorously, until the benzoic acid precipitate has been dissolved and the solution is alkaline to litmus paper. Extract the phenol with two 5 ml portions of ether, combine these and dry them over anhydrous sodium sulphate. Filter the solution into a small distilling flask, and distil off the ether, taking the usual precautions. When nearly all the ether has distilled off, pour the hot solution into an evaporating dish and the phenol will crystallise (but do not spill any hot phenol on the hands).

To obtain the benzoic acid, add dilute hydrochloric acid to the aqueous solution which was extracted with ether; filter off the white crystals of benzoic acid, wash them with a little cold water, and recrystallize from hot water.

$$Ph\text{-}O\text{-}CO\text{-}Ph \xrightarrow{\text{NaOH}} Ph\text{-}O^{\ominus}Na^{\oplus} + Ph\text{-}CO_2^{\ominus}Na^{\oplus}$$

Phenyl benzoate   Sodium phenoxide   Sodium benzoate

Dil. Acid

$$Ph\text{-}OH + Ph\text{-}COOH + NaCl$$

Phenol   Benzoic acid

(ii) N-Benzylamides of acids from esters

Esters are converted into the N-benzyl amides of the corresponding acids by heating with benzylamine in the presence of a little ammonium chloride as catalyst.

$$R^1\text{·}COOR^2 + Ph.CH_2NH_2 \longrightarrow R^1\text{·}CONHCH_2\text{·}Ph + R^2OH$$

The reaction proceeds readily when R^2 is methyl or ethyl.

Procedure: Place a mixture of the ester (200 mg), benzyl amine (1 ml) and 30 mg of powdered ammonium chloride in a pyrex test tube fitted with a small condenser. Heat the mixture for 1 hour. Cool the reaction mixture and wash with water to remove any excess benzylamine and to induce crystallisation. Often the addition of a little dilute HCl will help crystallisation (an excess of acid must be avoided since it dissolves N-benzylamide). Filter the solid amide, wash with a little petroleum and recrystallise it from aqueous ethanol, ethyl acetate or acetone.

3.5.8 Amides (Hydrolysis)

Amides may be hydrolysed by boiling with 20 percent sodium hydroxide solution to the corresponding acid after acidification of the sodium salt with dilute hydrochloric acid.

Primary amides: Aliphatic amides are acetamide, urea etc.

Derivatives: Nitrates, oxalates, picrates

(i) **Nitrates:** Dissolve 1 g of the solid in 3 ml of water by heating, cool and add concentrated nitric acid carefully. Collect the crystals, washed with cold water and crystallise from a suitable solvent. Note the melting point.

$$CO(NH_2)_2 + HNO_3 \longrightarrow CO(NH_2)_2 \cdot HNO_3$$

(ii) Oxalate: Mix concentrated aqueous solutions of urea and oxalic acid scratch the sides of the test tube with a rod. On cooling collect urea by filtration and note the melting.

Urea　　　　　　Oxalic acid　　　　　　Oxalyl urea
(or) parabanic
acid

(iii) Picrates: Refer the procedure under amines.

Aromatic amides: Benzamide ($C_6H_5CONH_2$).

Derivative procedure: Heat a mixturte of 0.2 gm of given solid and 4 ml of 10% aqueous sodium hydroxide solution, in a 50 ml flask for about 15 minutes. Cool and acidify with dil. sulphuric acid. Filter the solid, wash with water and purify by recrystalisation from hot water.

$$C_6H_5CONH_2 + NaOH \rightarrow C_6H_5CO_2Na + NH_3 \xrightarrow{H^+} C_6H_5COOH$$

3.5.9 Anilides (R-CO-NHPh)

Derivatives:

(i) p-Bromo-acetanilide

(ii) p-Nitroacetanilide.

(i) p-Bromoanilide: Dissolve by heating 0.2 g of the given compound in 1 ml of glacial acetic acid. Add to it 0.5 ml of bromine, dissolved in 3 ml of glacial acetic acid gradually, with constant shaking. Set aside for about 10 minutes. Dilute the mixture with 10 ml of water, shake vigorously. Filter the solid and purify by crystallisation from aqueous alcohol. Record the melting point.

$$Ph\text{-}NHCOCH_3 + Br_2 \longrightarrow p\text{-}Br\text{-}C_6H_4\text{-}NHCOCH_3 + HBr$$

Acetanilide　　　　　　　　　　　　p-Bromoacetanilide

(ii) p-Nitroanilide: Add 0.25 g of the given compound to 1 ml of glacial acetic acid contained in a boiling tube; introduce into the well stirred mixture 0.5 ml of concentrated sulphuric acid. The boiling tube is cooled in a mixture of ice and salt and then add drop by drop a cold mixture of 0.2 ml of concentrated nitric acid and 0.1 ml of concentrated sulphuric acid, while the temperature is maintained below 10°C. After the mixed acid has been added, remove the tube from the freezing mixture, and allow it to stand at room temperature for half an hour. Pour the contents into 5 ml of cold water, whereby the crude nitro derivative is at once precipitated. Filter the solid with a suction on a Buchner funnel, wash with water to remove acids and purify from industrial spirit. (o-nitroanilide remains in the mother liquour). Record the m.p.

$$Ph\text{-}NH\cdot COCH_3 \xrightarrow[H_2SO_4]{HNO_3} p\text{-}O_2N\cdot C_6H_4\cdot NHCOCH_3$$

3.5.10   Derivatives for Carbohydrates

(i) Osazone

(ii) acetates

(iii) 2,4-dinitrophenyl hydrazone.

The melting points of sugars and some of their derivatives, namely **osazones.**, are not reliable identification as those of a other classes of organic compounds.

(i) **Osazones:** Take 0.2 g of carbohydrate, 0.4 g of pure phenyl hydrazine hydrochloride, 0.6 g of crystalline sodium acetate and 4 ml of water in a boiling tube. Stopper the tube loosely with cork and place it in a beaker containing boiling water for few minutes. Shake the tube occasionally and note the time of immersion and the time of osazone first separates. The precipitate separates quite quickly. Collect the crystals by filtration and note the melting point.

Time of formation of osazone-d-fructose 2 minutes, d-glucose 5 minutes, d-glactose 17 minutes and sucrose 30 minutes.

In case of lactose and maltose osazones are obtained on cooling only.

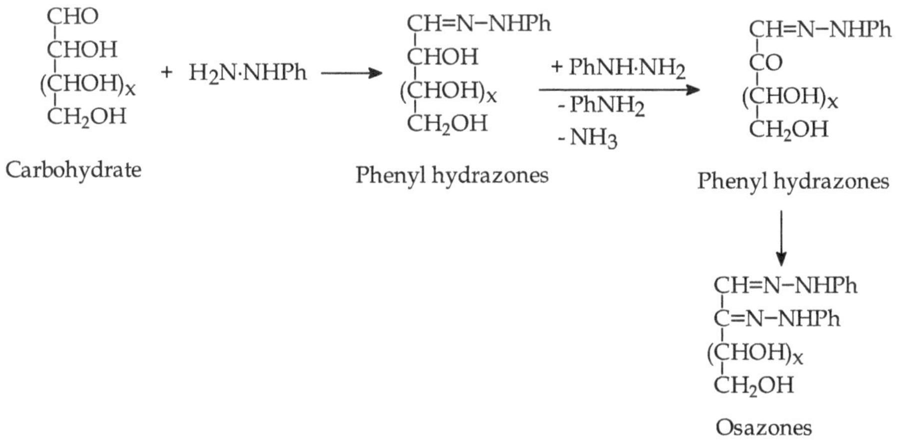

(ii) Acetates: Free hydroxyl groups are acetylated with acetyl chloride or acetic anhydride in the presence of a catalyst (H_2SO_4, fused sodium acetate or anhydrous zinc chloride). Number of acetyl groups introduced would depend upon the hydroxyl groups present.

Procedure: In a dry 50 ml of round bottomed flask fitted with a condenser, place catalyst (0.5 g of anhydrous zinc chloride or 1 g of fused sodium acetate) and 4 ml of acetic anhydride. Heat the contents in a boiling water bath until nearly all solid is dissolved. To this add 0.5 g of the carbohydrate to be acetylated and continue heating for 45 minutes to 1 hour. Pour contents in 30-40 ml of ice cold water. In case oil separates scratch with a glass rod. Filter and wash few ml of ice cold water. Crystallise from dilute alcohol (2:1). Dry and determine its melting point.

$$
\begin{array}{ccc}
\text{CHO} & & \text{CHO} \\
| & & | \\
\text{CHOH} & & | \\
| & \xrightarrow{\text{Ac}_2\text{O}} & \text{CHO Ac} \\
(\text{CHOH})_3 & \text{ZnCl}_2/ & | \\
| & \text{CH}_3\text{COONa} & (\text{CHOAc})_3 \\
| & & | \\
\text{CH}_2\text{OH} & & \text{CH}_2\text{OAc} \\
\text{Glucose} & & \text{Penta acetate of glucose}
\end{array}
$$

(iii) 4-Nitrophenyl hydrazones: This reagent has been used in the characterisation of a number of monosaccharides. e.g., Glucose, Fructose etc.

Procedure: Place 0.25 g of the compound in a test tube containing 3 ml of ethanol, add 0.25 g of p-nitrophenylhydrazine and heat the contents until the reaction appears complete. The p-nitrophenyl hydrazone slowly separates. Filter, wash with a **little cold** ethanol and recrystallise from ethanol.

3.5.11 Amino Acids

Derivatives:

(i) Benzoyl,

(ii) Toluene-p-sulphonyl and

(iii) phthaloyl derivatives and

(iv) 3,5-dinitrobenzoyl derivative.

(i) **Benzoyl derivative procedure:** A mixture of 0.25 g of the unknown amino acid, 5 ml of 10% sodium hydrogen carbonate solution and 0.5 ml of benzoyl chloride is shaken vigorously in a tightly stoppered test tube, remove the stopper from time to time since carbon dioxide is evolved.

When the oily, dense benzoyl chloride has disappeared, acidify with a dilute hydrochloric acid to congo red and filter. Extract the solid with a **little cold water** to remove any benzoic acid that may be present. Recrystallise the benzoyl derivative from hot water or from dilute ethanol. (**Note** : Any unreacted benzoyl chloride should be destroyed by pouring the contents of the test tube into a dilute ammonium hydroxide solution prior to discarding).

(ii) Toluene-p-sulphonyl derivative: Amino acids react with toluene-p-sulphonyl chloride to yield, in many cases crystalline derivatives.

A mixture of 0.1 g of the amino acid, 10 ml of 1 M sodium hydroxide solution and 0.5 g of toluene-p-sulphonyl chloride in 10 ml ether is shaken vigorously for 25-30 minutes. Separate the ether layer; acidify the aqueous layer to congo red with dilute hydrochloric acid. The derivatives crystallise out on standing in ice water. Filter the crystals and recrystallise from 4-5 ml of aqueous ethanol.

$$H_2N-C_6H_4-COOH \ + \ CH_3-C_6H_4-SO_2Cl$$

Glycin   Toluene-p-sulphonyl chloride   NaOH

$$H_3C-C_6H_4-SO_2-NH-C_6H_4-COOH$$

Toluene-p-sulphonyl derivative

3.5.12 Quinones

Derivatives

(i)   Reduction of the hydroquinone

(ii)   Semicarbazones

(iii) Reductive acetylation.

(i)   **Reduction of the hydroquinone procedure:** Suspend 0.2 g of the quinone in 3 ml of ether or benzene and shake vigorously with a solution of 0.5 g of sodium dithionite ($Na_2S_2O_4$) in 5 ml of 30% sodium hydroxide until the colour of the quinone has vanished. Separate the alkaline solution of the hydroquinone, cool it in ice-water and neutralise with concentrated hydrochloric acid. Collect the product (extract with ether, if necessary) and recrystallise it from ethanol or hot water.

3.5.13 Aromatic Hydrocarbons and Aryl Halides

Derivatives of Benzoenoids

Benzenoid hydrocarbons and halogeno hydrocarbons may be identified by their physical properties (m.p., b.p., etc.) and by derivatives resulting from ring-substitution (nitration bromination, sulfonation Friedel-Crafts acylation) and by side-chain oxidation to carboxylic acid. In addition, polycyclic hydrocarbons form crystalline complexes with picric acid.

(i) Nitrohydrocarbons: Most of them are solid. Procedures vary with the reactivity of the ring to electrophilic substitution so that some indication of the probable identity of the unknown substance is necessary.

Mix carefully and cool to room temperature 2 ml of each of concentrated nitric and sulfuric acids. Add the sample (0.2 ml), shake the mixture, and when any initial reaction has subsided, warm the mixture only to 40-50° for 2 to 5 minutes. Pour the mixture into 10 ml of ice cold water; rub the oily product with a glass rod until it solidifies.

$$Ar\text{-}H \; + \; H_2SO_4 \; \xrightarrow{\;HNO_3\;} \; Ar\text{-}NO_2$$

(ii) Side-chain oxidation (substances having alkyl side-chains only)

$$Ar\text{-}R \; + \; KMNO_4 \; \longrightarrow \; Ar\text{-}COOH$$

Add potassium permanganate (1 g) slowly to a heated (boiling water bath) mixture of sample (0.25 g) and 5% sodium hydroxide or sodium carbonate solution (3 ml). Heat the mixture under reflux until the purple colour of permanganate has disappeared (1-4 hrs). Cool the mixture, filter off manganese dioxide, acidify the filtrate with dilute (10%) sulfuric acid. If the acid precipitates on cooling, collect it. Otherwise, extract the solution with minimum amount of ether or CCl_4 and recover the acid by evaporating the solvent layer to dryness. Recrystallize it with boiling water.

(iii) Picrate: The picrates of monocyclic hydrocarbons (benzene, toluene) are in general only formed in solution and are not used as derivatives.

Dissolve naphthalene (0.15 g) in the minimum of hot ethanol and add to 2 ml of a saturated solution of picric acid in ethanol or acetone. Warm the mixture and then cool to allow the product to crystallise. Recrystallize the picrate from minimum of ethanol, ethyl acetate or benzene.

(iv) Sulfonamides

$$Ar\text{-}H \; + \; 2Cl\text{-}SO_3H \; \longrightarrow \; Ar\text{-}SO_2\text{-}Cl \; \xrightarrow{\;(NH_4)_2CO_3\;} \; Ar\text{-}SO_2\text{-}NH_2$$

Add chlorosulfonic acid (1 ml, caution) to an ice-cooled solution of the sample (0.3 g) in chloroform (.15 ml). When reaction subsides, warm the mixture to room temperature for 20-30 minutes, add ammonium carbonate (1 g) cautiously, mix thoroughly, and evaporate the mixture to dryness at 100°C on a boiling water bath. Recrystallise the sulfonamide from aqueous alcohol.

(v) Aroylbenzoic acids

Place a mixture of 0.5 g of the hydrocarbons, 5 ml of dry dichloromethane, 1.25 g of anhydrous aluminum chloride (powdered) and 0.6 g of pure phthalic anhydride in a 25 ml round-bottomed flask, and heat the mixture under reflux for 30 minutes. Cool in ice, dilute the mixture with dilute hydrochloric acid (5 ml). When the reaction has subsided, add 10 ml of water and shake thoroughly (All the solid material should pass into solution). Transfer the solution into a separating funnel and extract with ether. Drive off the ether by warming on a water bath and isolate the crude aroyl benzoic acid. Recrystallise the aroylbenzoic acid from aqueous alcohol or hot water.

(vi) **Aromatic halogen compounds - sulphonamides:** Many aryl halides in chloroform solution, when treated with chlorosulphonic acid affords the corresponding sulphonyl chlorides, which in turn readily converted into the aryl sulphonamides by reaction with conc. ammonia solution or with ammonium carbonate (use the experimental details given above).

(vii) **Aromatic nitro compounds**: Aromatic mononitro compounds may some times be characterised by conversion into the corresponding dinitro or trinitro derivatives.

Caution: The nitration of benzenoid hydrocarbon must be conducted with great care, preferably behind a safety screen fume cup-board, since many aromatic compounds react violently. Add about 0.5 g of nitrobenzene to a mixture of 1 cc each of concentrated nitric acid and sulphuric acid in a test tube. Warm the mixture gently on a water bath (50-60°C) for 5 minutes; shaking the test-tube continuously to ensure good mixing. Now pour the mixture on to about 50 ml cold water, collect the yellow solid which separates by suction filtration. Recrystallise from aqueous ethanol.

(viii) **Reduction to primary amines:** Place a small piece of granulated zinc in a test tube and add 0.25 g (four or five drops) of nitrobenzene followed by 2 cc of conc. hydrochloric acid. Shake the test tube well to ensure thorough mixing during the addition of the acid. Leave the mixture for about 10 minutes and warm the contents on a boiling water-bath with vigorous shaking until the hydrogen produced to reduce the nitrobenzene to aniline i.e., the odour of nitro compound is no longer present. Cool the reaction mixture and make it alkaline with dilute sodium hydroxide solution. Isolate the liberated amine by ether extraction. Characterise the amine by the preparation of a suitable crystalline derivative (see primary and secondary amines).

NO$_2$ → (Zinc / HCl) → NH$_2$ → Acetyl or Benzoyl derivatives

3.5.14 Nitriles : Hydrolysis to the Crboxylic Acid

(i) **Hydrolysis with alkali** : It works best for aliphatic members, while the acid hydrolysis is preferred for aromatic nitriles.

$$R-CN \longrightarrow R-COOH + NH_3$$

Reflux the unknown (1 g) with 50-70% sulphuric acid (3 ml of conc. acid added slowly to 3 ml of water) : hydrolysis will take 45 minutes to 1 hour. It may crystallize out on cooling in ice and filter off. Otherwise extract the acid with little ether, and disti. off the ether and isolate the acid. Conversion to the amide often leads to a satisfactory derivative.

3.6 Separation Techniques in the Purification of Organic Compounds

Solvent Extraction

Isolation of benzoic acid from aqueous solution of sodium benzoate by acidification followed by extraction.

Solvent extraction is an operation very frequently used in practical organic chemistry. In this experiment we apply this technique to the purification of crude organic acid.

Many organic acids (including the one under investigation) are sparingly soluble in water but are readily soluble in organic solvents such as toluene or methylene chloride. In contrast the sodium salts of these acids, which are obtained by reaction of the acids with aqueous sodium hydroxide, are readily soluble in water and insoluble in organic solvents. This is because the salts are ionic while the free acids are relatively undissociated.

$$Ar-COOH \; + \; NaOH \; \longrightarrow \; ArCOO^{\ominus} Na^{\oplus} \; + \; H_2O$$

$$ArCOO^{\ominus} Na^{\oplus} \; + \; HCl \; \longrightarrow \; Ar-COOH \; + \; NaOH$$

This experiment makes use of these differences in solubility. You will be supplied with a solution of an impure* organic acid (ex. benzoic acid) dissolved in toluene.

Chemicals: Benzoic acid, sodium hydroxide, methylene chloride or chloroform, dil HCl.

Experimental: Place the given solution (20 ml) in a separating funnel and add 20 ml of dilute sodium hydroxide (5%) solution. Close the funnel with a stopper and shake the contents thoroughly. Clamp the funnel vertically and allow the two layers to separate. The upper layer is toluene, the lower layer is sodium hydroxide solution.

This treatment converts the organic acid into its water-soluble sodium salt which will be present in the lower alkaline solution. Non-acidic impurities will remain in the upper toluene layer.

Remove the stopper from the funnel and run the lower layer into a 200 ml beaker. To make sure that all the organic acid is removed from the toluene layer, re-extract this layer with a further 20 ml of dilute sodium hydroxide solution. Combine the two alkaline solutions in the beaker and discard the toluene solution in the funnel.

Regeneration of the Free organic acid

Add dilute hydrochloric acid (5%) with stirring to the alkaline solution in the beaker until the mixture is just acid (test with congo red paper; it is blue in acid solution). This treatment will cause most of the organic acid to be precipitated as a white solid. Pour the resulting suspension of the organic acid into a clean separating funnel.

Re-extraction of the organic acid

Add methylene chloride or chloroform (15 ml) to the suspension in the separating funnel, stopper the funnel and shake the contents thoroughly. Allow the layers to separate. The organic acid will now have dissolved in the lower methylene chloride layer.

N.B.: Methylene chloride/chloroform, unlike toluene is more dense than water and hence forms the lower layer at this stage. Run off the lower layer into a small beaker. Re-extract the upper layer remaining in the funnel with a further 15 ml of methylene chloride. This should remove the last traces of the organic acid from the aqueous layer. Combine the two methylene chloride solutions in the beaker and discard the aqueous layer in the funnel.

Purification of the extract

Pour the methylene chloride solution of the organic acid into a clean separating funnel, add 20 ml of water and shake the mixture thoroughly. Traces of inorganic salts will dissolve in water. Remove the stopper, allow the layers to separate and run the lower methylene chloride solution into a small, clean, dry conical flask. Add a few grams of solid anhydrous sodium sulphate to the methylene chloride extract and after swirling the flask, allow it to stand for 10 min. This removes traces of water from the organic solution.

Isolation of the pure organic acid

Filter the clear methylene chloride solution, to remove the sodium sulpahte, into a small round-bottomed flask. Rinse the remaining sodium sulphate with a small additional amount of methylene chloride and filter the washings into the round-bottomed flask. Add a small piece of carborundum (boiling stone) and remove the methylene chloride solvent by distillation on the steam bath using the distillation apparatus. A residue of the purified organic acid (benzoic acid) will remain in the round-bottomed flask.

Recrystallise the material from minimum amount of boiling water. Collect, dry and weigh the crystals. Determine the melting point of the pure dry organic acid and place the crystals in a labelled sample tube.

* A mixture of an organic acid (benzoic acid) and a small amout of a coloured neutral compound dissolved in toluene. The same procedure to be adopted to any other acid.

3.7 The Possible Combination of Binary and Ternary Mixtures

Examples of binary mixtures:

1. Strong acid + Neutral including Hydrocarbons
2. Strong acid + Weak acid
3. Weak acid + Neutral including Hydrocarbons
4. Base + Hydrocarbon
5. Carbohydrate + Weak acid
6. Carbohydrate + Strong acid (other than p-nitro benzoic acid)

Organic Mixture Analysis

Examples of ternary mixtures:

1. Strong acid + Weak acid + Neutral
2. Carbohydrate + Weak acid + Neutral
3. Base + Hydrocarbon + Weak acid
4. Carbohydrate + Strong acid (other than p-nitro benzoic acid) + Neutral

3.8 Separation of a Mixture of Organic Compounds

Object: To separate the components of a mixture of two of the following:

* An organic base,

* an organic acid,

* an organic neutral substance.

Background

The simplest organic mixtures to separate are those in which one component can be dissolved in water, or can't be converted into a water-soluble salt, while the other components remain in solution in a solvent which does not mix with water.

Thus a mixture of an organic acid, such as benzoic acid, and a neutral compound, such as ethyl benzoate, can be separated by dissolving the mixture with dilute aqueous sodium hydroxide. The benzoic acid passes into the aqueous layer as its sodium salt while the ethyl benzoate remains in the dichloromethane solution. The benzoic acid can be recovered by separating the two layers and acidifying the upper aqueous layer with concentrated hydrochloric acid.

Similarly an amine can be separated from a neutral compound by shaking a solution of the mixture in dichloromethane with dilute hydrochloric acid. The amine goes into the aqueous layer as its hydrochloride salt and can subsequently be recovered, after separation of the aqueous layer, by basifying with aqueous sodium hydroxide.

The concepts involved in these separations are of fundamental importance, and find application in nearly every organic synthesis.

You will be required to separate a two-component mixture of one of the following types:

(i)   Carboxylic acid + neutral compound.

(ii)   amine + neutral compound,

(iii)   amine + carboxylic acid

Experimental *Priliminary Test*

Read carefully before carrying out the separation

* The procedure for separating a mixture into basic, acidic and neutral components is detailed below and is summarised into the scheme depicted in previous page. By following this scheme you should be able to isolate the two components of your mixture without difficulty.

* Before you proceed you should undertake some test-tube experiments to ascertain whether your mixture contains an acid or a base, or both, by the following methods.

* Dissolve a small amount of your mixture in about 2 ml of dichloromethane in a test tube. Add about 2 ml of dilute sodium hydroxide solution. Shake! Pipette off the sodium hydroxide layer into another clean test tube, using a dropper, and make just acid by adding a few drops of conc. hydrochloric acid. The formation of a precipitate or an emulsion indicates that the mixture contains an acidic compound.

* The presence of a base can be deduced by shaking a solution of the mixture (about 2 ml) with dil. hydrochloric acid. Separate off the acid layer with a dropper and basify with conc. Sodium hydroxide. The formation of precipitate or an emulsion shows the presence of a base.

   N.B. If you are uncertain which types of components are present after several tests seea demonstrator for assistance.

* Weigh out a sample (2.0 g) of the mixture and dissolve it in dichloromethane (50-100 ml). Pour the solution into a separating funnel.

Separation of the base

* **To extract a basic component** from the mixture, add dilute HCl (50 ml), stopper the funnel and shake it thoroughly. Remove the stopper and draw off the lower (CH_2Cl_2) layer into a clean conical flask. Pour the upper (aqueous) layer into another clean conical flask labelled **"B"** or "Basic component".

* Replace the CH_2Cl_2 layer in the separating funnel, and repeat the extraction with a second portion of dil. HCl (50 ml). (Several extractions with small volumes of solvent are more efficient than a single extraction with a large volume).

* Again separate the two layer; the lower (CH_2Cl_2) layer is returned to the separating funnel, and the upper aqueous layer is added to the flask labelled **"B"**.

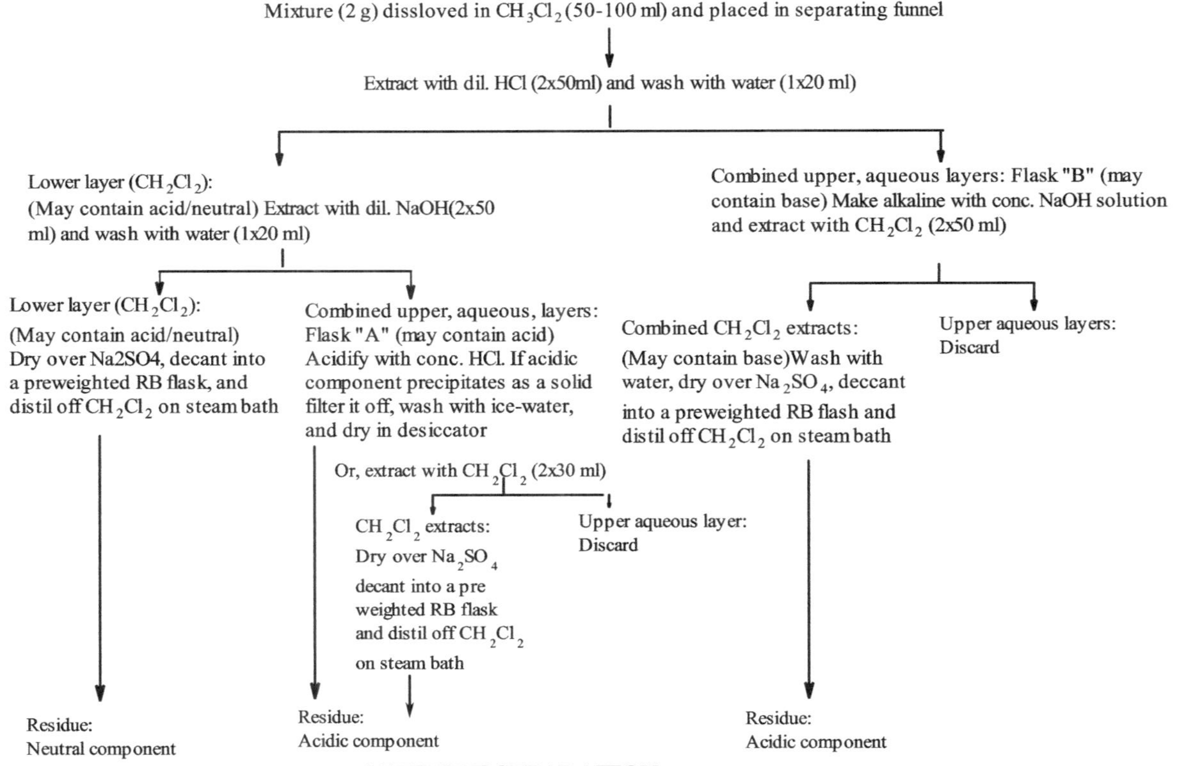

Mixture (2 g) dissloved in CH₃Cl₂ (50-100 ml) and placed in separating funnel
Extract with dil. HCl (2x50ml) and wash with water (1x20 ml)
Lower layer (CH₂Cl₂):
(May contain acid/neutral) Extract with dil. NaOH(2x50 ml) and wash with water (1x20 ml)
Combined upper, aqueous layers: Flask "B" (may contain base) Make alkaline with conc. NaOH solution and extract with CH₂Cl₂ (2x50 ml)
Lower layer (CH₂Cl₂):
(May contain acid/neutral) Dry over Na2SO4, decant into a preweighted RB flask, and distil off CH₂Cl₂ on steam bath
Combined upper, aqueous, layers: Flask "A" (may contain acid) Acidify with conc. HCl. If acidic component precipitates as a solid filter it off, wash with ice-water, and dry in desiccator
Combined CH₂Cl₂ extracts: (May contain base)Wash with water, dry over Na₂SO₄, deccant into a preweighted RB flash and distil off CH₂Cl₂ on steam bath
Upper aqueous layers: Discard
Or, extract with CH₂Cl₂ (2x30 ml)
CH₂Cl₂ extracts: Dry over Na₂SO₄ decant into a pre weighted RB flask and distil off CH₂Cl₂ on steam bath
Upper aqueous layer: Discard
Residue: Neutral component
Residue: Acidic component
Residue: Acidic component
MIXTURE SEPARATION
73

* Finally, wash the CH_2Cl_2 in the funnel with water (20 ml) to remove residual hydrochloric acid, add the aqueous washings to the basic component flask (**"B"**) and return the CH_2Cl_2 layer to the separating funnel.

* Set aside flask **"B"** for the time being.

* **To separate an acidic component,** extract the remaining CH_2Cl_2 solution in the separating funnel twice with 50 ml portions of dil. NaOH and once with 20 ml of water, combining the three alkaline layers in a conical flask labelled **"A"** or "Acid component".

* Set aside flask **"A"**.

* **To isolate a neutral component,** Place the CH_2Cl_2 solution, from which base or acid components have now been removed, in a clean dry conical flask, add some anhydrous sodium sulpahte (1-2 g), and swirl the mixture occasionally for 10 minutes to remove residual water.

* Decant off the CH_2Cl_2 solution (or filter it) into a 100 ml round bottom flask of known weight, and either distil off the solvent on a steam bath, or remove it on a rotary evaporator.

* Weight the residue, recrystallise it from a suitable solvent and determine its m.p. Store it in a stoppered sample tube labelled "Neutral component".

* **To isolate the basic component,** carefully neutralise the combined HCl-extracts in Flask **"B"** by adding conc. NaOH until the solution is alkaline to pH paper. If solid substance is present it should separate as a solid or otherwise an oil .

* Pour the whole mixture into a separating funnel and extract twice with 30 ml portions of CH_2Cl_2. Combine the CH_2Cl_2 extracts and dry them over anhydrous Na_2SO_4.

* Decant or filter the CH_2Cl_2 solution into a clean, dry, preweighed 100 ml round bottom flask, and either distill off the solvent or remove it on a rotary evaporator.

* **To isolate the acidic component,** carefully neutralise the combined NaOH extracts in flask **"A"** by adding cone. HCl drop by drop with stirring until the solution is just acid to pH paper. If a carboxylic acid is present the solution will become cloudy and / or a solid will precipirate.

* Allow the mixture to cool and then isolate the solid acidic component by either of the following two procedures.

1. Filter off the solid acid using a Buchner or Hirsch funnel, wash it with a little ice-cold water and dry it in a desiccator.

Or

2. Transfer the mixture to a separating funnel and extract the acidic component with two 30 ml portions of CH_2Cl_2. Combine the extracts and dry them ever

anhydrous Na_2SO_4. Decant (or filter) the CH_2Cl_2 solution into a clean, dry, preweighed 100 ml round bottom flask and distill of the solvent or remove it on a rotary evaporator.

* In either case determine its weight, recrystillise it from a suitable solvent and take its m.p. Store it in a stoppered sample tube labeled "Acidic component".

Record the infrared spectrum of each component as a Nujol mull, and identify significant infrared peaks in your spectrum; for at least one of the components you should be able to identify the infrared bands corresponding to the functional group.

Results Record your findings by filling in the appropriate table below.

Acidic component	Present*/Not presnet*
Weight isolated: Crude Purified	
Recrystallisation solvent	
M.P.	
Diagnostically significant bands	

Basic component	present*/not present*
Weight isolated: Crude Purified	
Recrystallisation solvent	
M.P.	
Diagnostically significant infrared bands	

Neutral component	present*/not present*
Weighted isolated: Crude Purified	
Recrystallisation solvent	
M.P.	
Diagnostically significant infrared bands	

3.9 Organic Qualitative Analysis

Early in your career you will meet a sample of two (may be more) of the following functional groups; alchohol, phenol, aldehyde, ketone carboxylic acid, amine (primary, secondary, tertiary). Later the list of functional groups you are expected to recognize will be expanded, and the next series includes: ester, amide, nitrile, salt (of carboxylic acid or of amine).

Quantities: In a text-book the quantities of materials and reagents needed are usually given by weight. However, the instructions define amounts that are convenient rather than stoichiometric; and with practice, quantities of solid can be estimated with sufficient accuracy by volume rather than by weight. For your first two samples, weigh the quantities required, crush any solid to powder, and measure the volumes as depths (in cm) in a 5 mm tube. Thereafter, judge quantities by volume rather than by weight; include in your report a statement of the quantities used.

In the absence of other information, assume:

If two organic liquids are to react, they do in equal volumes.

If an organic liquid is to react with a organic solid, the apparent volume of the solid should be twice that of the liquid, to allow for the spaces between the grains.

Bench Reagents

These have approximate concentrations:

Dilute sodium hydroxide & All are about2M; for sodium hydroxide and dilute hydrochloric acid hydrochloric acid the concentrations are about 10% by weight:

Dilute ammonia for ammonia, about 5%.

Dilute sulphuric acid 1M; about 10% by weight

Conc. hydrohloric acid above 10M

'50% sulphuric acid' used for hydrolysis of amides means 50% by weight. The reagent is prepared by adding concentrated sulphuric acid slowly and with stirring to twice its volume of water.

3.10 Analysis of Unknown Compounds

3.10.1 Sequence of Preliminary Tests

1. Note the appearance and, if appropriate, smell (Caution). Determine the m.p. or b.p.

2. Lassaigne test (if appropriate).

3. Test the solubility in water (cold, then hot)

Whether or not the material dissolves completely, test whether the pH of the mixture is the same as that of the tap water, or has changed. Universal indicator paper is better than litmus.

4. Test the solubility with dilute HCl. If the material is soluble in water, then acidify the aqueous solution.

5. Test the solubility in bench dilute sodium hydroxide solution. If then material is soluble in water, then make the aqueous solution alkaline.

6. Test the solubility in cold dilute sodium hydrogen carbonate (sodium bicarbonate). The material may dissolve slowly; look for small bubbles on particles of a solid, and the disappearance of the smallest grains first.

7. Test the material with Brady's reagent.

These tests reveal the groups : Carboxylic acid

> Phenol
>
> Amine
>
> Aldehyde / Ketone.

If no groups is yet identified, examine the infrared spectrum and look for evidence of the following functional group.

> Ester
>
> Amide
>
> Alcohol
>
> Nitrile.

When you have identified the functional group present check your conclusion with demonstrator before proceeding to make derivatives.

Choose derivatives with care, in order to distinguish clearly between possible alternative compounds of similar m.p. / b.p

3.10.2 Interpretation of Preliminary Tests

Test No.

1. If the substance is clearly more soluble in dilute hydrochloric acid than in water, you can expect an amine.

    If the material dissolves in water and the addition of dilute hydrochloric acid causes a ppt., then you can expect a salt of an organic acid.

2. If the substance shows enhanced solubility in sodium hydroxide solution, then you can expect a phenol, or a carboxylic acid.

3. Effervescence with aqueous sodium hydrogen carbonate shows the likely presence of a carboxylic acid function, or a salt of an amine.

4. Brady's reagent is used for testing for the presence of an aldehyde or ketone. To prepare a 2,4-dinitrophenyhydrazone as a derivative, make up your own reagent using solid 2,4-dinitrophenythydrazine.

It is important: To test for an amine function before testing for an aldehyde or ketone.

Addition of an amine to Brady's reagent may give a precipitate of 2,4-dinitrophenylhydrazine.

Tollens' reagent: To 2 ml of 2% silver nitrate solution add one drop of bench dilute sodium hydroxide solution shake, and then add bench dilute ammonia dropwise with shaking till the ppt. just dissolves.

Aldehydes: React with in two minutes producing a black ppt. - if the test-tube is clean – a silver mirror. Ignore a black colour, which does not settle.

5. If the material dissolves in conc. sulphuric acid, or reacts with it (e.g., turns to tar), then it contains - and / or non-bonding electrons that can accept a proton. Compounds that remain unchanged are alkanes; benzene, naphthalene, and their alkyl-, and halogenated derivatives.

6. Insolubility in ether indicates strong intermolecular forces. The commonest examples of such are a salt, e.g., aniline hydrochloride, or multiple hydrogen-bonding, as a sugar.

Amines: Reaction with nitrous acid.

Dissolve about 0.1g of the amine in 5 ml of dilute hydrochloric acid in a test-tube, then half fill the tube with ice. To this solution add a solution of sodium nitrite (0.1 g) dissolved in the minimum quantity of water (about 1 ml); you can expect one of the following:

A. Effervescence, reveals a primary aliphatic amine.

B. A yellow precipitate, either a solid or an oil. Make the solution alkaline with bench dilute sodium hydroxide; you will see either no change. You have a secondary amine with one or both substituent's aliphatic.

C. No apparent reaction. Either an aliphatic tertiary amine or a primary aromatic amine

Distinguish between these as follows: Add solid urea, a quantity at least as great as the amount of sodium nitrite added. In a separate vessel dissolve a few grains of 2-naphthol in about half a test-tubeful of dilute sodium hydroxide solution. Mix the two solutions. A red, orange, or yellow precipitate indicates that the starting material was a primary aromatic amine.

3.10.3 Form of Report

Your report should include the following:

A title e.g., unknown No.3

A record of the tests you have done, including any that gave a negative result and of the derivatives that you prepared, stating the solvent used for the recrystallization, and the m.p. found. (If the material crystallises with solvent of crystallisation, then the m.p. depends on the solvent used).

The structures of the unknown, and of the derivatives you made.

A summary of the evidence for your conclusion that has the form of three paired thermometer readings; for example:

Sample is N-methylaniline

Derivatives

Found: b.p.   194;   Acetyl, m.p.   103;   Benzoyl,   m.p.   60

Lit.   b.p.   196;   m.p.   102   m.p.   63

3.11 Specimen Report on the Identification of Single Substance

Test	Observation	Inference
Appearance	Pale yellow crystalline Solid – faint smell of almonds	-----
M.P	$57 - 58^0$	----
Ignition test	Smoky flame no residue	Aromatic/ No metallic
Lassaigne test	Nitrogen – positive Sulphur – negative Halogens – negative	N present
Solubility		
H_2O	Slightly soluble in hot	
Ether	Soluble	
Dil. NaOH	Insoluble	Not acid or phenol
Dil. HCl	Insoluble	Not amine
Conc. H_2SO_4	Brown colour develops	Nnconclusive
Unsaturation		
Br_2 / CCl_4	No reaction evident	No alefinic C=C
Br_2 / H_2O	No reaction	
$KMnO_4$	Decolorisation occurred after a few seconds	Easily oxidised group present

Test	Observation	Inference
Carbonyl tests:		
2, 4-dinitrophenyl-hydrazine solution	Yellow-orange ppt	>C=O present (aldehyde or ketone)
Schiff's reagent	Pink colour	Aldehyde
Tollent's test	Silver mirror	Confirms – CHO
Sodium metal	No reaction with a solution of unknown in dry ether	No OH group
$FeCl_2$	No colour change	No phenol or enols
Ester test	Negative	
$TiCl_2$	Solution decolorised on boiling With unknown	$-NO_2$ group

Infrared Spectrum

Nujol mull spectrum contained peaks at

1709 cm^{-1} (S)	-	Aromatic CHO
2760 cm^{-1} (W)	-	CHO
1615 cm^{-1} (m)	-	Aromatic C=C
1535 cm^{-1} (m)	-	$Ar-NO_2$
1350 cm^{-1} (s)	-	Ar-NO2

From preliminary tests the substance is an aromatic nitro-aldehyde. From tables, it could be m-nitrobenzaldehyde (lit., m.p. 58 °C).

Preparation of Derivatives

1. 2,4-Dinitrophenylhydrazone, m.p. 280-285 °C.

Recrystallised from ethanol/CH_2Cl_2, m.p. 288-290 °C (lit., mp 292 °C).

1. **Oxidation to m-nitrobenzoic acid:** Suspension of substance in 5% aq. Na_2CO_3 boiled under reflux with aq. $KMnO_4$ until homogenous; cooled, acidified (dil. HCl), and added solid $Na_2S_2O_5$ to dissolve MnO_2. Filtered off pale yellow solid, m.p. 136-138 °C. Recrystalized from ethanol to crystals, m.p. 140-141 °C (Lit., m.p. m-nitrobenzoic acid 141 °C).

O_2N—C$_6$H$_4$—CHO $\xrightarrow{(O)}$ O_2N—C$_6$H$_4$—COOH

Conclusion

The unkown substance is probably m-nitrobenzaldehyde.

3.12 Infrared Spectrum

The sample is handled either as a liquid film if the material is a fluid, or as a Nujol mull that is, a suspension of a fine powder in Nujol; the latter is a mixture of aliphatic hydrocarbons. Nujol produces strong signals:

About 2900 cm^{-1}	Aliphatic C-H stretch
1460 cm^{-1}	Aliphatic C-H bend
1380 cm^{-1}	Aliphatic C-H bend

On the record of the spectrum you should state the phase (liquid film, or Nujol mull, as appropriate), and identify signals due to nujol.

Additionally, you should examine the regions of the spectrum specified below:

3300-3200 cm^{-1} O-H	or N-H stretch, hydrogen-bonded. Usually a broad signal of medium or weak intensity. A sharp peak on the high-frequency side indicates a free OH or N-H group.
3300-2500 cm^{-1} O-H	Stretch due to a carboxylic acid. Very broad, and may not be easily seen.
2250 cm^{-1} CN	Stretch, sharp peak of medium or weak intensity.
1740-1640 cm^{-1} CO	Stretch, singles usually strong or medium intensity, and usually sharp. Look particularly at the signal of highest frequency. For free (that is, non-conjugated) groups:

1735 ± 15	Ester
1725 ± 15	Aldehyde

1715 ±15	Ketone
1710 ± 15	Carboxylic acid
1660 ±30	Amide
	Conjugation lowers the frequency by about 30 cm^{-1}
900-730 cm^{-1}	C-H (arene) bend: usually strong, signals. The position(s) of the signal(s) may reveal the pattern of substitution in derivative of benzene.

3.13 Calculation of Percentage Yield

In any preparative experiment you will be asked to calculate the yield. This requires that you first write a balanced equation for the synthesis e.g

$$ROH + R^{1}COOH \rightarrow R^{1}CO - OR + H_2O$$

Then identify the minority reactant i.e., this example requires equal numbers of moles of alcohol and acid. Let us suppose that your calculations, from the masses of each used and their molecular weights, show that there are fewer moles of alcohol than of acid. i.e., the alcohol is the minority reactant, thus we base our yield on this reactant (ROH).

Calculate moles ROH taken = mass (ROH / M.W.(ROH)

Calculate moles of product = mass (R^1 CO-OR)/M.W. (R^{1}CO-OR)

Since only 1 mole of ester is produced from 1 mole of alcohol.

Example A

At the end of this and several of the following experiments you are asked to calculate the % yield of product. This is the correct procedure:

Firstly ascertain the correct stoichiometry for the reaction yielding the desired product,

Preparation of n - butyl bromide the overall reaction is

$$C_4H_9OH + Br^{-} + H_2SO_4 \rightarrow C_4H_9Br + OH^{-}$$

Which indicates that 1 mole of n–butanol reacts with 1 gm^{-} ion of bromide (as 1 mole of NaBr).

Inspection of the experimental details indicates that 0.17 mole of NaBr is required but only 0.135 mole of n-C$_4$ H$_9$ OH is consumed. It is normal practice to calculate the percentage of a theoretical yield based upon the reactant present in smallest molar amount (i.e. in this case the yield would be based upon n- $C_4 H_9 OH$).

If this particular experiment gave a 100% yield of crude product, from 10 g. (0.135 mole) of n-$C_4 H_9 OH$ one would expect 0. 135 mole n-$C_4 H_9 Br$.

Since the molecular weight of n-$C_4 H_9 Br$ is 137,

0.135 mole $= 0.135 \times 137$ g. $= 18.5$ g.

(i.e. theoretically 10 g. $C_4H_9OH \rightarrow 18.5g. \ C_4H_9Br = 100\% \ yield)$.

In fact the experiment yielded only 12.4 g. $C_4 H_9 Br$.

The actual yield is therefore $12.4/18.5 \times 100 = 67\%$ of the theoretical.

Example B: Preparation of ethyl acetate (calculation of % yield)

Amount used:Ethanol 5 ml (40 g), acetic acid (52 g), sulphuric acid 1 ml, calsium chloride (2.5 g).

$$C_2H_5OH + CH_2COOH \rightarrow CH_2COOC_2H_5 + H_2O$$

The yield of the compound obtained in an organic preparation can be calculated as a percentage of the yield theoretically possible from the weights of the original compound taken. Thus the in the preparation of ethyl acetate, the equation shows that 60 g of acetic acid and 46 g of ethanol reacts to give 88 g of ethyl acetate.

If equimolecular quantities of acid and alcohol were taken in the preparation, the percentage yield of the ethyl acetate would clearly be the same if calculated on the basis of either reagent. Thus,

46 g ethanol should give **88 g** of ethyl acetate.

Therefore: 40 g of ethanol should give **76 g** of ethyl acetate.

Practical yield: 50 g ester was obtained. Hence the yield of ethyl acetate

$$50 / 76 \times 100 = 66\%$$

Similarly, since 60 g of acetic acid should theoretically give 88 g of ethyl acetate,

52 g (actual amount taken) should give **76 g** of the ester, and hence the yield calculated,

$$50 / 76 \times 100 = 66\%$$

On the basis of acetic acid is used.

CHROMATOGRAPHY

4.1 Introduction

- Chromatography is a dynamic process in which the mobile phase moves in definite direction.

- IUPAC definition: Chromatography is a physical method of separation in which the components to be separated are distributed between two phases, one of which is stationary while the other moves in a definite direction.

- The stationary phase may be a solid, or a liquid supported on a solid or gel, the mobile phase may be either a gas or a liquid.

- Mobile phase - phase that moves through chromatograph

- Stationary phase - column; phase that is stationary in chromatograph

- Bonded phase - reactive groups imparted to stationary phase in order to achieve selectivity

4.2 Classification of Chromatographic Techniques

Nature of the mobile phase	Nature of stationary phase	Mechanism of separation	Technique	Name of the chromatographic technique
Gas	Liquid	Partition	Column	Gas liquid chromatography (GLC)
	Solid	Adsorption	Column	Gas solid chromotography
Liquid	Partition		Column	Liquid liquid chromatography (LLC)
			Planar	Thin layer chromatography (TLC) Paper
Cromtography (PC) liquid	Bonded liquid	Modified partition	Column	High performance liquid chromatography (HPLC)

Table *contd..*

Nature of the mobile phase	Nature of stationary phase	Mechanism of separation	Technique	Name of the chromatographic technique
			Planar	High performance thin layer chromatography (HPTLC)
Solid	Adsorption	Column		Classical liquid, solid chromatography (LSC) & HPLC
			Planar	TLC HPTLC
	Ion Exchange	Column		Ion Exchange chromatography (IEC)
	Exclusion	Column		Exclusion chromatography (EC) or Gel permeation chromatography (GPC)

4.3 Thin-Layer Chromatography (TLC)

Chromatography, of which there are many forms, is a technique for qualitative analysis and is useful for detecting compounds in a mixture, even when they are present in very low concentrations. Thin layer chromatography (TLC) is a simple and convenient form of technique enabling both micro and preparative quantities of mixtures to be separated rapidly.

The advantage of TLC over paper chromatography are well known, and include the speed of migration of the eluent, the speed of separation of the mixture, the degree of resolution of the mixture and, most important, the relative ease with which the separation is achieved.

Principle based on adsorption phenomenon: Essentially a glass plate is covered with a thin coating of an adsorbent material (e.g., silicagel, alumina, cellulose)). The film is activated by drying and a solution of the mixture to be separated is spoted on the plate at one end. The plate is then placed upright in a closed vessel or in a bottle containing an appropriate solvent(s), so that the end that has been spotted comes into contact with the eluent (solvent). The solvent climbs the film by capillary action carrying with it the components of the mixture at different rates depending on their relative affinities for the adsorbent as compared with the migrating eluent.

Theory: Separation in TLC include adsorption and partition mechanism, and the technique can involve reversed phase partition. In practice, a combination of several mechanisms usually occurs, although one (usually adsorption) may predominate.

R_f value: A measurement (R_f value) of the movement of the spot with respect to the eluent front can be made by measuring the distance moved from the origin as follows.

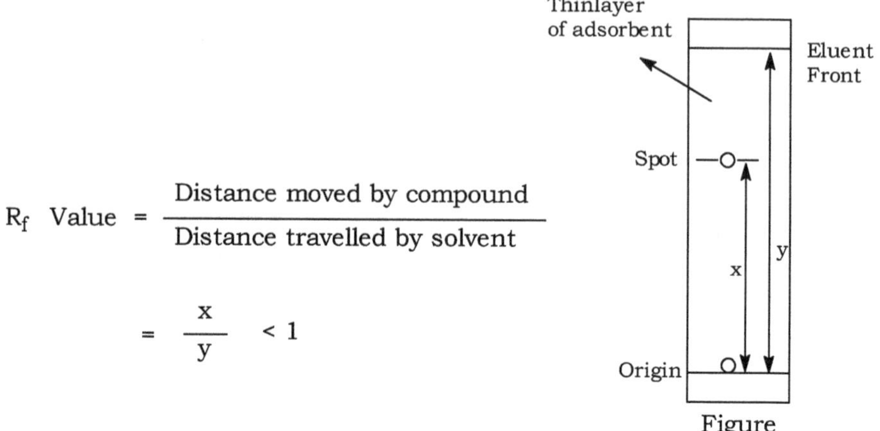

$$R_f \ \text{Value} \ = \ \frac{\text{Distance moved by compound}}{\text{Distance travelled by solvent}}$$

$$= \ \frac{x}{y} \ < 1$$

R_f values are extremely dependent on many factors and must therefore be considered approximate value only. To obtain reproducible values precaution should be taken to maintain the following conditions as much as possible.

1. Quality of the adsorbent
2. Thickness and uniformity of the adsorbent layer
3. Degree of activity of the adsorbent
4. Quality of the solvent
5. Temperature
6. Degree of chamber saturation
7. Amount of sample
8. Impurities present in the sample mixture.

Adsorption Chromatography

Adsorption is the most commonly operating mechanism in thin layer chromatography separation. The sample is continuously fractionated as it migrates through the adsorbent layer and the degree of separation achieved is strongly dependent upon the surface of the adsorbent. Competition for active adsorbent sites between components of the mixture to be separated and the eluent, produces continuous fractionation. As the process continues, the various components of the mixture adsorb and desorb with different ease and hence move at different rates, allowing separation to occur. Polar solvents effect a higher rate of movement of compounds over the adsorbent. It is usual to start with a relatively non polar solvent followed by solvent of increasing polarity till a good seperation is achieved.

The following is a list of solvents in increasing order of polarity.

1. n-Hexane
2. Petroleum ether
3. Toluene
4. Diethyl ether
5. Chloroform
6. Ethyl acetate
7. n-Butanol
8. n-Propanol
9. Acetone
10. Ethanol
11. Methanol
12. Water.

Adsorbents: Many solids have been used as adsorbents. Many adsorbents such as alumina, silica gel, active carbon, cellulose powder, kieselguhr and sephadex etc., can be obtained commercially. They often require activation before use; this can be achieved by heating, possibly in a vacuum, when the adsorbent loses water and other adsorbed materials. Three or four hours of heating at 200°C is usually sufficient for most solids. Two important properties of the adsorbent are its particle size and its homogeneity, because adhesion to the support largely depends on them. A particle size of 1-25 μm is usually recommended. Some examples of absorbents which have been used for representative separations by TLC are given below.

Table 4.1 Adsorbents for TLC

Solid	Used to separate
Silica gel	Aminoacids, alkaloids, sugars, fatty acids, lipids, essential oils, steroids and terpenoids.
Alumina	Alkaloids, food dyes, phenols, steroids, vitamins, carotenes
Kieselguhr	Sugars, dibasic acids, fatty acids triglycerides, aminoacids, steroids.
Celite	Steroids, inorganic cations
Cellulose powder	Amino acids, food dyes, alkaloids
Starch	Amino acids
Sephadex	Amino acids, proteins.

Preparation of plates: The adsorbent is spread on a suitable firm support, which may be quite rigid or flexible. Glass plates were the original support and are still the most usual. The size used depends on the type of separation to be carried out and the type of chromatographic tank and spreading apparatus available. Microscope slides are often very convenient for rapid qualitative separation. Most of the commercial apparatus is designed for plates of 20 × 5 or 20 × 20 cm, and those are now regarded STANDARD. It is important that the surface of the plate shall be flat, without irregularities or blemishes.

Glass plates are cleaned thoroughly before use, washed with water and a detergent, drained, and dried. it is important not to touch the surface of the cleaned plates with the fingers.

The first step is to make the adsorbent into a slurry with water, usually in the proportions Xg of adsorbent and 2x cm^3 of water. The slurry is thoroughly stirred, and spread on the plate by using commercial spreaders, which might be called the 'moving spreader'.

The film thickness is a most important factor in thin layer chromatography. The standard thickness is 250 μm. Thicker layer (0.5 to 2.0 mm are used for prepartive separations; with a loading of upto 250 mg on 20 x 20 cm plate. One difficulty with thick layers is their tendency to crack on drying. There are in principle three or four ways of applying the thin layer to its support; spreading, pouring, dipping and spraying.

Spreading: It is possible to spread the adsorbent in a number of ways, but the main objective is to produce an absolutely even layer with no lumps or gaps, which adheres evenly and securely to the support. It is not essential to use a commercial spreader, but it is very much easier, and unless cost is an important consideration.

Pouring: Many workers prefer not to use mechanical spreading method at all. If the adsorbent is very finely devided and of homogeneous particle size, and if no binder is used, a slurry can be poured on a plate and allowed to flow over it so that it is evenly covered. Preparation of plates by pouring is particularly easy with certain types of alumina, but water alone is not usually suitable for making the slurry; a volatile liquid such as ethanol (or an ethanol-water mixture) or ethyl acetate is preferable. The appropriate amounts of liquid and solid adsorbent needed to cover a plate have to be found by trial and error, and exactly those quantities should be used to ensure that the thickness of the layer is reproducible, but the thickness of the layer is not known by this method.

Dipping: Small plates such as microscope plates (slides) can be spread by dipping in a slurry of the adsorbent in chloroform or other volatile liquid. Again the exact thickness of the layer is not known, but this is a most convenient method for making a number of plates for rapid qualitative separations. After spreading, the plate is allowed to dry for 5-10 minutes, and if it has been made with an aqueous slurry, it is further dried and activated by heating at about 100°C for 30 minutes. Plates made by volatile organic liquids may not need this further drying. Plates may be kept for short periods in a desiccator, but long storage is not recommended.

Detection of the separated zones: This may be achieved by one or more of the following methods.

(a) **Direct observation:** This is only suitable if the spots are coloured.

(b) **Ultra-violet radiation:** This can be used to assist detection if the spots are not directly visible under normal lighting. If the sheets or plates containing a fluorescent indicator are used, exposure to radiation of 254A wavelength causes substances that absorb radiation of that wavelength to appear as dark spots against a glowing background.

(c) **Iodine vapour:** If a few crystals of iodine are placed in a beaker which is then covered by a watch glass, iodine vapour will soon fill the container (caution : iodine vapour is toxic). A dried eluted chromatogram is then placed inside the beaker and the iodine vapour will render visible the spots related to many colourless organic substainces.

(d) **Conventional chemical spray treatments:** Appropriate sprays containing chemicals reactive with the constitutents of the mixture under investigation produce colour spots. Thus, cold 10% solution of conc.sulphuric acid in ethanol can be used, but charing with strong acids is precluded, because of the presence of the organic binder.

Experimental Procedure

Initially, practice the use of TLC using mixture of dyes provided (red, yellow and blue). Use benzene as the developing solvent.

Application of sample: Make an applicator by softening the middle of a melting point tube over a very small flame (from a microburner), quickly draw the ends of the tube apart, make a mark at the middle of the capillary with the edge of a piece of broken tile, and snap the capillary in half.

Fill the applicator by dipping the capillary tip vertically into the dye mixture solution.

A pencil line to denote the origin is lightly drawn at least 1.5 cm from the bottom edge of the sheet. A solution contianing a 0.1-1.0% average concentration of the mixture to be separated is then spotted on to this line. About 0.5 micro litre (μl) of the solution is spotted on to the plate or sheet using one of the capillary prepared above. The tube when dipped into a solution of the mixture fills by capillary action. By touching the adsorbent surface gently with the filled tube, a spot of about 2 mm diameter can be produced without damaging the surface of the layer. If a larger quantity of deposited sample is desired, additional spots may be added over the first ones, provided each spot is dried before the next is added. Two spots per plate may be applied. Apply them so that they are 4 mm from one end of the plate, 8 mm a part and equidistance from the edges as indicated. The spots must be dry before the chromatogram is eluted.

Eluents: The following list contains common eluents which, alone or in various combinations, should effect a separation.

Acetic acid	Dioxane	Methylene chloride
Acetone	Ethanol	Pyridine
n-butanol	Ethyl acetate	
Chloroform	Hexane	
Diethyl ether	Methanol	

Development: Line the inside of a beaker with a strip of filter paper, then pour in the developing solvent to moisten the paper and give a depth of 2-3 mm on the bottom of the beaker. Place the chromatoplate upright in the beaker (with the end with the spots on at the bottom). Cover the beaker with a watch glass when the solvent has risen 5-6 cm up the plate, remove, the plate from the beaker and immediatly mark the solvent front, with a pin or pencil. It may be necessary to visualize the compounds after chromatography by exposing the plate to iodine vapour. Place the plate for 4-6 minutes in a beaker containing a few crystals of iodine, which is covered by a clock (watch) glass. Mark the position of the brown spots by pricking round them with a pin, and for each spot calculate the approximate R_f value of the compound.

This is given by

$$R_f = \frac{\text{Distance travelled by compound}}{\text{Distance travelled by solvent front}}$$

confirm your conclusions by running a further chromatogram using the unknown mixture as one spot and the suspected mixture as a second spot. (The R_f values may vary with the quality of the TLC plate and hence comparisons between the unknown and the standard must be made on the same plate).

While choosing a developing solvent to acheive good separation start with a relatively non polar solvent like petroleum ether. If the compound do not move on the chromatoplate use a more polar solvent, e.g., benzene, chloroform, ethyl acetate, methanol or ethanol, acetic acid in that order. Frequently a mixed solvent is valuable. Thus if the compound moves only slightly with benzene, try a 9:1 mixture of benzene : chloroform.

Experimental results

(a) Solvent used

(b) Mixture of unknowns x and y

  Distance moved by solvent front =

  Distance moved by x =

  Distance moved by y =

$$R_f(x) \quad =$$

$$R_f(y) \quad =$$

Experiment 1

Separation of dye mixture: This explains the principle of adsorption TLC.

Method: The dye test mixture (Azobenzene + Sudan red G + 1,4-Dimethylamino anthraquinone) is spotted on the chromatogram plate with the help of the capillary tube, keeping the spots as small as possible.

A Strip of filter paper is lined inside of a beaker, then pour in the developing solvent, toluene to moisten the paper and give a depth of 3 mm on the bottom of the beaker. Cover the beaker with a watch glass and the elution allowed to proceed for about 10 minutes to enable complete separation to be obtained.

Discussion: Three dyes will separate from the mixture. From top to bottom they are

Azobenzene (Yellow)

Sudan red G (Red)

1,4-Dimethyl amino anthraquinone (Blue)

Experiment 2

Separation of plant pigments

This illustrates the usefulness of TLC for separation of some natural products that are extremely difficult to separate by conventional chemical methods.

Equipment and reagents

Mortar and pestle

Acetone

Petroleum ether (40-60°)

Diethyl ether

Spinach, mint, grass (many other plant materials may be substituted)

Method: Approximately 1 g of grass is chopped into a mortar and ground. It is transferred to a boiling tube, 4 ml of acetone are added and the tube is corked and shaken. It is then allowed to stand for 10 minutes, 4 ml of water is then added and the mixture is shaken. This is followed by the addition of 4 ml of petroleum ether, after which the mixture is shaken once again. The green colour is extracted into the petroleum ether layer which is separated or decanted.

An activated chromatogram plate is spotted with this decanted solution and (eluted) in the beaker with the following (eluent) developer

Diethyl ether **1** part

Acetone **1** part

Petroleum ether (80-100°C fraction) **2** parts

The compounds that are separated are sensitive to light, so it is recommended that the apparatus should be covered with dark paper whilst the chromatogram is eluting (5-10 minutes).

Alternative eluents (developer)

Eluents A: Acetone+Petroleum ether (40-60°) (1:9)

Eluents B: Chloroform+Petroleum ether (40-60°) (3:7)

Results: Many jones should separate out. The main groups of compounds will fall into the following classes in decreasing order of R_f values)

Carotenes

Pheophytins

Chlorophylls

Xanthophylls

Experiment 3

Separation of some common indicators

Reagents : Fluorescein (**X**), Eosin (**Y**) and Methyl red (**Z**)

Method: A dye solution consists of the mixed indicators mentioned above, in methanol is spotted on the chromatogram plate and eluted with the following eluent.

Diethyl ether **1** part

Acetone **1** part

Petroleum ether (80-100°) **2** parts

Three spots will become visible. In decreasing order of R_f values they are

Methyl red (**Z**)

Fluorescein (**X**)

Eosin (**Y**)

Experimental results

(A) solvent used

(B) Mixture of common indicators **X,Y,Z**

Distance moved by the solvent front =

Distance moved by **X** =

Distance moved by **Y** =

Distance moved by **Z** =

$R_f(\mathbf{X})$ =

$R_f(\mathbf{Y})$ =

$R_f(\mathbf{Z})$ =

Conclusions

4.4 Preparative Thin Layer Chromatography (PTLC)

Aim: To separate the components present in the given mixture quantitatively by preparative thin layer chromatography.

Principle: Preparative thin layer chromatography can be used for relatively small preparative separations, that is, quantitative separations in the 10-100 mg scale. TLC plates of 20 × 20 cm size, with a layer (0.5-2 mm) adsorbent, the mixture is applied as a band, instead of spots, at 1 cm from one end of the TLC plate. This is mainly an adsorption chromatography technique. The sample applied as a band gets adsorbed onto the thin layers (stationary phase) & gets partitioned between the thin layers and mobile phase, when the mobile phase percolates through the thin layer. The percolation of mobile phase is by means of capillary forces. As adsorption and partition go on simultaneously, the sample will move along the mobile phase. When more than one component/analyte is present, movement of components is according to their differential migration rates. The component, which has greater affinity towards the stationary phase, will move later than the one which has lesser affinity towards the stationery phase.

Part I. Preparation of slurry

Mix 50 g of silica gel G in a mortar to a smooth consistency with 100 ml of distilled water and transfer the slurry quickly to the spreader/applicator.

Part II. Preparation of TLC plates

Clean the glass plates of size 20 X 20 cm thoroughly with chromic acid, rinse with distilled water and dry. Arrange these lengthwise, in a row on a thick plastic sheet. Mount the applicator on the glass plates at one end of the plastic sheet. Move the applicator slowly and steadily over the glass plates. A uniform thin coating of silica gel G on the glass plates is thus obtained. Allow the coated plates 1 to air dry for about 15 minutes and then transfer to the drying rack. Then keep in a hot oven at 100 $^{\circ}$C for 2 hours to remove water from the adsorbent and activate the absorbent. Cool the TLC plates and store in a desiccator over silica gel.

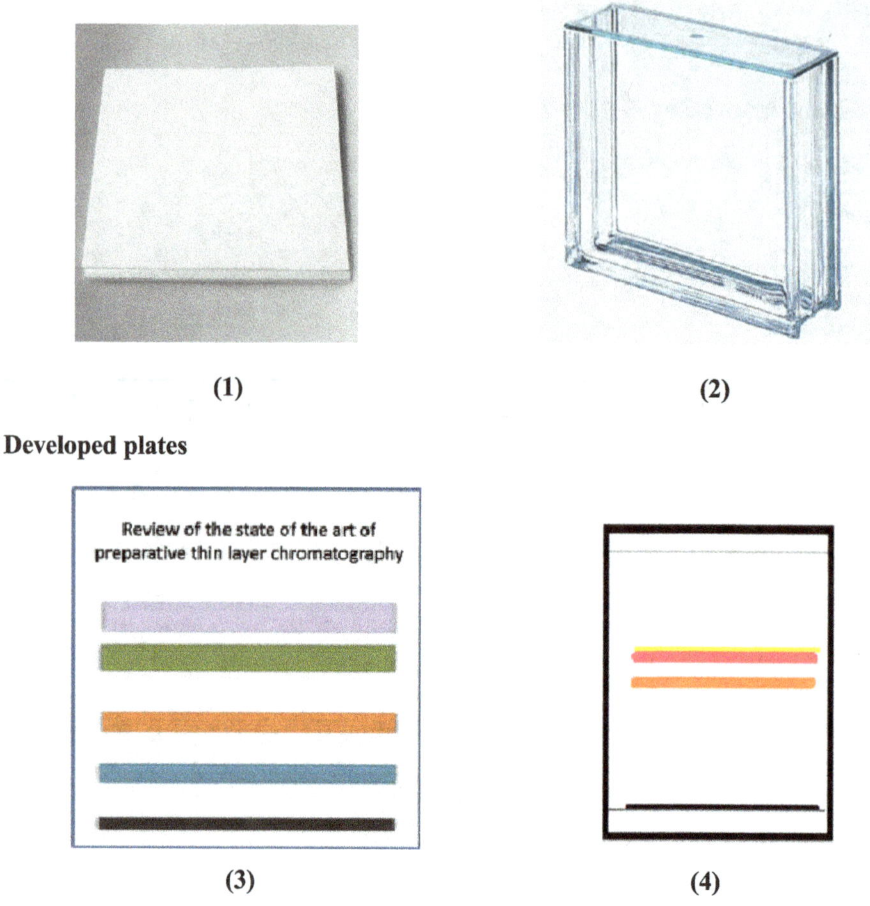

Fig. 4.1 Coated plates

Prepare a saturated solution of given compound in acetone/methanol so that the solution contains 10-100 mg. of compound. Place a capillary tube inside the solution; take out the capillary tube from the solution as soon as the solution enters into the capillary tube. Apply the solution gently on TLC plate just 5 mm above the bottom as a band. The sample solution gets adsorbed on silica plate. Let solvent evaporate on plate. Place gently the plate inside the development tank 2 in a slant position and cover with the lid. Observe that the loading point is above the solvent. Allow the solvent to ascend on the plate by capillary action, the compound applied as band moves upward direction along with the solvent. Remove the plate from the development tank after the solvent reaches to the maximum height. Then allow the solvent to evaporate completely.

Part III. Detection

To locate the position of the bands for separation into individual components, visualize at one end of developed plate (3, 4) by iodine vapour or UV light covering the rest of the plate.

Part IV. Recovery of the Components

After locating the position of the bands, mark the region of bands at one end of plate. Then scrap the bands from the plate and collect the components in separate flasks. Then leach the compound from the adsorbent using an organic solvent. Filter off solution and distill off the solvent to get a concentrated solution. Then transfer into a china dish and allow the solvent to evaporate completely to get dry compound

4.5 Paper Chromatography

The technique was such a remarkable advance upon previous methods, and of such wide application, that its discoverers, Martin and Syringe, were awarded the Nobel prize for chemistry in 1952.

This is usually regarded as a typical partition system, where the stationary phase is water, held by adsorption on cellulose molecules, which in turn are kept in a fixed position by the fibrous structure of the paper. It is now realized, however, that adsorption of components of the mobile phase and of solutes, and ion exchange effects, also play a part and that the role of the paper is by no means merely that of an inert support.

One characteristic feature of the paper technique is that there is nothing corresponding to the column effluent of the usual gas or liquid system. The separated substances are located and identified on the paper and there is thus a reasonably permanent record of the separation. There is no collecting of a series of fractions and sophisticated continuous monitoring devices are not needed. Quantitative determination of the separated substances may be carried out on the paper, but it is desired to remove them by cutting out the part of the paper containing each substance and elute each one separately.

Paper chromatography is a form of partition chromatography especially adapted to separating very small quantities of material. The cellulose of filter paper contains adsorbed water (about 20%). This combination is the stationary phase. A drop of a solution containing the mixture to be separated is applied in solution as a spot on the filter paper, and the developing solvent (which is immerscible with water) passes through a filter paper by capillary flow. The component of the sample will move at different rates through the paper. The rates being proportional to the partition co-efficients between the adsorbed water and the particular solvent used for developing, although adsorption by cellulose is also important.

When the position of the separated jones have thus been detected, it is essential to identify each individual component. In the normal course each one gives a characteristic colour with the locating agent, as is often found with inorganic substances, but less commonly with organic compounds.

The simplest method of identification is based on the R_f value, the ratio of the distance travelled by the spot and that travelled by the solvent front.

Paper chromatography is an invaluable tool in biochemistry, especially for separating and identifying mixtures of amino acids.

Experiment

Separation of synthetic food dyes : Using the apparatus shown below, a spot of a dye solution consisting of the mixed dyes in methanol is applied near the centre of a piece of filter paper punctured at the centre. A filter paper wick is placed through the central hole; the filter paper is placed over the solvent contained in the lower half of a petridish, which is covered to reduce evaporation. As the solvent travels up the wick and through the paper, the separate components appear as concentric rings.

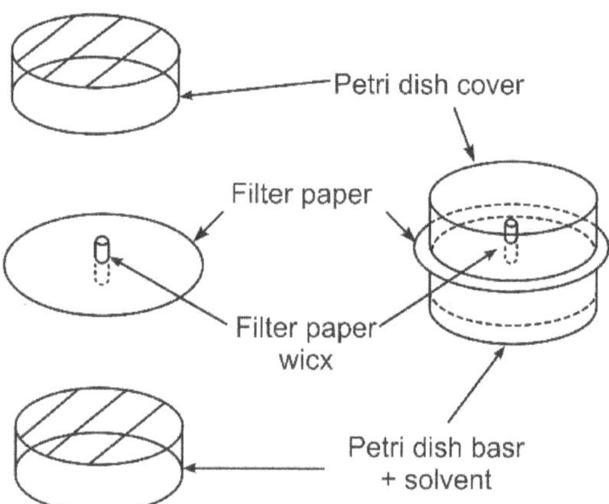

Fig. 4.2 Apparatus for paper chromatography of food dyes

Apparatus for Paper Chromatography for Food Dyes

Use Whatman No.1 filter paper (12.5 cm). Make a small hole at the centre of the paper. Prepare the wick from another piece of the filter paper (about 1 × 1.5 cm) and roll tightly to give a wick 1 cm long. Partially fill the lower half of the petri dish with the developing solvent (propan-1-ol : distilled water, 2:1).

Apply a spot of dye (about 1 cm in diameter) to the centre of the filter paper using a capillary tube, insert the wick, and rest the filter paper on the petri dish so that the wick dips into the solvent. Cover with the second petri dish and allow the chromatogram to develop for 10-15 minutes. Carefully remove the wick and allow the chromotogram to dry.

Report on the constituents of the three dyestuff mixtures provided and retain the chromatograms in your lab. book.

Separation of Amino acids

Mixtures of amino acids resulting from protein hydrolysis can be separated by paper chromatography. In this experiment a mixture of three amino acids (glycine, alanine and leucine) will be separated using the developing solvent-methanol : pyridine : water (20:1:5).

Since amino acids are colourless, their migration on paper cannot be followed visually. If the dried chromatogram is sprayed with ninhydrin reagent, each amino acid will appear on the paper as a blue spot. This is possible because ninhydrin reacts with amino acids to give a blue dye whose colour is so intense that the reaction can be used to detect very small amounts of amino acid.

Amino acids can be characterised by their R_f (rate of flow) values. The R_f value is defined as the ratio of the distance travelled by the amino acid to the distance travelled by the solvent.

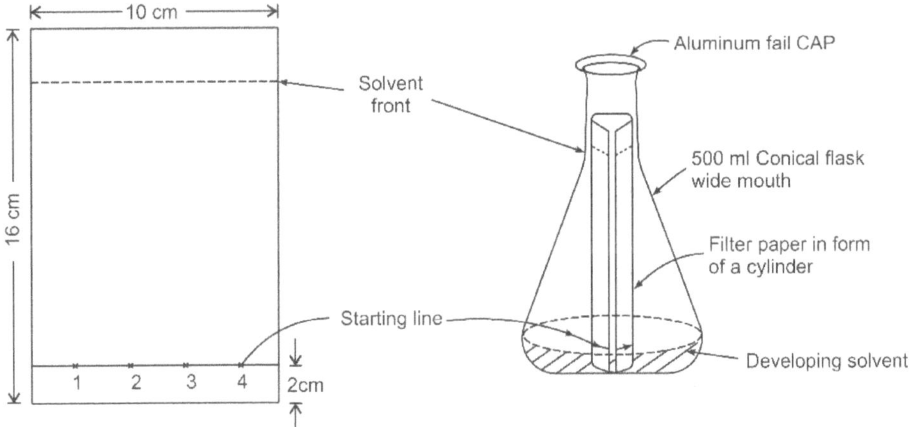

Fig. 4.3 Paper chromatography

On a strip of filter paper (Whatman No.1) 10 cm in width, draw a faint line 2 cm from the lower end (use a pencil). This is the starting line. Mark off the points 1-4 along this line at 2 cm intervals.

N.B. Protect the paper from the bench by means of an ordinary sheet of paper, and touch the paper with the fingers as little as possible as contact with skin protein can give rise to false spots.

Using a fresh capillary for each application, apply a spot of the individual amino acids at points 1, 2 and 3. At point 4 apply a similar spot of the mixture. Allow the spots to dry.

In a clean, dry conical flask (500 ml) place 20 ml of the developing solvent. Position the filter paper in the flask as shown in the diagram and make sure that the paper cylinder stands vertically and the starting line is above the surface of the liquid. Cover the mouth of the flask with aluminium foil, and allow the solvent to rise up the paper for about 1½ hr. Remove the paper, mark the upper limit of the solvent front with a pencil and hang up the paper to dry.

The spots are developed by spraying the dried chromatogram with the ninhydrin reagent.* The amino acids will appear as blue or purple spots.

Calculate the R_f values of each of the pure amino acids and for the components of the mixture using the relationship.

$$R_f = \frac{\text{Distance travelled by spot}}{\text{Total distance travelled by solvent}}$$

Identify the components of the mixture by comparison of R_f values and comment on the effectiveness of separation. Fix the dried chromatogram into your laboratory book.

* *Spray the dried chromatogram with the ninhydrin reagent in a Fume Cupboard. Use rubber gloves when spraying the chromatogram.*

4.6 Column Chromatography

This procedure utilizes differences in the strength of adsorption of given compounds to the surfaces of solids. These attractions are very dependent upon the nature of the solid adsorbent as well as the substances adsorbed. Since common adsorbents, such as alumina, silica gel, silicic acid and cellulose, are reasonably polar in character, the more polar compounds are more strongly adsorbed.

In this type of chromatography the sepration is effected by adsorption on a solid stationary phase. Several types of forces that cause interaction of the substance with the solid are hydrogen bonding, vander waals forces, electrostatic interaction etc. A cylinderical glass column of the type shows below is employed to separate a mixture by column chromatography. The size of the column which should be used depends on the volume of the sample to be separated. For many purposes a bed volume of about five times the sample volume should be suffice, with a ratio of diameter to hight of about 1:15, although for very intricate separations the ratio may need to be as much as 1:100.

To pack a column, first clean the glass column thoroughly and dry. Now position it vertically on an iron stand with the help of a big clamp. Close the stop cock and fill the column with an appropriate solvent. Insert a small plug of glass wool to the bottom of the column. Pour some fine dry sand so that it forms a 1 cm layer on the top of the glass wool. of the adsorbent; while the liquid is slowly draining from the pipette the tip is run round the inside of the column, care being taken not to touch the packing; any sample remaining on the walls can be care fully washed on to the column in a similar way, using pure solvent. Then a suitable solvent (mobile phase) is allowed to percolate through the column. The process is known as elution. It is common to use a relatively non-polar solvent in the beginning. The adsorption depends on both the nature of the solvent and the adsorbent. For effective separation the eluting solvent must be significantly less polar than the components of the mixture. The rate of the movement of a given component is related to the strength of its adsorption to the solid. Different compounds vary in this respect and if the rates of movement on the column are somewhat different, a separation will occur. Polar solvents are used for eluting strongly adsorbed components while non-polar solvents are used for weakly adsorbed components of a mixture.

An approximate order of increasing polarity of some common solvents is as follows. **Petroleum ether, hexane, benzene, toluene, chloroform, ethanol and water.** A mixture of two solvents of different proportions may be more useful for separation than a single solvent in many instances.

The most widely used column packing for adsorption chromatography is alumnium oxide, (alumina) other than silica gel. The particle size of the commercially available grade is in the range 50-200 µm (70-290 mesh). Alumina may be obtained in basic (pH 10), neutral (pH 7) and acidic (pH 4) forms and it is important to ensure the correct type is employed. For example, basic alumina may lead to hydrolysis of esters, acidic alumina may lead to dehydration of tertiary alcohols; in these circumstances neutral alumina is to be recommended.

Experiment 1

Separation of a mixture of o- and p-nitroaniline

Aim: To separate o-nitro aniline and p-nitro aniline in the given mixture

Materials needed: Chromatography column (3 cm internal diameter, 30 cm length), silica gel, (column grade, 60-120 mesh, 200 g.), conical flasks, mixture of o-nitro-aniline, and p-nitro aniline (100 mg each), cotton, t.l.c. micro slides (for monitoring the column chromatography separation).

Procedure: Clean the column with chromic acid. Rinse with distilled water and dry it. Loosely plug the bottom end of the column with a small piece of cotton using a glass rod. Fill the column with hexane/petroleum ether to half level. Prepare uniform slurry of 75 grams of silica gel in about 300 ml of benzene. Carefully transfer the slurry into the column with a stopcock slightly open. The slurry slowly settles down. Drain out any excess solvent. When all the solvent is drained out and the slurry level in the column is unchanged, close the stopcock. The packed column height should be 2/3 of the length of the column. Now, clamp the column vertically and carefully. Load the mixture of o-nitro aniline and p-nitro aniline as discrete plug by adsorbing it over silica gel. Then insert a further small loose plug of cotton over the sample. Moisten the column by running in 25 ml hexane from the dropping funnel; the column must be kept covered with solvent to a depth of about 5 cm. throughout the experiment to avoid the formation of air bubbles. Open the stopper and fill the column with

10% ethyl acetate/hexane. Feed the solvent continuously from the top. After complete elution, initially an orange band appears at the top of the column.

After 30 minutes two bands, a clear orange band (due to o-nitro aniline) moving faster than the pale yellow band (due to p-nitro aniline) with a white gap in between them appear. Collect the column eluate in serially labeled conical flasks. In each conical flasks collect about 150 ml eluate. Concentrate these fractions separately to a small volume. Examine their TLC. Combine the fractions that show the same t.l.c. behavior and evaporate the solvent to get pure o-nitro aniline and p-nitro aniline. Record the yield and their melting points.

Experiment 2

Aim: To purify the given commercial anthracene containing carbazole and naphthacene sample by column chromatography

Materials needed: Chromatography column (3 cm internal diameter, 30 cm length), silica gel, (column grade, 200 mesh), commercial anthracene (1 g), conical flasks, cotton, t.l.c. micro slides (for checking the purity of anthracene).

Procedure: Clean the column with chromic acid. Rinse with distilled water and dry it. Loosely plug the bottom end of the column with a small piece of cotton using a glass rod. Fill the column with petroleum ether to half level. Prepare uniform slurry of 75 grams of silica gel in about 300 ml of n-hexane or petroleum ether. Carefully transfer the slurry into the column with a stopcock slightly open. The slurry slowly settles down. Drain out any excess solvent. When all the petroleum ether is drained out and the slurry level in the column is unchanged, close the stopcock. The packed column height should be 2/3 of the length of the column. Now, clamp the column vertically and carefully. Then load the technical sample of anthracene (1 g) as discrete plug by adsorbing it over silica gel. Then insert a further small loose plug of cotton over the sample. Develop the chromatogram with 200 ml of petroleum ether. Over a time three bands will be seen.

a. Top - a narrow blue fluorescent band (carbazole)

b. Middle - a yellow non-fluorescent band (naphthacene)

c. Bottom - a broad blue violet fluorescent band (anthracene)

Continue the development with 75 ml of n-hexane/petroleum ether Change the receiver when fluorescent material commences to pass into the eluate. Stop the elution when the naphthacene band almost reaches the cotton wool plug. Distill off the solvent from the eluate containing anthracene. Pure anthracene (m.p. $215^{\circ}C$) visibly fluorescent in day light will be obtained. Check for the purity by means of t.l.c.

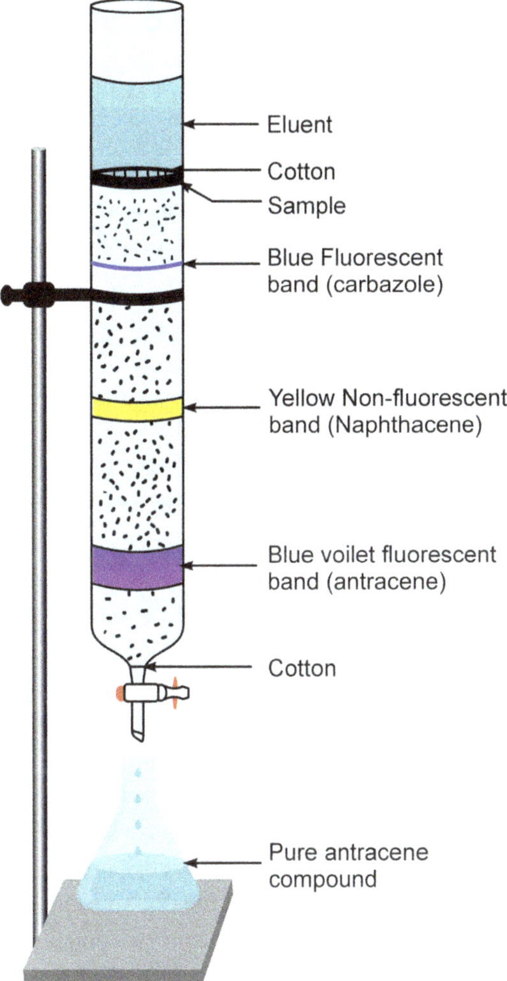

Fig. 4.4 Column chromatography

Experiment 3

Separation of *cis* and *trans*-azobenzene

Ordinary samples of azobenzene contain about 1% of the cis-isomer; irradiation with ultra-violet light increases this proportion.

Dissolve 0.5 g of azobenzene in 50 ml of light petroleum ether (b.p. 60-80), put the solution in a loosely stoppered test tube and stand it under an ultra-violet lamp for about 30 minutes.

Prepare a column 1.5 cm in diameter and 10 cm long, using light petroleum ether (b.p. 60-80°C) as the solvent and alumina (about 50 g) as the adsorbent. Allow solvent to run down until it is just above the top of the column, and cautiously add the azobenzene solution, letting the solvent flow out at a rate of 1-2 ml per minute. Elute with light petroleum ether, using the same flow rate. The *cis*-azobenzene remains as a compact band at the top, and a more diffuse band of *trans*-azobenzene moves down the column. Change the receiver flask when this band is just about to leave the column, and collect the yellow solution of the *trans*-isomer. Evaporation of the solvent gives *trans*-azobenzene, m.p. 68°C.

When the eluate becomes colourless change the eluting solvent to light petroleum ether containing 1% of methanol to elute the cis-azobenzene. Collect the fraction containing the *cis*-isomer, wash with water to remove the methanol, dry over anhydrous sodium sulphate and evaporate the solvent under reduced pressure to obtain cis-benzene, m.p. 61°C.

Experiment 4

Separation of plant pigments of spinach extract by column chromotography

Aim: It is well suited for a student lab exercise and a lecture demonstration of the nonquantitative extraction by column chromatography of plant pigments from spinach extract.It involves extraction, separation and identification.

Materials

Disposable (pasteur) glass pipet, (14.5 × 0.5 cm) or a small column, cotton, sterilized sand, neutral alumina (200-400 mesh), mortar and pestle, porcelain dish, spinach, petroleum ether, (40-65°), acetone, paper towels or clean dry cloth, test tubes and test tube rack.

Procedure: Preparing the extract

Take 50 g of dry spinach and place the spinach in a mortar to grind it. Transfer the pulp into a 600 ml beaker and add 100 ml of a mixture of 8:2 volume, petroleum ether (40 - 65^0 C) and acetone; stir it with the help of a spatula or glass rod until the solvent is of a green colour. Decant the liquid into a 100 ml conical flask.

To prepare the column for separation of the plant pigments (carotenoids, b-carotene and chlorophyll) in the spinach extract: Secure a disposable (pasteur) glass pippet or any other small column to a ring stand with a clamp or a rubber band.

Soak a small cotton piece in petroleum ether and stuff it from the top of the pippet column into the bottom end, gently, but firmly with a piece of wire or a thin glass rod. Next, transfer petroleum ether to the column until it is filled two thirds. Now add alumina slowly tapping the column with a pencil or hand finger to ensure even packing.

Be careful to maintain the level of petroleum ether well above the top of the alumina to avoid the formation of air bubbles in the column.

Bring the alumina to about 6 cm high or two thirds of column height, followed by another piece of cotton on top of the alumina, so that it does not disturb the alumina. Now transfer some of the spinach extract prepared earlier to the column. Wait until spinach extract falls just on top of the cotton, then immediately transfer petroleum ether to the column keeping the petroleum ether level well above the top layer of cotton during the separation procedure.

A definite yellow band of the carotenoid fraction will begin descending at this point, while green band of the chlorophyll fraction will ramain at the top of the alumina column.

Within a few minutes the carotenoid fraction is collected in a test tube and is sufficient for concentration to analyse in a spectrophotometer, scanning from 380 to 500 nm, yielding the characteristic visible spectrum for b-caroteine.

The chlorophyll fraction can be eluted from the column by changing the eluting solvent from petroleum ether to acetone. Continue adding acetone to the column until the chlorophyll fraction is collected in another test tube.

In conclusion, this procedure is an easy way of illustrating the concept of extraction, separation, and analysis of a biochemical mixture, in less than an hour, without cumbersome and expensive glass ware and material. It can be adapted to suit the needs of a student laboratory excercise or a lecture demonstration and one can encourage experimentation with different plant extracts and solvent mixtures by taking suitable columns of different sizes using the same procedure. See the purity by TLC.

4.7 Gas Liquid Chromatography

Gas-liquid chromatography is an extremely powerful method for separating mixture of relatively volatile organic compounds. The method is particularly suitable for the analysis of small quantities of material.

G.L.C is a type of partition chromatography in which the mixture of compounds is separated by partitioning the components between a flowing gas and stationary liquid phase. It is dependent for success on small differences in the partition coefficients of the substances to be separated.

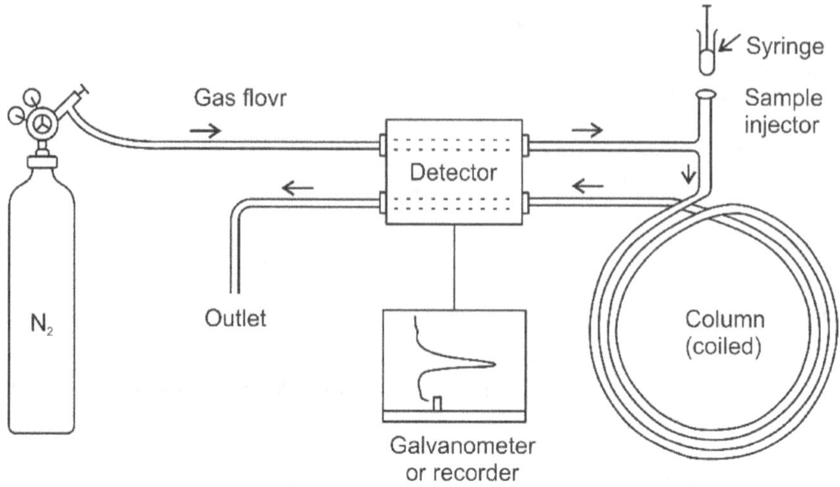

Fig. 4.5 Gas liquid chromatography

The material to be separated is volatilized into a carrier gas (nitrogen in this case) which flows at a constant rate over a non-volatile liquid (silicone oil in this case) supported on an inert powdered solid (celite in this case) packed into a column. The components of the mixture move through the column at different rates, depending on their partition coefficients between the gas and liquid phases, and emerge separately into the gas stream to be detected.

The G.L.C apparatus will be set up by the demonstrator and is ready for use. **Do not use the instrument unless you have received instructions in its use from the demonstrator.** Note the details of column length, column packing, carrier gas and flow rate, and temperature which are given on the card attached to the instrument.

Using the micro-syringe provided (the plunger of this syringe is extremely delicate and should be handled with care) inject 5 µl of the original mixtrue of methylene chloride and carbon tetrachloride into the apparatus. Simultaneously start the stop-watch.

A peak on the recorder tracing will indicate the emergence of the first component from the column, and this should be followed by a second peak for the second component. Note the times at which these peaks occur. (The areas under the peaks correspond approximately to the amount of each component present).

Examine in similar fashion 5 µl sample of the pure components as obtained by you in the fractional distillation. Only **two** fractions should be examined and these should be selected by reference to the graph. Report on the purity of these samples. The purity of the components of the mixture is reported in terms of retension times (Rt).

4.8 GLC Separation of Fatty Acid Methyl Esters

Qualitative analysis of a mixture is possible because, for a particular and reproducible set of instrument conditions, every compound has a known retention time on the column. Two compounds with identical retention times under identical conditions may therefore be the same, but this cannot be proved by GLC alone.

Quantitative analysis is possible when the identity of the compound is known. The magnitude of the response of the instrument's detector is proportional to the amount of substance. If the detector is coupled to a suitable strip-chart recorder, then this response is displayed as a peak on the chart. The area enclosed by the peak is then proportional to the amount of substance, and previous calibration of the instrument allows quantitative estimation.

Experimental procedure

The components of homologous series of esters can be separated quite readily on the Perkin-Elmer F11 instrument fitted with a flame ionisation detector. Ask for assistance from a demonstrator before using the instrument which should have an oven temperature of $210°$ and contain a 2 metre column of support material coated with 5% Apiezon 'L'. An injection setting '4' corresponds to an injection temperature of $260°$. Check these temperatures using the 'Read' button to give a recorder deflection, and set the 'Sensitivity' scale to 20×10^4; ensure that the instrument is temperature stabilised and the recorder baseline is horizontal before proceeding.

Inject 2 µl of the mixture of methyl esters of the monobasic esters $CH_3(CH_2)_x CO_2CH_3$ (x = 6,8,10,12 and 14). If necessary adjust the attenuation of the instrument to give peaks which are wholly on the chart, and repeat the injection. When a reasonable chromatogram has been obtained proceed immediately with the analysis of the unknown two-component mixture which will be supplied by the demonstrator. As far as possible, the instrument conditions for the two runs should be identical. The components of the unknown mixture can be identified by comparison of their retention times R_t against each component of the mixture.

For an homologous series $C_nH_{2n+1}CO_2CH_3$.

$$\text{Log } R_t \mu\ n$$

Demonstrate this relationship graphically.

4.9 Flash Chromatography

Introduction: Since its introduction by Still in 1978, Flash Chromatography has proved extremely popular in the laboratory for the rapid isolation of organic and organometallic compounds from crude reaction products and/or their purification. Its advantages include rapid separation times, typically 10-15 minutes, and the ability to separate compounds with an R_f value difference of 0.15.

The technique basically consists of forcing a chosen solvent system, using compressed air or nitrogen, through a closely packed column of inert support material in order to separate the components of a sample mixture initially placed at the top of the column.

Unlike medium and high pressure liquid chromatography which operates between 50-300 p.s.i., using mainly pre packed columns. Flash Chromatography operates at pressure of typically 5-10 p.s.i., and with columns packed in situ.

Silica is the most widely used column support material and is dry or slurry packed depending on the column diameter. Uniformly graded to a particle size of 40 – 63 im, this packing is able to provide several advantages over medium and high pressure liquid chromatography with respect to speed of operation and economy of solvent use.

Other support materials used include alumina, Polygosil 60, amine-bonded and special chiral columns.

During operation, the sample to be separated is normally added in solution to the top of the column support material and covered with a thin layer of acid washed sand to prevent disturbance when the column is filled with solvent. Relatively insoluble substances, on the other hand, are normally submitted to a prior absorption stage involving a five-fold weight of deactivated chromatography grade silica before column addition.

Using flash chromatography, a wide variety of substances have been separated including benzenoid derivatives, bypyridyls, imidazoles, peptides, pyrazoles, semi-synthetic antibiotics, sulphones and thiazolidines.

1. Column Packing (note*)

(a) In a fume cupboard pour dry silica (or alternative column packing) into the column to give a depth of 15 cm.

(b) Turn column tap to 'ON' position and with the aid of a funnel pour solvent on to the column, continue filling until solvent level is 7.5 cm from the top of the column. (ensure a receptacle for the expelled solvent is in place).

(c) Fit column head and screw down cap to give finger-tight fit.

(d) Connect column head to compressed air line and ensure air-bleed is open.

   * Turn compressed air on and adjust air-bleed to give a pressure of approximately 5-10 p.s.i. This will force the solvent through the silica displacing any trapped air. When the solvent level has reached the top of the silica, turn the column tap to the 'OFF' position then turn the compressed air off and ensure air bleed is fully open before removing the column head.

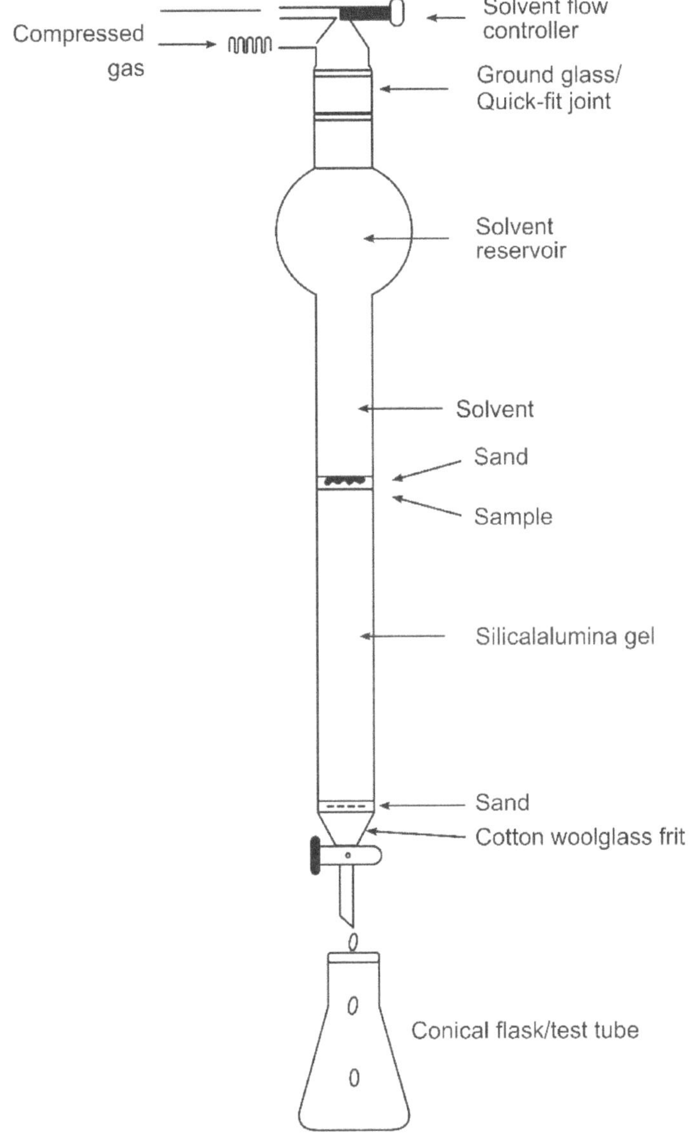

Fig. 4.6 Experimental set-up for flash column chromatography

2. Sample Loading

(a) Remove column head (**NB:** do not use pressure tube connection, grip securely round the joint).

(b) Apply sample onto the top of the silica (or alternative column packing) either (i) in the minimum amount of solvent or (ii) coated on to silica

(c) For (i) start here. Turn column tap to 'ON' position, fit column head and turn compressed air on to load sample onto the silica. When the solvent level has reached the top of the silica turn column tap to 'OFF' position then turn compressed air off and ensure air bleed is fully open before removing the column head.

(d) Wash the column walls with a little solvent to remove any traces of sample and repeat c), before proceeding to 4 a)

3. Sand layering and column elution

(a) For (ii) start here. Pour sand (acid washed) onto the silica until a layer approximately 1.2 – 2.5 cm has been formed.

(b) Carefully add solvent to column with the aid of a funnel, (tapping the sides of the column to allow any trapped air to displace from sand) until the solvent level is 2.5 to 5 cm from the top of the column.

(c) Fit column head, turn column tap to 'ON' position, turn compressed air on and collect fractions. The size of the fraction collected will depend on the size of column used. (The ideal flow rate is approximately 5 cm solvent height per minute).

Assay of materials eluted from column

Eluent can be monitored continuously by connecting the column in series with a UV or IR detector, or fractions can be examined by thin layer chromatography.

Reference: W.C. Still, M.Khan and A. Mitra, J. Org. Chem. 1978, 43, 2923;

W.C. Still U.S. Pat. 429 3422 (1979).

4.10 High Performance Liquid Chromatography (HPLC)

Principles of Chromatography:

" Chromatography is a physical process.

- Any Chromatography   system is composed of three Components:

1. Stationary phase

2. Mobile phase

3. Mixture to be separated

Why use HPLC

- Simultaneous analysis
- High resolution
- High sensitivity (ppm-ppb)
- Good repeatability
- Small sample size
- Moderate analysis condition
- No need to vaporize the sample just as in GC
- Easy to fractionate the sample and purify

HPLC types-applications

SCALE	CHROMOTOGRAPHIC OBJECTIVE
Analytical	Information : Compound identification and concentration
Semi-preparative	< 0.5 gram : Compound purification (small)
Preparative	> 0.5 gram : Compound purification (large)
Process (Industrial)	Grams to kilograms : Manufacturing quantities

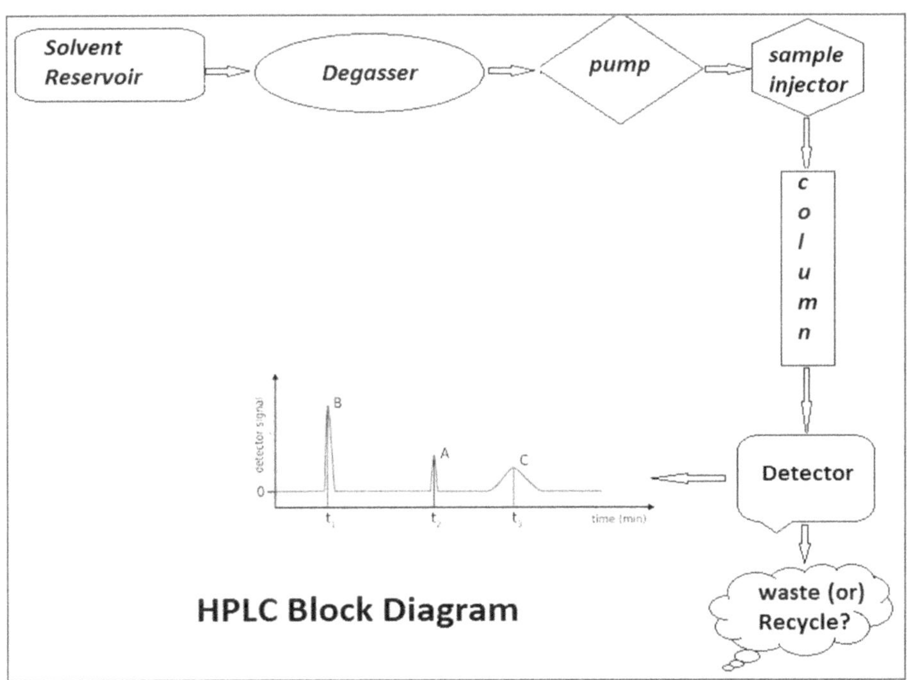

Mobile Phase:

The mobile phase should:

- Not alter column and its characteristics
- Be compactable with detector
- Dissolve the sample
- Have a low viscosity
- Permit easy sample recovery, if desire
- Pure no contaminants
- Commercially available at reasonable price

All the solvents should be degassed for smooth operation.

Mobile phase elution types:

- Isocratic
- Gradient

Isocratic analysis

- Isocratic: Same mobile phase throughout the elution of the sample.
- Most pharmaceutical assays are isocratic
- Common in QC applications
- Simpler HPLC & premixed mobile phases.

Disadvantages

- % Limited peak capacity
- % Problems with samples containing analytes of diverse polarities.
- % Difficult to quantitate due to excessive band broadening with long RTs.

***Gradient Analysis*:**

- Gradient: Strength of the mobile phase is increased with time during sample elution.
- Suitable for complex samples.
- Samples containing wide polarities.
- Amenable for high-throughput screening applications and for impurity testing.
- It yields better separation for early peaks and sharper peaks for late eluters.

Disadvantage

- More complex instrumentation.
- Greater skills in method development
- Difficulties in method transfer.

Gradient Elution example:

Pump A: Water

Pump B: Methanol

Mobile phase elution types:

- ¨ Isocratic
- ¨ Gradient

Isocratic analysis

- Isocratic: Same mobile phase throughout the elution of the sample.
- Most pharmaceutical assays are isocratic
- Common in QC applications
- Simpler HPLC & premixed mobile phases.

Disadvantages:

- % Limited peak capacity
- % Problems with samples containing analytes of diverse polarities.
- % Difficult to quantitate due to excessive band broadening with long RTs.

Gradient Analysis:

- Gradient: Strength of the mobile phase is increased with time during sample elution.
- Suitable for complex samples.
- Samples containing wide polarities.
- Amenable for high-throughput screening applications and for impurity testing.
- It yields better separation for early peaks and sharper peaks for late eluters.

Disadvantage

- More complex instrumentation.
- Greater skills in method development
- Difficulties in method transfer.

Gradient Elution example:

Pump A: Water

Pump B: Methanol

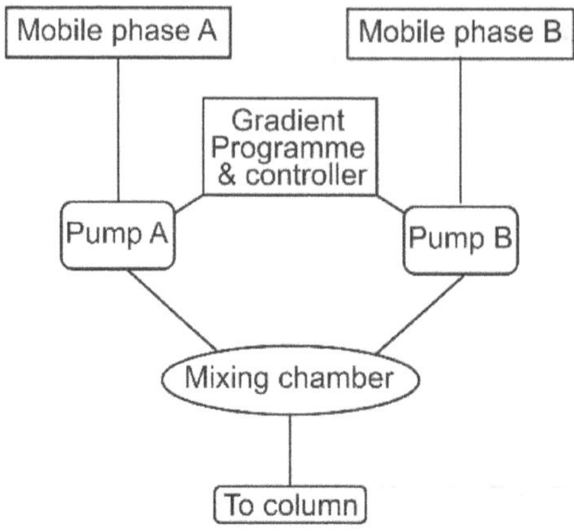

Time	Module	Action	Value
0.01	Pumps	B.conc	15
5.00	Pumps	B.conc	15
10.00	Pumps	B.conc	90
20.00	Pumps	B.conc	90
25.00	Pumps	B.conc	15
25.01	Controller	Stop	

Column:

- Types of Columns

- Normal phase (historically first): Stationary phase is highly polar-least polar comes off first.

- Reverse phase: Stationary phase is non-polar more polar comes off first.

Columns chemical interactivity with solute is mainly depends on the following interactions in different columns. C18 or C8 columns: hydrophobic interactions, Phenyl: π - π interactions, amino column: basic interactions, Cyano: largely dipolar, Chiral columns: Hydrogen bonding, π - π interactions, dipole induced dipole attractions in polysaccharide type columns (amylose based-CHIRALPAK IA, CHIRALPAK IG and cellulose based-CHIRALPAK IB and CHIRALPAK IC) suitable for enantiomeric separations.

4.10.1 HPLC Detectors

" A detector is required to sense the presence and measure the amount of a sample component in the column effluent.

" Good detectors exhibit: high sensitivity, low noise, a wide linear response range and unaffected by changes in conditions, response to all types of compounds or selective, non- destructive, cheap, reliable and easy to use.

" The function of the detector in HPLC is to monitor the analyte emerging from the column.

·· The output of the detector is an electrical signal that is proportional to property of the mobile phase and/or the solutes

4.10.2   Detectors Commonly used in HPLC

·· UV-Visible absorbance detector (UV-VIS)

·· Fluorescence detector

·· Photo-diode array detector (PDA)

·· Electrochemical (ECD)

·· Refractive index (RI)

·· Mass detectors (MS)

·· Conductometric detector

·· Chiral detector (Polarimetric & circular dichroism)

·· Evaporative light scattering detector (ELSD)

·· Radiochemical detection

4.11   Quantitative Analysis by HPLC

Area normalization: The area percentage calculation procedure reports the area of each peak in the chromatogram as a percentage of the total area of all peaks. Area percentage does not require prior calibration and does not depend upon the amount of sample injected within the limits of the detector. If all components respond equally in the detector and are eluted, then area percentage provides a suitable approximation of the relative amounts of components. In this method no response factors are used.

There are two important methods for quantification: (i) the external and (ii) internal standard methods; both the methods are performed using a calibration curve.

The external standard method creates a calibration curve for a standard sample and unknown samples are quantified using the calibration.

In the internal standard method, a fixed amount of an internal standard substance is added to an unknown sample when creating a calibration curve using a standard sample, and a calibration curve is created with the concentration ratio vs. peak area ratio for quantification.

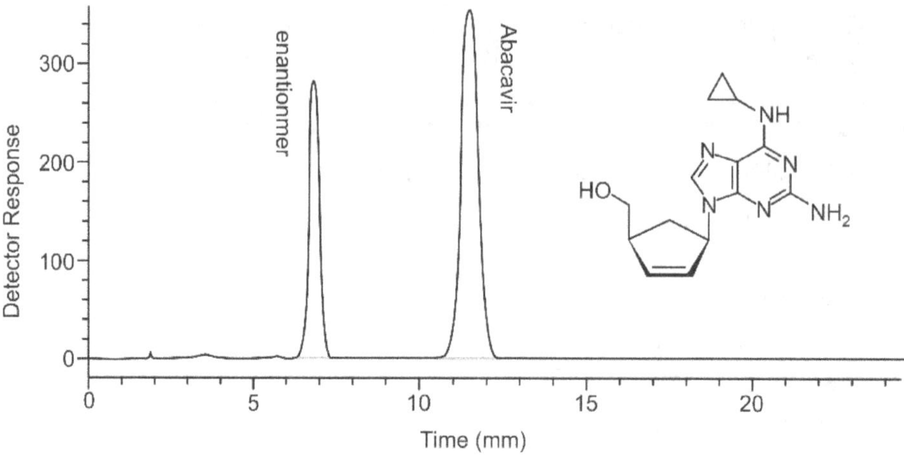

Figure. HPLC Chromatogram of Abacavir and its enantiomer.

Drug Name   :   Abacavir

Column   :   CHIRALPAK IG-3(150 × 4.6, 3ìm)

Mobile phase   :   10 mM Ammonium bicarbonate in water [(water: ACN)
                    (05/95 v/v)]

Quantitative analysis calculation:

Response factor of Abacavir in standard = peak area/standard amount

Amount of Abacavir in sample = Abacavir peak area/response factor

An injection containing Abacavir at a concentration of 1000 µg/mL is resulted area 100000

Response factor = 100000/1000 = 100

An injection of the sample with unknown concentration of Abacavir has a peak area of 50000

The amount of Abacavir present in the given sample = 50000/100 = 500 µg/m

HPLC Trouble-Shooting

If you have a problem, use trouble-shooting guide. Locate the table for type of problem you have, find the possible cause and use short description of the solution-corrective action.

Trouble-shooting reference table

SYMPTOM	POSSIBLE CAUSE	CORRECTIVE ACTION
No peaks	Detector lamp off	Turn lamp on
No flow	Pump off	Start pump
No pressure	Leak	Check pump for leaks
High back pressure variable	Obstruction in the column	Flush the column
Retention time	1. Change in the mobile composition 2. Air trapped in the pump (or) column	1. Check, make up of mobile phase, 2. Purge air from pump head and use column oven to avoid temperature fluctuations.
Split peaks	Contamination in guard or analytical column	Clean (or) replace the analytical column
Base line drift	Temperature fluctuations	Control column and mobile phase temperature
Board peaks	Mobile phase flow rate too low	Adjust flow rate
Negative peaks	Recorder leads reversed	Check polarity
Ghost peak	Contamination in injector or column	Flash injector between analysis and follow column washing guidelines.

ESTIMATION OF ASSAY

5.1 Estimation of the Amount of Aspirin

Principle : Estimation of aspirin depends upon the reaction given below:

$$\text{(aspirin, } -COOH, -OCOCH_3) \xrightarrow{2NaOH} \text{(salicylate, } -COO^{\ominus}Na^{\oplus}, -OH) + CH_3COO^{\ominus}\ Na^{\oplus}$$

A known amount of powdered tablets of aspirin is treated with a known excess of 0.5 N solution of Aq. NaOH. By back titration we will find out how much sodium hydroxide solution is left out after the hydrolysis with the help of standard HCL (By volumetric procedure)

If the known volume of excess of 0.5N NaOH taken is $= x$ cc

After the hydrolysis of aspirin the volume of NaOH left out is $= y$ cc

Then the volume of NaOH reacted with asprin $= x - y$ cc

As per the stoichiometry 2 moles of NaOH are needed to hydrolyze 1 mole of Asprin :. 80 gms of NaOH are needed to hydrolyze 180 gms of Aspirin.

By calculation it is possible to calculate the amount of aspirin present in the tablet if X-Y cc of NaOH are consumed.

Chemicals required:

1. A.R.Grade anhydrous Na_2CO_3 solid

2. 0.5 N HCl

3. Methyl orange indicator

4. 0.5 N NaOH Solution

5. Phenolphthalein indicator

6. Aspirin tablets (Disprin)

Procedure:

(a) Preparation of standard solution ~ 0.5N Na₂CO₃ solution: Weigh accurately around 2.65 gms of anhydrous Na_2CO_3 into a clean 100 ml standard flask. First dissolve in small quantity of distilled water and later make up the solution up to 100 ml mark with distilled water.

$$Na_2CO_3 \longleftrightarrow 2\,Na^+ = CO_3^{2-}$$

$$CO_3^{2-} + 2\,H_2O \longrightarrow H_2CO_3 + 2[OH]^-$$

$$\text{Eq. wt of } Na_2CO_3 = \frac{\text{Mol. wt.}}{2} = \frac{106}{2} = 53$$

Calculation

If 2.65 gms of Na_2CO_3 are dissolved in 100 ml of H_2O. We get 0.5N Na_2CO_3 solution If W gms of Na_2CO_3 are dissolved in 100 ml of H_2O. What is the concentration.

$$\frac{W \times 0.5}{2.65} = N_{Na_2CO_3}$$

(b) Standardization of 0.1 N HCl: Pipette out 20 ml of std Na_2CO_3 solution into a clean conical flask and add 20 ml distilled water. Add 2 or 3 drops of methyl orange indicator. Titrate this solution with approx 0.1 N solution of HCL till one drop changes the colour from yellow to orange.

$$2HCl + Na_2CO_3 \rightarrow 2NaCl + H_2O + CO_2$$

Calulation:

$$V_{HCl} \times N_{HCl} \rightarrow V_{Na_2CO_3} + N_{Na_2CO_3}$$

$$N\,HCl = \frac{V\,Na_2CO_3 \times N\,Na_2CO_3}{V\,HCl}$$

(c) Standardization of approx 0.1 N NaOH solution: Fill the burette with standard HCL. Pipette out 20 ml of approx 0.1 N NaOH solution into a clean conical flask. Add 2 drops of phenolphthalein indicator. Titrate this solution with HCL until the solution becomes colourless. Repeat the titrations a number of times to get at least two constant volumes for HCl.

$$NaOH + HCl \rightarrow NaCl + H_2O$$

$$V_{HCl} \times N_{HCl} = V_{NaOH} \times N_{NaOH}$$

Calculation:

$$N\ NaOH = \frac{V\,HCl \times N\,HCl}{V\,NaOH}$$

(d) Assay of Aspirin: Weigh accuratly 5 tablets of aspirin. Powder these tablets in a dry mortar. Now weigh accurately a quantity of the powder equivalent to 0.35 gms of aspirin and transfer it into 250 ml conical flask. Then add to it 20 ml of standard NaOH solution by means of a pipette. Boil gently for 10 minutes. Cool it to room emperature. Titrate the excess of NaOH solution present in the flask with standardized HCl solution using phenolphthalein indicator, till the pink colour of the solution becomes colourless. Note the volume of HCl solution *(y* ml) required. Repeat this experiment as many times as possible. With these values we calculate the amount of aspirin present in the given tablets of aspirin.

Model Calculation:

S.No.	1	2	3	4	5
Wt. of tablet	A gms	B gms	C gms	D gms	E gms

$$\text{Average wt. of table} = \frac{A+B+C+D+E}{5} = x\ gr$$

x gms of tablet contains 0.35gm of aspirin

S.No.	Aspirin + Vol. NaOH	Vol. Of HCl	Indicator
1	20 ml	y ml	Phenolphthalein
2	20 ml	y ml	
3	20 ml	y ml	

20 ml of 0.5 N NaOH consumed **x** ml of HCl

? ml of 0.5 N NaOH consumed yml of HCl

$$\frac{20 \times y}{x}\ \text{ml of NaOH (This is the vol. NaOH left in the flask after Hydrolysis)}$$

Therefore volume of NaOH solution consumed for the Hydrolysis of

$$\text{aspirin} = 20ml - \frac{20 \times y\ ml}{x} = zml$$

20ms NaOH in 1000 ml -   0.5 N

1ml of 0.5 N NaOH contains how many grams of NaOH ?

20 × 1 / 1000 = 0.02 gram

This means if we prepare exactly 0.5 NaOH solution - then 1 ml of this solution contains 0.02 g NaOH by weight.

80 gms of NaOH can hydrolyse 180 gms of aspirin

0.02 gms of NaOH can hydrolyse? gms of aspirin

$$\frac{0.02 \times 180}{80} = 0.045 \text{ gms of aspirin}$$

1 ml of 0.5 NaOH hydrolyse 0.045 g aspirin

1 ml of xN NaOH hydrolyse?

$$\frac{x\, N \times 0.045}{0.5\, N} = \text{Wt of aspirin (wg)}$$

which undergoes hydrolysis with 1 ml of **x** N NaOH Wt of Aspirin present = Wg × Z ml.

$$\%\text{Purity} = \frac{\text{Actual aspirin}}{\text{Theoretical wt of aspirin}} \times 100$$

$$= \frac{\text{wg}}{\text{x gms}} \times 100$$

5.2 Estimation of Ibuprofen

Principle: Estimation of Ibuprofen depends on the reaction given below.

Structure:

As per the stiochiometry

1 mole of ibuprofen reacts with 1 mole of NaOH

Mol. Formula of ibuprofen = $C_{13}H_{18}O_2$; Mol. weight = 206.28.

1 Mole of ibuprofen reacts with 1 mole of NaOH.

i.e. 206.8 gms of ibuprofen reacts with 40 gms of NaOH.

Chemicals required:

1. A.R grade anhydrous sodium carbonate solid

2. ~ 0.1 N HCl solution

3. ~ 0.1 N NaOH solution

4. Methyl orange indicator

5. Phenolphthaline indicator

6. Distilled chloroform

7. 15% v/v Methanol or ethanol

Procedure:

1. **Preparation of standard solution of Na_2CO_3:** Weight accurately about 0.53 gms of AR Na_2CO_3 into a 100 ml standard flask. Dissolve in minimum amount of distilled water

   Make it upto 100 ml mark by adding distilled water. Shake well to get uniform concentration.

$$N_{Na_2CO_3} = X \text{ gms}/53 \times 1000/100$$

2. **Standardisation of ~ 0.1 N HCl solution:** Pipette out 20 ml of standard Na_2CO_3 solution into a clean conical flask and add 20 ml distilled water. Add 2 to 3 drops of methyl orange indicator. Titrate this solution with approximately 0.1 N solution of HCl taken in a burette till one drop changes the colour from yellow to orange.

$$N_{HCl} = V_{Na_2CO_3} \times N_{Na_2CO_3}/V_{HCl}$$

3. **Standardisation of approximately 0.1N NaOH solution:** Fill the burette with standardized HCl solution. Pipette out ~ 0.1 N solution of NaOH into a conical flask. Add 2 to 3 drops phenolphthaline indicator. Titrate this solution with HCl solution until the solution becomes colourless. Repeat the titrations a number of times to get at least two constant volumes for HCl.

$$N_{NaOH} = V_{HCl} \times N_{HCl}/V_{NaOH}$$

4. **Estimation of ibuprofen:** Weigh and power 5 tablets of cambicam. Weigh the quantity of powered cambicam powder equivalent to the average weight of the tablets. Extract the ibuprofen with 60 ml of distilled $CHCl_3$ for 10 to 15 minutes in a conical flask. Filter and wash the residue 3 times with 10 ml of $CHCl_3$ each time & filter. Add the washings to the filtrate. Gently evaporate the $CHCl_3$ on a hot water bath. Dissolve the residue (ibuprofen) in 100 ml of 95% MeOH. Titrate the methanolic solution (whole of 100 ml solution) with standard 0.1N NaOH using phenolphthaline indicator. Note the volume of NaOH required for the titration.

Model Calculatios:

Normality of Na_2CO_3 solution	=	0.09811 N (assume)
Normality of HCl solution	=	0.10606 N (assume)
Normality of NaOH solution	=	0.101287 N (assume)

Titre value of ibuprofen Vs NaOH = 19.0 ml

40 gms of NaOH in 1000ml	1N
4.0 gms of NaOH 1000 ml	0.1 N
? gms of NaOH 1 ml of	0.1 N

$4 \times 1/1000$ =   0.004 g (this is the weight of NaOH present in exactly 1 ml of 0.1 N NaOH)

According to stoichiometry

40 gms of NaOH are equivalent to 206.28 gms of ibuprofen

0.004 gms of NaOH is equivalent to how many gms of ibuprofen?

$0.004 \times 206/40 = 0.0206$ gms of ibuprofen

1 ml of 0.1 N NaOH is equivalent to 0.0206 g of ibuprofen

1 ml of 0.101287 N NaOH is equivalent to how many gms of ibuprofen?

$0.101287 \times 0.0206/0.1 = 0.020865$ g of ibuprofen

Ibuprofen present in 100 ml solution prepared in step 4

$= 0.020865 \times 19$ ml $= 0.396435$ gms.

% Purity $=$ Weight of ibuprofen found/weight of ibuprofen claimed $\times$ 100.

$= 0.396435/0.4 \times 100$

$= 99.10875$ %

5.3 Estimation of Analgin

Principle: Analgin is a well known analgesic. It is known as Dipyrone and Metamisole. It's aqueous solutions are neutral. It is soluble in water (1 gm in 1.5 ml of water) and also in MeOH.

$C_{13}H_{16}N_3NaO_4S.5H_2O$

$\therefore$ Mol. Wt = 351.36

Analgin undergoes oxidation quantitatively with iodine. Therefore it can be estimated iodometrically by titrating a known quantity of a commercial drug containing analgin with standard solution of iodine. This is a redox reaction.

As per the stoichiometry: 127 gms of I_2 oxidizes $351/2 = 175.5$ gms of analgin. Then if x gms of I_2 are consumed in the oxidation, what would be the weight of analgin which undergoes oxidation.

$$(X \text{ gms of } I_2 \times 175.5) / 127$$

Iodine solution can be standardized by using standard hypo solution. In turn hypo solution can be standardized by titrating with a standard solution of potassium dichromate which is a primary standard.

Chemicals required:

1. Analar $K_2Cr_2O_7$ solid

2. 0.1 N Hypo solution

3. 0.1 N Iodine solution

4 1% Starch solution

5. Dil H_2SO_4

6. Potassium iodide solution (10%)

7. Uncoated analgin tablets (Novalgin) - 5 each.

Procedure: Preparation of 0.1 N standard solution of potassium dichromate

Weigh accurately about 0.49 gm of A.R.Grade $K_2Cr_2O_7$ solid into a 100 ml standard flask. Dissolve first in small quantity of H_2O and then make it up to 100 ml mark with distilled water. Shake thoroughly to ensure uniform concentration.

$$\text{Normality of } K_2Cr_2O_7 = \frac{Wg}{49} \times \frac{1000}{10}$$

(a) **Standardization of approx 0.1 N Hypo:** Pipette out 20 ml of standard solution of $K_2Cr_2O_7$ into a clean iodination flask. Add 15 ml of 10% KI solution and 6 ml of Conc. HCl followed by a pinch of $NaHCO_3$ solid.

Keep the flask in the dark 5 minutes for the completion of the reaction. Titrate the liberated iodine with hypo solution till a faint yellow colour appears in the solution. Add 1 ml of freshly prepared starch indicator (5gm in 25 ml water) where by the whole solution acquires blue colour. Continue the titration till the blue colour disappears and gives way to green colour. Note the volume of hypo needed for titration. Repeat the titrations a number of times to get at least two constant volumes for hypo.

$$N_{hypo} = \frac{V\,K_2Cr_2O_7 \times N\,K_2Cr_2O_7}{V_{hypo}}$$

(b) Standardization of approximate 0.1 N iodine solution: Pipette out 20 ml of iodine solution into a clean iodination flask and add 20 ml of distilled water. Then titrate this solution with hypo till a faint yellow colour remains in the solution. At this stage add 1 ml of starch indicator where by the whole solution aquires blue colour. Continue titrations till the blue colour disappears. Note the volume of hypo. Repeat the titration number of times to get at least three constant volumes for hypo.

$$N\,I_2 = \frac{V_{hypo} \times N_{hypo}}{V\,I_2}$$

Estimation of Analgin

Weigh separetly 5 tablets of Novalgin. Determine the average of weight of each tablet. Powder the 5 tablets of novalgin in mortar. Weigh accurately the novalgin powder equivalent to the average wt. of the 5 tablets and transfer into 100 ml standard flask. In the beginning dissolve the novalgin powder in 10 ml of distilled water. Then make up the solution upto 100 ml mark with 95% ethyl alcohol. Shake the solution well to get uniform concentration in the solution. Pipete out 25 ml of this solution into a clean iodination flask. Add slowly iodine solution from a burette to the iodination flask with shaking till a faint pink colour remains in the solution, at least for a minute. Note the volume of I_2 consumed.

Repeat this titration to get at least two constant volumes for I_2 solution (say it as X ml).

Model Calculation:

127 gms of I_2 in 1000 ml of water - IN

12.7 gms of I_2 in 1000 ml of water - O.lN

? gms of I_2 in 1 ml - 0.1N

$$= \frac{1 \times 12.7}{1000} = 0.0127g \text{ of } I_2$$

1 ml of 0.1 N I_2 corresponds to 0.0175 gms of analgin

1 ml of × N I_2 corresponds to?

$$\frac{xN \times 0.0175}{0.1} = Z \text{ gms}$$

Amount of analgin present in 25 ml solution = Z gms × **x** ml = w gms

25 ml of solution contains w gms of analgin

100 ml of solution contained ? gms of analgin

$$\frac{100 \times W}{25}$$

$$\% \text{ Purity} = \frac{\text{wt. of analgin determined}}{\text{wt. of analgin printed on the foil}} \times 100$$

5.4 Estimation of Codeine

Principle: Estimation of Codeine depends on the reaction given below

Codeine phosphate $\xrightarrow{NH_3}$ Codeine : $C_{18}H_{21}NO_3$ Molecular weight: 299 $\xrightarrow{HCl}$

Many cough syrups contain codeine as codeine phosphate. It is an expectorant. A known quantity or volume of syrup containing codeine phosphate will be taken and it is treated with ammonia solution; where by free codeine will be liberated from its quaternary salt. Then a known excess of standard HCl solution *(x ml)* will be added to free codeine where by codeine forms its quaternary salt with HCl. The excess of HCl that was left out after the formation of quaternary salt can be found out by titration with a standard solution of NaOH as per the following relationship:

$$V_{HCl} \times N_{HCl} = V_{NaOH} \times N_{NaOH}$$

$$V_{HCl} = \frac{V_{NaOH} \times N_{NaOH}}{N_{HCl}}$$

Volume of HCl used for quaternization of codeine $= \text{x} - V_{HCl} = \text{y ml}$

Then we find out weight of HCl (in grms) present in y ml. Let us say it was W gms.

As per the stoichiometry

36.5 gms of HCl is capable of quarternizing 299 gms of codeine

W gms of HCl is capable of quarternizing ? gms of codeine.

$$\frac{W \times 299}{36.5} = Z \text{ gms of codeine}$$

Procedure

(a) **Preparation of standard solution of 0.05 N Na₂CO₃:** Weigh accurately 0.265 gms of Na_2CO_3 and transfer it into 100 ml standard flask. Dissolve it in minimum quantity of water and then make up the solution upto 100 ml mark with distilled water. Shake the solution well to get uniform concentration.

$$N\,Na_2CO_3 = \frac{Wt.\,of\,Na_2CO_3}{53} \times \frac{1000}{100}$$

(b) **Standardization of 0.05 N HCl:** Fill the burette with the given HCI solution. Pipette out 20 ml of standard Na_2CO_3 solution into a clean conical flask. Add 20 ml of water and 2 drops of methyl orange indicator and shake well. The solution aquires yellow colour. Titrate this solution with HCI until one drop of HCI changes the colour of the solution from yellow to orange colour. Note the volume of HCl. Repeat the titrations a number of times to get at least two constant volumes for HCl.

$$N_{HCl} = \frac{V\,Na_2CO_3 \times N\,Na_2CO_3}{V_{HCl}}$$

(c) **Standardization of 0.05 N NaOH solution:** Fill the burette with 0.05 N HCI solution. Pipette out 20 ml of NaOH solution into a clean conical flask. Add 1 or 2 drops of phenolphthalein indicator. The solution acquires pink colour. Titrate this solution with 0.05 N HCI till the solution becomes colourless. Note the volume of HCl. Repeat the titration a number of times to get at least two constant volumes for HCl. Find out the normality of sodium hydroxide solution.

$$N_{NaOH} = \frac{V_{HCl} \times N_{HCl}}{V_{NaOH}}$$

(d) **Estimation of codeine:** Measure 5 ml of codeine phosphate syrup (Mitlinctus or Corex) into a dry clean conical flask. Dilute with dilute NH_3 solution until the solution is just alkaline to litmus paper. Afterwards extract 4 times with 25 ml portions of $CHCl_3$ each times in a separating funnel. Wash the Chloroform extract with 10 ml portion of water, three times. Combine the chloroform extract and evaporate the solvent on a water bath. The residue that is obtained is mixed with about 5 ml alcohol and again it is evaporated. Dissolve the remaining residue in 5 ml of 0.05 N HCI and shake well till the residue is completely soluble. Titrate this solution with 0.05 N NaOH using methyl red indicator to find out the unreacted HCl. Note the volume of NaOH needed for the titrations.

Model Calculations:

S.No.	Vol. codeine phosphate syrup (5 ml) and 0.05 N HCl (5 ml)	Vol. NaOH (SAY)	Indicator
1.	5 + 5 ml	3.2 ml	Methyl red
2.	5 + 5 ml	3.2 ml	- do -

$$V_{HCl} \times N_{HCl} = V_{NaOH} \times N_{NaOH}$$

3.2 ml is the volume of 0.05 N NaoH needed to neutral is the excess of HCl left after quaternizing 5 ml of codeine phosphate.

Then what is the volume of HCl corresponding to 3.2 ml NaOH. This we can find out by using the above relationship

$$V_{HCl} = \frac{V_{NaOH} \times N_{NaOH}}{N_{HCl}}$$

$$V_{HCl} = \frac{3.2 \times 0.059\,N}{0.0425\,N} = 4.42\,ml$$

Then volume of HCI used by codeine phosphate.

Eg: = 5.0 ml- 4.442 ml = 0.558 ml

The amount of HCI present in 1 ml of 0.05 N HCI quaternizes 0.0232 gms of codeine phosphate

0.558 ml of 0.0425 N HCI quaternizes ? of codeine phosphate.

$$= \frac{0.558 \times 0.0425 \times 0.0232}{0.05} = 0.011003\,gms$$

$$= 11.003\,mg$$

OR

1 ml of 0.0425 HCl quaternizes how much codeine

$$= \frac{0.425 \times 0.0232}{0.05} = 0.01972\,gms$$

Amount of codeine present in the given amount

$$= 0.01972 \times 0.558$$

$$= 0.011003\,gms = 11.00\,mg$$

5.5   Estimation of Chlorides in Ringer's Lactate Solution

Principle: Ringer's lactate solution is administered intravenously to patients suffering from dehydration. It contains three electrolytes namely NaCI, KCI and $CaCl_2$. The total chloride contents in the solution are 0.403 g per 10 ml.

A known quantity (10 ml) of Ringer's lactate solution is treated with a known excess of 0.1 N solution of $AgNO_3$. Silver nitrate reacts with chlorides and form the corresponding amount of AgCl. The excess (unreacted) of $AgNO_3$ will be determined by titrating with a standard solution of NH_4/KSCN. The volume of NH_4/KSCN consumed in the titration corresponds to the volume of $AgNO_3$ left unreacted.

In a similar manner, a blank titration is performed by taking 20 ml of 0.1 N $AgNO_3$ solutions with NH_4/*KSCN* solution.

Volume of thiocyanate solution required for titrating

20 ml of 0.1 N $AgNo_3$ solutions: y ml

Volume of thiocyanate required for titrating

the excess 0.1 N solution of $AgNO_3$ x ml

Volume of $AgNO_3$ reacted with chlorides

present in Ringer's lactate solution   y - x ml

(a) Preparation of standard solution of sodium chloride (0.1N): Weigh accurately around 0.58 g of AR sodium chloride and transfer it into a 100 ml standard flask. Dissolve NaCI in small quantity of distilled water and then make up the solution up to 100 ml mark.

$$N_{NaCl} = \frac{Wt.\, of\, NaCl}{Eq.\, Wt\, of\, NaCl} \times \frac{1000}{10}$$

(b) Standardization of approximately 0.1N solution of $AgNO_3$: Neutral solution of NaCI when titrated with $AgNO_3$ solution, silver chloride precipitate will be formed.

$$AgNO_3 + NaCl \rightarrow AgCl + NaNO_3$$

In the above titration, potassium chromate indicator is used to detect the end point. When potassium chromate indicator method is used, it is called Mohr's method.

Pipette out 20 ml of standard sodium chloride solution into a conical flask. Add 1 ml of potassium chromate indicator and then titrate with about 0.1 N $AgNO_3$ solution. During the titration, red colored precipitate of silver chromate is formed temporarily and imparts red color to the solution. But on stirring the solution present in the conical flask, the red color disappears. At the end point, the red color persists. Note the volume of AgNO$_3$ consumed in the titration. Repeat the titrations till two concurrent values are obtained for AgNO$_3$ solution.

$$N_{AgNO_3} = \frac{V_{NaCl} \times N_{NaCl}}{V_{AgNO_3}}$$

Note:

1. It will be good to carry out the titration in a porcelain container. It will help us to identify the colour change at the end point.

2. In this titration, a large quantity (1 ml) of indicator is used. Therefore, at the end point the volume of AgNO$_3$ which reacted with the indicator must be determined and has to be subtracted from the titer value. This is called indicator correction.

$$2AgNO_3 + K_2CrO_4 \rightarrow Ag_2CrO_4 + 2KNO_3$$

Determination of indicator correction: Pipette out 1 ml of indicator solution into a conical flask and dilute with distilled water. As done in the case of chlorides estimation, titrate the solution with standard AgNO$_3$ solution till the solution acquired red color. The volume of AgNO$_3$ consumed is known as indicator correction.

(c) **Standardization of potassium thiocyanate solution:** Potassium thiocyanate is a hygroscopic substance. Hence it cannot be used as a primary standard. That means we cannot prepare a standard solution of KSCN by weighing method.

Approximately, 0.1 N solution of KSCN is made and then its concentration is determined by titration with a standard solution of 0.1 N AgNO$_3$. (dissolve about 9.7 g of KSCN in I L of distilled water. We get approximately 0.1N Solutions).

Pipette out 20 ml of standard solution of AgNO$_3$ into a conical flask. Add 5 ml of AgNO$_3$ into a conical flask. Add 5 ml of dil HNO$_3$ (6 M) solution and 1 ml of ferric nitrate indicator solution. Take KSCN solution in the burette. Add slowly KSCN solution from the burette to AgNO$_3$ solution. During the titration white precipitate of AgSCN forms. As soon as thiocyanate solution reaches the solution in the conical flask, the indicator undergoes temporarily a chemical reaction and forms a red colored solution. But when the flask is shaken, the red color disappears. But at the end point, the solution in the conical flask acquires a permanent red color. Note the volume of thiocyanate solution. Repeat the titrations a number of times to get two concurrent readings for the volume of thiocyanate solution consumed.

$$AgNO_3 + KSCN \rightarrow AgSCN \downarrow + KNO_3$$

$$N_{KSCN} = \{V_{AgNO3} \times N_{AgNO3}\} / V_{KSCN}$$

(d) Estimation of chloride in Ringer's lactate solution: Pipette out 10 ml of Ringers Lactate solution into a clean conical flask containing 20 ml of distilled water. To this, add 20 ml of 0.1 N AgNO$_3$ solution, 1.5 ml of conc HNO$_3$, 2 ml of nitrobenzene and 1 ml of ferric ammonium sulfate solution. Shake well, titrate this solution with 0.1 N/KSCN solution until the solution acquires reddish yellow color. Note the volume of thiocyanate needed for the titration. Repeat the titrations to get at least two concurrent values for thiocyanate solution.

Model Calculations:

S.No.	Vol of AgNO$_3$	Vol of KSCN	Indicator
1	20 ml	22 ml	
2.	20 ml	22 ml	
3	20 ml	22 ml	

Therefore 20 ml of AgNO$_3$ are equivalent to 22 ml of thiocyanate

S.No.	$\left(\dfrac{\text{Volume of}}{AgNO_3 + \text{Ringers lactone}} \right)$	Vol of KSCN	Indicator
1.	10 ml + 20 ml	9.7 ml	
2.	10 ml + 20 ml	9.7 ml	
3.	10 ml + 20 ml	9.7 ml	

22 ml of KSCN solution is equivalent to 20 ml of 0.1 N AgNO$_3$

9.7 ml   "   ?   "

$$\frac{9.7 \times 20}{22} = 8.818 \, ml \text{ of AgNO}_3 \text{ (this is the volume of AgNO}_3 \text{ left unreacted)}$$

Volume of 0.1 N AgNO$_3$ required to react with total Cl$^-$ ions present in 10 ml of RL solution:

20ml - 8.818 = 11.1819 ml

$$AgNO_3 + Cl^- \rightarrow AgCl + \overset{\ominus}{N}O_3$$

1000 ml of 1N AgNO$_3$ reacts with 35.5 g of Cl$^\ominus$ ions

1000 ml of 0.1 N AgNO$_3$ reacts with 3.55 g of Cl$^\ominus$ ions

1 ml of 0.1 N AgNO$_3$ reacts with 00.00355 g of Cl$^\ominus$ ions

1 ml of xN $AgNO_3$ reacts with how many gms of $Cl^{\ominus}$ ions

$$\frac{x\,N\times0.00355}{0.1} = Z\,g\,of\,Cl^{\ominus}ions$$

If 11.1819 ml of standard $AgNO_3$ is consumed, means how many grams of $Cl^{\ominus}$ ions might have reacted:

$$11.1819 \times Z = W\,g\,of\,Cl^{\ominus}\,ions$$

Therefore, W g is the total weight of total chloride ions present in 10 ml of RL solution.

Then W × 10 gives the total weight of total chlorides present in 100 ml of RL solution

$$Percent\;purity = \frac{Total\,weight\,of\,chlorides\,found}{Total\,weight\,of\,chlorides\,\,claimed}\times100$$

5.6 Estimation of Ascorbic Acid

Aim: To determine the amount of ascorbic acid present in CELIN (Vitamin C) tablets

Principle: Ascorbic acid undergoes oxidation to dehydro ascorbic acid. Quantitatively with one electron oxidizing agents such as Cerric ammonium sulphate (CAS)

Therefore a known quantity of ascorbic acid can be titrated directly with standard solution of CAS using ferroin indicator. From the volume of the CAS we calculate the Wt. of CAS present in that. Then as per the stoichiometry.

632 gms of CAS oxidizes 88 gms of Ascorbic Acid

X × gms of CAS oxidizes ? gms of Ascorbic Acid.

$$\frac{x\times88}{632} = Z\,g\,wt.of\,ascrorbic\,acid$$

Ferroin Solution: Dissolve 0.695 gms iron (II) sulphate hepta hydrate and 1.485 gms of 1,1O-phenanthroline monohydrate in 100 cm^3 of distilledwater.

Procedure: Weigh and powder 5 tablets of Celin (Vitamin-C). Weigh accurately powdered celin around 0.15 gms into a clean conical flask. Dissolve it in a mixture of 30 ml of distilled water and 20 ml dil. H_2SO_4. Add 2 drops of ferroin indicator where by the whole solution acquires red colour. Titrate this solution with a standard solution of 0.1 N cerric ammonium sulpahte till the red colour of the solution gives way to blue colour. Note the volume of CAS required for the titration.

Model Calculation:

	1	2	3	4	5
Wt of tablets	0.600	0.602	0.598	0.621	0.59

Average Wt. = 0.604 gms

Wt. of Celin powder taken = 0.15 ms

632 gms of CAS in 1000 ml – I N

63.2 gms of CAS in 1000 ml - O.I N

? gms of CAS in 1 ml - O.1N **soultion**

$$\frac{63.2 \times 1ml}{1000\ ml} = 0.0632\,gms$$

(This is the wt. of CAS present in 1 ml of 0.1 N solution of CAS)

Volume of CAS consumed in the titration = 16.2 ml (say)

According to stoichiometry :

632 gms of CAS oxidizes 88 gms of Ascorbic acid

0.0632 gms of CAS oxidizes? gms of Ascorbic acid

$0.0632 \times 88/632 = 0.0088$ g

1 ml of 0.1 N CAS oxidises 0.0088 g of vit-C

16.2 ml of 0.1 N CAS oxides how many gms of vit-C

$16.2 \times 0.0088/1 = 0.14256$ g

0.15 g of CELIN contain – 0.14256 g of Vit-C

0.604 g of CELIN – do – how many gms of vit-C

$0.604 \times 0.14256/0.15 = 0.57$ g of vit-C

5.7 Estimation of the Amount of Diazepam

$$C_{16}H_{13}ClN_2O$$
Mol wt : 284

Aim: To determine the amount of Diazepam in Zepose (5mg tablet) by U.V. spectroscopy method.

Branded name for diazepam (zepose); (usage-anticonvulsant and sedative).

Principle: The amount of diazepam in zepose tablets can be determined by making use of specific absorbance value ($A_{1cm}^{1\%}$) of diazepam. According to Beer's Lambert law the absorbance is directly proportional to concentration. After measuring the absorbance of appropriate dilute solution of test drug diazepam at 280 nm, concentration of test solution can be calculated by using the specific absorbance of value of Diazepam ($A_{1cm}^{1\%}$ is 446 nm).

$$OD = A_{1cm}^{1\%} \times b \times c$$

$$C = \frac{OD}{A_{1cm}^{1\%}} \times 1 = \frac{OD}{446} \times 1$$

Procedure: Weigh accurately 5 tablets of zepose and find out their average weight. Powder the 5 tablets in a clean and dry mortar. Weigh accurately around 80 mg. of the zepose powder into a clean 100 ml beaker. Add 15 ml of water and allow it to stand about 15 minutes with occasional shaking. Now add 25 ml of 0.5% w/v solution of sulphuric acid in methanol (about 1.0 ml of conc. sulphuric acid in 300 ml of methanol). Shake the solution for about 10 minutes and filter the solution and dilute to 100 ml in 100 ml standard flask. Find out the absorbance of this solution using UV visible spectrophotometer at 280 nm.

Model Calculation:

Wt. of one tablet = 0.211 gm

Amount of powdered tablet weighed = 0.081 gm

Optical density recorded = 0.856

$$\text{Concentration} = \frac{OD}{A_{1cm}^{1\%} \times b}$$

$A = 0.856\ ;\ A^{1\%} = 446;\ b = 1\ cm$

$$C = \frac{0.856}{446 \times 1} = 0.0019192 \text{ g/100 ml}$$

0.081 g of zepose contained 0.0019192 gm of diazepam

0.211 g of zepose contains? gm og diazepam

$$= \frac{0.211 \times 0.0019192}{0.081} = 0.0049987 \text{ gm}$$

$$\% \text{ Purity} = \frac{\text{wt. determined by assay}}{\text{wt. printed on the foil}} \times 100$$

$$\% \text{ Purity} = \frac{0.0049987}{0.005} \times 100$$

$$= 99.97\ \%$$

5.8 Estimation of Riboflavin

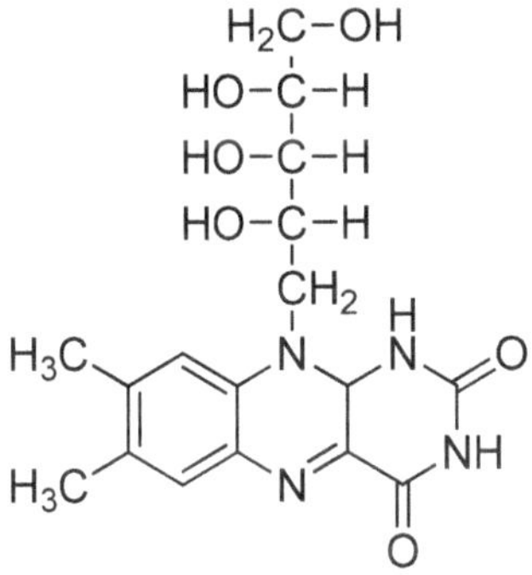

Principle: Riboflavin is a yellow coloured crystalline compound with some odour. Its saturated solution is neutral to litmus. It is very slightly soluble in water, alcohol and NaCl solution but soluble in dilute alkalies. It is not soluble in chloroform and ethers.

λ_{max} : 220; 266; 371; 444; 475

It can be estimated by using specific absorbance method.

Specific absorbance is the absorbance of a specified concentration in a cell of specified path length.

The most common form in pharmaceutical analysis is the A(1 %; 1 cm) which is the absorbance of a 1 g in 100 ml (1% W.V.) solution in a 1 cm cell. The Beer-Lambert equation therefore takes the form

$$A = A_{1cm}^{1\%} \times b \times c$$

$$C = \text{grams per 100 ml}$$

$$B = \text{cell length (1 cm)}$$

The units of A(1 %; 1 cm) are $dlg^{-1} cm^{-1}$

Therefore $\qquad \varepsilon = \dfrac{A_{1cm}^{1\%} \times \text{Mol.wt}}{10}$

Where ε = molar extinction coefficient

In the equation

$$A_{(O,D)} = A_{1cm}^{1\%} \times b \times c$$

C (Gms/100 ml) can be calculated if A (optical density) and the value of $A_{1\,cm}^{1\%}$ are known as indicated below

$$C = \dfrac{OD}{A_{1cm}^{1\%}} \qquad\qquad \text{..... (Eq. – A)}$$

That means after measuring the absorbance (OD) of appropriate dilute solution of the test compound at the λ (wave length) where $A_{1cm}^{1\%}$ is known (i.e. by using standard specific absorptivity value of the 1% solution of the test compound at a particular absorption maxima) (λ), we can calculate the concentration of the test compound by using equation - A.

Procedure: Weigh accurately 5 tablets of riboflavin. Combine the tablets and powder them in a clean and dry mortar.

Weigh exactly around 25 to 30 milli grams of powdered riboflavin into a clean beaker (100 ml capacity). Add 1.5 ml of glacial acetic acid and 25 ml of distilled water and heat on a water bath for one hour with occasional stirring. At the end of the heating period dilute the solution with another additional portion of 25 ml of distilled water and cool. Add 7.5 ml of 1 molar NaOH solution with continuous stirring and then transfer into a 250 ml standard flask. Make up the solution up to 250 ml mark with distilled water. Shake the solution well to get uniform concentration. Then filter the solution through a Whatmann filter paper. Discard the first 50 ml portion of the filtrate and collect the remaining filtrate. Find out the absorbance of the filtrate using UV -Visible spectrophotometer at the wavelength (absorption maxima = 444mn) where the specific absorptivity value ($A_{1cm}^{1\%}$) of riboflavin is reported.

Model Calculation:

Wt. of 5 tablets

$0.0918 + 0.0924 + 0.0912 + 0.0914 + 0.0912$

Average Wt. $(0.458 / 5) = 0.0916$

Amount of riboflavin powder weighed $= 0.025$ gms

Optical density recorded $= 0.352$

$$C = \frac{OD}{A_{1cm}^{1\%} \times b}$$

where $OD = 0.352$

$$A_{1cm}^{1\%} = 323 \text{ nm}$$

$$b = 1 \text{ cm}$$

$$= \frac{0.352}{323 \times 1} = 0.0010897 \text{ gms/100 ml}$$

Wt. of riboflavin present in 100 ml $= 0.0010897$ gms

Wt. of riboflavin present in 250 ml $=?$ gms

$$= \frac{250 \times 0.0010897}{100} = 0.0027242 \text{g}/250 \text{ ml}$$

0.025g of riboflavin powder contains 0.0027242 gms of the drug

0.0916 gms of powder contains?

$$= \frac{0.0916 \times 0.0027242}{0.025} = 0.00998 \text{ gms}$$

$$\%\text{Purity} = \frac{0.00998}{0.01} \times 100$$

$$= 99.8\%$$

5.9 Estimation of Calcium Ions

Aim: To estimate volumetrically the amount of calcium ions present in the tablets of calcium sandoz.

Principle: Calcium ions present in Calcium Sandoz ($CaCO_3$) tablets are estimated by using volumetric method. A solution containing a known weight of the drug is titrated with a solution of EDTA (Di sodium salt of ethylene diamine tetra acitic acid) using Eurochrome black-T as indicator.

Procedure: Weigh accurately the given tablet of Calcium Sandoz. ($\sim$ one tab) Powder it in a clean and dry mortar. Weigh accurately around 0.5 gms of

Calcium Sandoz powder and dissolve in 50 ml of distilled water by heating on low flame (some white compound remains undissolved. It may be an additive like binder, starch etc). Cool the solution to R.T and add 5 ml of 0.1 M $MgSO_4$ solution and 10 ml of ammonia - ammonium chloride solution (It is a buffer solution and maintains a pH of 10). Add match head size of Euro Chrome Black-T indicator where by the whole solution turns into wine red colour. Titrate with 0.1 M EDTA till the solution acquires blue colour. Note the volume of EDTA consumed in the titration.

Model Calculations:

Wt. of Calcium Sandoz tablet   = 1.1952 gms

Wt. of Calcium Sandoz powder = 0.5 gms

S.No.	Solution to be Titrated	Vol. EDTA	Indicator
1.	$CaCO_3$ solution 50 ml dist H_2O 10 ml Buffer solution 5 ml 0.1 M $MgSO_4$ solution	25.4 ml	Eurochrome black-T wine red to blue

Vol. of EDTA required = 25.4 ml

Vol of EDTA required to react with calcium ions

   = Total Vol of EDTA required for titration $-$ 5.0 ml of $MgSO_4$

i.e. 25.4 $-$ 5 = 20.4 ml

(Gram Mol. Wt of EDTA) 372 gms of EDTA in 1000ml $-$ 1.0 M

37.2 gms of EDTA in 1000 ml- O.lM

? gms of EDTA in 1 ml = 0.1 M $= \dfrac{37.2 \times 1}{1000} = 0.03728g$

$\therefore$ Wt. of EDTA present in 1 ml of 0.1 M Soln = 0.03728g

Wt. of EDTA present in 20.4 ml of 0.1 M EDTA

Solution = 20.4 × 0.03728

          = 0.7588 gms of EDTA

As per stoichiometry

372 gms of EDTA equivalent to 100 (gram mol. Wt of $CaCO_3$)

0.7588 gms of EDTA equivalent to? gms of $CaCO_3$

$$= \frac{0.7588 \times 100}{372} = 0.20397 \text{g}$$

0.5 gms of calcium Sandoz contained 0.20397 gms of $CaCO_3$

1.1952 gms of calcium Sandoz contained? gms of $CaCO_3$

$$= \frac{1.1952 \times 0.20397 \text{ g}}{0.5} = 0.4875 \text{ g of Ca } CO_3$$

100 gms of $CaCO_3$ contain 40 gms of calcium ions

0.4875 of $CaCO_3$ contain? gms of calcium ions

$$= \frac{0.4875 \times 40}{100} = 0.19502 \text{ g of calcium ions}$$

SYNTHESIS OF ORGANIC MOLECULES

6.1 Preparation of *t*-Butyl Chloride

Discussion: Tertiary alcohols react very readily with concentrated hydrochloride acid to produce the corresponding chloride. The mechanism is S_N1 type involving formation of an intermediate carbonium ion. Using tertiary butyl alcohol most of the carbonium ion combines with the chloride ions to form *t*-butyl chloride, although there is a side reaction where elimination of a proton from the intermediate carbonium ion (E_1 reaction) gives isobutene which escapes from the reaction mixture.

Tertiary alcohols react very readily with concentrated hydrochloric acid to produce the corresponding chloride.

$$(CH_3)_3C-OH+HCl \rightarrow (CH_3)_3C-Cl$$

Experimental: Mix t-butanol (25g., 31.5ml., 0.34 mole) and conc. hydrochloric acid (85 ml.) in a 150 ml separation funnel., Shake the mixture periodically during 20 minutes, loosening the stopper after each shaking to relieve any internal pressure. Allow the stand for a few minutes until the layers separate, and remove and discard the lower acid layer. Wash the upper layer with saturated sodium bicarbonate solution (10ml.) (caution: CO_2 pressure). Dry the organic layer with anhydrous sodium sulphate, filter and distil the filtrate. Collect the fraction of b.p. $49\text{-}51^{O}$

$$H_3C-\underset{\underset{CH_3}{|}}{\overset{\overset{CH_3}{|}}{C}}-OH \;+\; HCl \;\rightleftharpoons\; H_3C-\underset{\underset{CH_3}{|}}{\overset{\overset{CH_3}{|}}{C}}-Cl \;+\; H_2O$$

Mechanism: It is S_N^{1} type reaction preceded by formation of an intermediate carbonium ion. Using t-butyl alcohol most of the carbonium ions combines with the chloride ion to form t-Butyl Chloride, although some eliminates a proton (E_1 reaction) to give isobutene which escapes from the reaction mixture.

Questions

1. The chlorides of tertiary aliphatic alcohols are readily prepared by the action of conc. hydrochloric acid upon the alcohol at room temperature, whereas primary aliphatic alcohols require sulphuric acid and the corresponding sodium halide. Explain the reason for this difference.

2. Predict the order of reactivity of the following compounds towards reaction with sodium iodide in anhydrous acetone:

$$CH_2 = CH \; CH_2 \; Cl, \qquad CH_2 = CH \; CH_2 \; CH_2 \; Cl,$$

$$CH_2 = CH \; Cl, \qquad CH_3 \; CH(Cl) \; CH_3$$

Give reasons for your predictions

3. Indicate, with reasons, whether the following compounds would be expected to undergo predominant S_N^1 or S_N^2 reaction.

(a) $C_6 \, H_5 \, CH_2 \, Br$ (b) $C_6 \, H_5 \, CH(Br) \, C_6 \, H_5$ (c) $C_6 \, H_5 \, CO \, CH_2 \, Br$

Reference: N Allinger etal 'Organic Chemistry" Chapter 17

6.2 Preparation of Ethyl Acetoacetate

α-Dicarbonyl compounds such as acetylacetone and ethylacetoacetate are very important precursors for many heterocyclic compounds.

Under acidic conditions, α-dicarbonyl compounds can be used in the synthesis of oxygen containing heterocyclic compounds.

Experimental Procedure: All reagents and glass apparatus must be dry. Sodium hydride itself is spontaneously inflammable in moist air but can be safely handled in air when dispersed in heavy mineral oil. Nevertheless, the dispersion in oil reacts extremely vigorously with water: any small quantities of excess reagent should be destroyed with ethanol.

Fit a 250 ml round bottomed flask with a reflux water condenser and a $CaCl_2$ drying tube. Place dry ethyl acetate (50 ml excess) in the flask and add sodium hydride dispersed in oil (calculate the weight required to give 3.84 g, 0.16 mole NaH). A vigorous reaction will ensure with the evolution of hydrogen and reaction is continued by heating on an oil bath at 100^0 for 2 hour during which time the mixture may become partially solid.

Allow the mixture to cool to room temperature and cautiously add dilute sulphuric acid with through mixing until the solution becomes acid to litmus. During acidification, the mixture may become pasty but any solid will redissolve as acidification progresses. Pour the solution into cold, saturated NaCl solution (50 ml) in a separating funnel and shake vigorously. Run off the lower aqueous layer, transfer the upper layer of ethylacetoacetate to a dry flask using ethyl acetate to rinse out the flask and dry the solution over anhydrous sodium sulphate.

Vacuum Distillation

A demonstrator will assist in setting up equipment. Assemble a vacuum distillation apparatus using a 100 ml distilling flask, a Claisen head and pig receiver. Fit the claisen head with an air leak drawn from capillary tubing through a screw cap adopter and with a thermometer. Filter the crude ethylacetoacetate into the distillation flask using a little ethyl acetate for rinsing. Turn on the water pump fully and consult demonstrator for beginning heating to ensure that the air leak is functioning properly. The first fraction come over will be the ethyl acetate washings. Ethyl acetoacatate itself boils at $73^0/15$ mm, $82^0/20$ mm, $87^0/30$ mm or $92^0/mm$ yield 10 g.

Record IR spectrum of your product as a liquid film.

Questions

1. Why is it necessary to neutralize the reaction mixture with acetic acid in order to isolate the products?

2. What is the function of the NaCl solution in the work-up?

3. What information on the structure of ethyl acetate is evident from the IR spectrum?

6.3 Preparation of Succinic Anhydride

Discussion: This method of dehydration is the only useful procedure for forming the anhydride of succinic acid, a dibasic acid. This reaction enables the preparation of the anhydride in a state of high purity.

Experimental: A 50 ml round-bottomed flask is fitted with a reflux condenser protected by a calcium chloride drying tube, place 5 g of succinic acid and 9 ml of redistilled acetic anhydride. Reflux the mixture gently on a water bath for 1 hour with occasional shaking until the solid is completely dissolved. Heating is continued for some more time to ensure the completeness of the reaction. The water bath is removed. The reaction mixture is allowed to cool slowly until the formation of crystals. The flask is now cooled in an ice bath, and the crystals are transferred to a Buchner funnel or a centered glass funnel. The product is washed twice with 5 ml of portions of cold ether and dry in a vacuum desiccator. The yield is about 3.5 g. m.p. 119-120^0 C.

6.4 Preparation of Maleic and Fumaric Acids

Discussion: The two isomeric unsaturated dibasic acids, maleic and fumaric acids are called cis-trans or geometrical isomers. The two compounds exhibit markedly different properties.

Maleic acid is easily formed from an anhydride by hydrolysis.

An isomeric anhydride related to fumaric acid does not exist. The trans isomer fumaric acid cannot form a cyclic anhydride because of the distance between the carboxyl groups.

Maleic acid is easily converted into fumaric acid by the action of acids, halogens, and sunlight. The trans isomer is more stable compound, and thus this exothermic reaction proceeds nearly to completion.

The separation of the mixture of the two compounds is easily accomplished since maleic acid is very water soluble while fumaric acid has but very limited water solubility.

Experimental: In a 125 ml Erlenmeyer flask containing 15 ml of boiling water, 10 ml of maleic anhydride is introduced. Heating the contents until the solid disappears, and then the solution is kept warm for ten minutes longer. The flask is cooled in ice water. The crystalline maleic acid is collected on the Buchner funnel. Its melting point is 130° C.

6.4.1 Conversion of Maleic Acid into Fumaric Acid

Dissolve 5g of maleic acid obtained above, without purification, in 5ml of warm water, add 10 ml of concentrated hydrochloric acid in a 50 ml round-bottomed flask fitted with a reflux condenser. Heat the mixture gently under reflux for 30 minutes. Crystals of fumaric acid separates from the hot solution. The mixture is cooled and filtered. It is recrystallised from hot water. m.p. $286\text{-}287^o$ C.

Ref: Laboratory Exercises Organic Chemistry by James M Sugihara 1969 Burger publishing company.

6.5 Hydroboration

Reference: Hydroboration : H.C. Brown, W. A. Benzamin, 1962 (Angewandte chemie)

One of the most valuable synthetic organic chemistry procedures developed within recent years is that of hydroboration, the addition of diborane (B_2H_6) across isolated carbon-carbon double bonds. The resulting alkyl boranes may be hydrolysed to alkanes, oxidised with alkaline hydrogen peroxide to alcohols, or oxidised with acidic dichromate to ketones.

Diborane is relatively selective and exhibits predominantly anti-Markownikov addition to olefins. For example, 1-hexene yields 94% 1-hexanol and 6% 2-hexanol, after hydroboration and oxidation.

$$3\ C_4H_9CH{=}CH_2\ +\ B_2H_6\ \longrightarrow\ (C_4H_9CH_2CH_2)_3B$$

$$C_4H_9CH_2CH_2OH\ (94\%) \qquad\qquad C_4H_9(CHOH)CH_3\ (6\%)$$

The overall addition of the elements of water is therefore also predominantly anti-Markownikov, and differs from the strong acid-catalysed hydration of olefins.

$$C_4H_9CH = CH_2\ +\ H_3O^+\ C_4H_9\ (CHOH)\ CH_3 \qquad 100\%$$

Furthermore hydroboration is highly stereospecific, the appropriate reaction sequence with 1-methylcyclopentene giving the *trans*-alcohol - i.e., overall *cis*-hydration.

Although many of the alkyl boranes, like B_2H_6 itself, are spontaneously inflammable, the reagents may be safely generated and used *in situ*. Thus diborane is conveniently prepared from sodium borohydride and the etherate of boron trifluoroide:

$$3\ NaBH_4\ +\ BF_3\ \longrightarrow\ 2B_2H_6\ +\ 3\ NaBF_4$$

6.5.1 Two-stage conversion of Norbornene into *exo*-Norborneol.

Experimental Procedure

This experiment must be performed in a fume cupboard. It may be performed by 2 persons working together; see note 1 for time required.

A 250 ml three-necked flask is fitted with a condenser, a short gas inlet tube and a pressure equalised dropping funnel. The flask is stood inside a large glass basin on a magnetic stirring hotplate. The top of the condenser is fitted with a diglyme-filled bubbler, and the exit tube of the bubbler is connected to a glass tube just dipping into a conical flask (25 ml) of acetone [Note 2].

In the flask is placed a Teflon-coated stirring bar, sodium borohydride (1.7 g) and 40 ml dry diglyme. The mixture is stirred until all the borohydride has dissolved and norbornene (14.1 g) is added; boron trifluoride diethyl etherate (11.5 ml) [Note 3] and diglyme (10 ml) are placed in the dropping funnel.

The apparatus is flushed with nitrogen and the gas flow is reduced to a minimum using the 'bubbler' as a guide. The mixture is stirred gently and the boron trifluoride solution is added dropwise over about 45 min. The reaction mixture is stirred for an additional period of 1 hr and water (15 ml) added cautiously to destroy any diborane. When evolution of hydrogen gas has ceased the bath surrounding the flask is half-filled with warm water which is maintained at 40-50 °C. Dilute sodium hydroxide solution (20 ml; 2N) is added to the reaction mixture followed by dropwise addition of 30% solution (100 volumes) hydrogen peroxide (20 ml). The warm reaction mixture is stirred for an additional period of 1 hr., and may, if necessary, be left to cool overnight at this stage [Note 1].

Ether (50 ml) and sodium chloride (50 g) are added to the cooled, stirred mixture and the slow N_2 flow is stopped. The ether phase is separated, the aqueous salt-saturated layer is re-extracted with ether (2 × 25 ml) and the combined ether extracts washed twice with 25 ml volumes of saturated sodium chloride solution and dried over anhydrous $MgSO_4$. The ether is removed under reduced pressure and the product is crystallised from 60-80° petroleum-ether. Submit your total yield in a labelled tube.

Notes:

1. The total time required for this experiment is longer than average. It is recommended that part of one laboratory period be spent in assembling the oven dried apparatus which may be left overnight with a slow stream of dry N_2 passing through it. The $BF_3.Et_2O$ should be distilled and the other reagents measured out. The experiment [approx. 3½ hr duration] may then be commenced as early as possible on the following laboratory period, and taken at least to the point after adding the H_2O_2/NaOH solutions. If it is interrupted before this stage the yields of product will be severely reduced since the trialkyl borane will suffer possible oxidation and hydrolysis.

2. The purpose of the acetone is to trap any diborane which may be swept through the system with a too-vigorous gas flow. It forms di-isopropoxyborane, $[(CH_3)_2CHO]_2BH$.

3. The $BF_3.Et_2O$ must be freshly distilled in clean, dry, apparatus since it rapidly deteriorates on exposure to moist air. Add 2 ml sodium-dried ether to 25 ml. etherate and distil this from approx. 0.5 g calcium hydride. B.p. 46 °/10 mm. Store the distillate for a minimum period in a stoppered flask.

6.6 Catalytic Hydrogenation

Catalytic hydrogenation is one of the most important techniques in preparative organic chemistry. In essence it allows the addition of hydrogens across a multiple bond $(C \equiv C, C = C, C \equiv O, C \equiv N$ etc.) using a catalyst which may be in a homogeneous or heterogeneous reaction phase. Without the catalyst the reaction will not proceed. The most common heterogeneous catalysts are derived from the noble metals [Ni, Pd, Pt, Rh and Ru] and are used in the pure state as a finely divided powder, or suspended on an inert substrate such as carbon. While the reaction conditions of solvent, temperature and pressure can be used to control degrees of hydrogenation, much greater selectivity is obtained by employing different catalysts. Partial 'poisoning' of catalysts is another method of promoting selective reduction. The best known example is the use of lead salts [Lindlar catalyst], or a base like quinoline, to inhibit the activity of palladium catalysts and thusallow the partial hydrogenation of an acetylenic to a *cis*-ethylenic group.

$$
-\!\!-C \equiv C-\!\!- \quad \xrightarrow[\text{Catalyst}]{\text{H}_2/\text{Lindlar}} \quad -\!\!- \underset{\underset{H}{|}}{C} = \underset{\underset{H}{|}}{C}-\!\!- \quad \xrightarrow{\text{H}_2/\text{Pd}} \quad -\!\!- CH_2-\!\!- CH_2-\!\!-
$$

Hydrogenation is widely employed in industrial processes. The 'hardening' of oils involves the hydrogenation of a portion of the liquid glycerides of unsaturated acids into higher melting saturated analogues, using a nickel catalyst, and gives edible fats.

6.6.1 Reduction of Maleic to Succinic Acid

$$
\begin{array}{c} H-C-CO_2H \\ \parallel \\ H-C-CO_2H \end{array} \quad + \quad H_2 \quad \longrightarrow \quad \begin{array}{c} CH_2-CO_2H \\ | \\ CH_2-CO_2H \end{array}
$$

Experimental procedure

Dissolve maleic acid (20 g) in ethanol (40 cc) in a hydrogenation flask (250 cc) and add the catalyst (100 mg. of palladized charcoal). Fill both burettes with hydrogen after evacuating air from the system, but do not commence stirring the mixture until accurate volumes have been noted. Calculate the expected uptake of hydrogen under prevailing ambient conditions and be prepared to interrupt the experiment to refill the burettes.

Once agitation of the mixture has commenced there may be an induction period before gas absorption. Plot a graph of gas uptake (cc) against time (minutes). When uptake of hydrogen ceases measure the final volume of hydrogen used and compare this with the calculated value.

Filter the solution through a sinter-funnel packed with a layer of 'Celite' or Kieselguhr, wash it with a little ethanol, *but avoid allowing the catalyst to become dry* or having air drawn over it. After filtration the catalyst may be destroyed with conc. hydrochloric acid. Evaporate the ethanol filtrate under reduced pressure and determine the weight and m.p. 185°C. If necessary recrystallise the succinic acid from water.

Compare the i.r. spectra of maleic and succinic acids as Nujol mulls.

Experimental results

Weight of maleic acid   =

No. of moles   =

Atmospheric temperature   =   °C = °K

Atmospheric pressure   =   mm.Hg

Expected uptake $H_2 = x\dfrac{22,400}{1} \times \dfrac{760}{1} \times \dfrac{1}{273} ml$

Actual volume H_2 =

Burette reading   Volume absorbed   Time

0   0

Reference: Catalytic Hydrogenation. Techniques and applications in organic synthesis, R.L. Augustine, Arnold, 1965.

6.7 Preparation of α-D Glucose Pentaacetate from Glucose (Acetylation)

The acetylation is in almost all cases effected with acetic anhydride using either the alkaline or an acidic catalyst.

A reducing sugar exists as an equilibrium mixture of two isomers. The formation of hemiacetal structure creates a new center of asymmetry, permitting the existence of the two isomers, α-D-glucose and β-D-glucose.

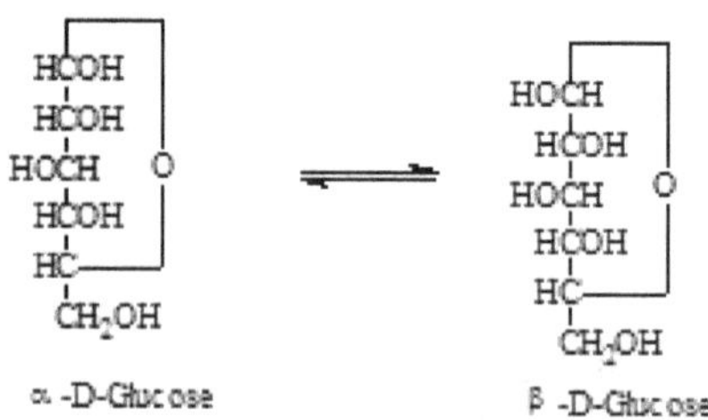

The acetylation with acetic anhydride of a reducing carbohydrate can yield both possible isomeric pentaacetates, depending mainly upon the catalyst and temperature. The acetates, once formed, may also be isomerized.

α-D-glucose pentaacetate

An excess of acetic anhydride is hydrolyzed by cold water, from which the product crystallizes.

Experimental

Place 1 g of anhydrous D-glucose in a 50 ml round-bottomed flask, add 0.8 g of anhydrous sodium acetate and 6.0 ml of acetic anhydride. A reflux condenser is attached, and the flask is continuously heated with a flame to initiate the reaction as the unit is swirled to distribute the solid particles of the sugar continue heating for a further 2 hours till a clear solution is obtained and then pour the reaction mixture into 100 ml of ice water with vigorous stirring. The product separate as an oil, but it crystallises readily upon rubbing the gummy solid against the sides of the beaker with a stirring red. Filter off the crystals, wash well with cold water and recrystallise from industrial spirit or methanol. The melting point is 132°C. The yield is 1.25g.

6.8   2,4,6-Tribromoacetanilide

Dissolve 1 g of 2,4,6-tribromoaniline in 15 ml of acetic anhydride and add 2 drops of concentrated sulphuric acid in a dry boiling tube and shake it occationally. After 15 minutes, pour the reaction mixture into excess of water. Filter the solid, wash and dry. Recrystallised from alcohol. m.p. 231°C.

Mechanism

6.9 Aliphatic Electrophilic Substitution

6.9.1 Preparation of Iodoform
from Ethanal (Acetaldehyde)

Chemicals: Ethanal, iodine, potassium carbonate.

Experimental: Pour ethanal (5 m) into a conical flask (100 ml) containing a solution of potassium carbonate (3 g) in water (10 ml). To this solution, add crystals of iodine (3 g) in small portions, mixing well until all the iodine is dissolved. Heat the resulting solution gently to 60° on a steam bath and maintain at this temperature with occasional swirling until the iodine colour has been discharged. Cool the mixture to room temperature and collect the yellow crystals of iodoform by filtration at the water pump using a small Buchner flask and funnel. Wash the crystals with water and suck dry. Do not dry the crystals in the oven. Determine melting point and weight. From the above equations it can be seen that 4 moles of iodine give 1 mole of iodoform. Calculate the theoretical yield of iodoform from 3 g of iodine and hence your percentage yield for the experiment.

Mechanism

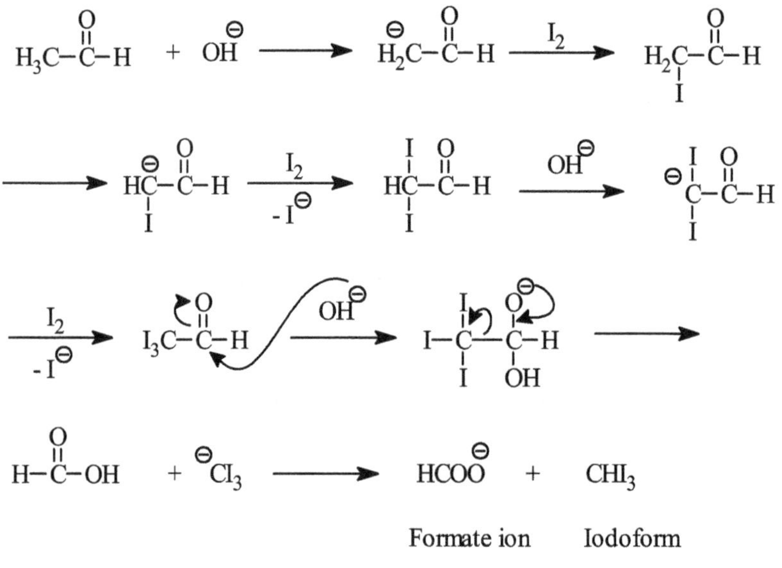

$$H_3C\text{-}\overset{\overset{O}{\|}}{C}\text{-}H \;+\; \overset{\ominus}{O}H \longrightarrow H_2\overset{\ominus}{C}\text{-}\overset{\overset{O}{\|}}{C}\text{-}H \xrightarrow{I_2} H_2\overset{\underset{I}{|}}{C}\text{-}\overset{\overset{O}{\|}}{C}\text{-}H$$

$$\longrightarrow H\overset{\ominus}{\underset{I}{C}}\text{-}\overset{\overset{O}{\|}}{C}\text{-}H \xrightarrow[-I^{\ominus}]{I_2} H\overset{\overset{I}{|}}{\underset{I}{C}}\text{-}\overset{\overset{O}{\|}}{C}\text{-}H \xrightarrow{OH^{\ominus}} \overset{\ominus}{\underset{I}{\overset{I}{C}}}\text{-}\overset{\overset{O}{\|}}{C}\text{-}H$$

$$\xrightarrow[-I^{\ominus}]{I_2} I_3C\text{-}\overset{\overset{O}{\|}}{C}\text{-}H \xrightarrow{OH^{\ominus}} I\text{-}\overset{\underset{I}{|}}{C}\text{-}\overset{\overset{O^{\ominus}}{|}}{\underset{OH}{C}}\text{-}H \longrightarrow$$

$$H\text{-}\overset{\overset{O}{\|}}{C}\text{-}OH \;+\; \overset{\ominus}{C}I_3 \longrightarrow HCO\overset{\ominus}{O} \;+\; CHI_3$$

Formate ion   Iodoform

Electrophilic addition to alkenes (Bromination of alkenes)

Reference: Roberts and Caserio, p.175 et seq. Sykes, chapter 6

Compounds containing multiple carbon-carbon bond react readily with bromine to give addition compounds; e.g. alkenes give 1, 2 – dibromoalkanes

$$R\,CH{=}CH\,R^1 + Br_2 \rightarrow R\,CH(Br)CH(Br)R^1$$

The vicinal dibromides obtained from alkenes are useful intermediates for the preparation of alkynes, when R and R^1 are capable of withstanding the dehydrohalogenation procedure.

6.9.2 Bromination of a Chalcone

$$\text{Chalcone} \xrightarrow{\text{Br}_2 \text{ in CCl}_4} \text{dibromide product}$$

Experimental: Dissolve the pure chalcone (0.02 mole) in carbon tetrachloride (100 ml) and cool the solution in an ice-bath. To this solution, carefully add slowly with stirring a solution of bromine (Danger*) in carbon tetrachloride (0.02 mole). When addition is complete, the precipitated dibromide is filtered off and washed with two portions of hot carbon tetrachloride. Crystallise the compound to constant melting point using a suitable non-hydroxylic solvent. Obtain the i. r. spectrum of your product as KBr disc or Nujol mull.

*A solution of bromine in CCl_4 of known concentration will be provided. Since this is extremely corrosive, great care should be exercised in its use. Rubber gloves should be used and the reaction should be carried out in a fume cupboard.

Questions

1. How could this reaction be used to convert 1, 2- diphenylethylene (stilbene) to 1,2 – diphenylacetylene ?

6.10 Aromatic Electrophilic Substitution Reactions

Nitration

Introduction: Nitration is one of the most important unit process used in industrial synthetic organic chemistry. Not only do nitration products fine wide applications as solvents, dye stuffs, pharmaceuticals and explosives, but they are also useful as intermediates for the preparation of other compounds particularly amines, which are prepared by the reduction of the corresponding nitro compounds. A variety of reagents are used for nitration. The following are some of the very important reagents, namely, fuming nitric acid, and aqueous nitric acid, and a mixture of nitric and sulfuric acids.

6.10.1 Preparation of *m*-Dinitrobenzene

$$\text{nitrobenzene} \xrightarrow{\text{H}_2\text{SO}_4 \text{ / HNO}_3} \text{m-dinitrobenzene} + \text{H}_2\text{O}$$

Chemicals: Fuming nitric acid, sulphuric acid, nitro benzene.

Experimental: This preparation of *m*-dinitrobenzene should be carried out in a fuming cupboard. Carefully mix together 5 ml of fuming nitric acid and 7ml of concentrated sulphuric acid in a 100 ml round-bottomed flask, and fit a reflux condensor to the flask. An apparatus with ground-glass joints should be used, fuming nitric acid rapidly attacks corks. Place a few pieces of pumice stone in the acid and then very slowly add 4.0 ml of nitrobenzene in 1 ml portions, pouring it down the condenser and shaking the flask well to ensure thorough mixing. Heat the mixture, with frequent shaking, on a boiling water bath for 30 minutes. After the flask has cooled, pour the contents slowly into abut 150 ml of cold water in a beaker stirring thoroughly. The residual acid will dissolve in the water and m-dinitrobenzene will collect as an yellow solid. Filter using a Buchner funnel and filter pump, wash thoroughly with cold water and allow it to drain. Finally recrystallize the product from 30 ml of methylated spirit, to free it from traces of the ortho and para-isomers. The yield is 4.8 g m.p. 89-90°C.

Mechanism

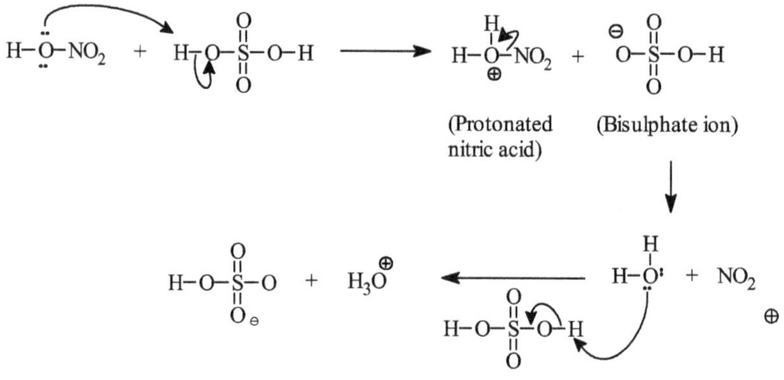

(Protonated nitric acid)   (Bisulphate ion)

Overall reaction:

$$HONO_2 + 2H_2SO_4 \longrightarrow NO_2^{\oplus} + H_3O^{\oplus} + 2HSO_4^{\ominus}$$

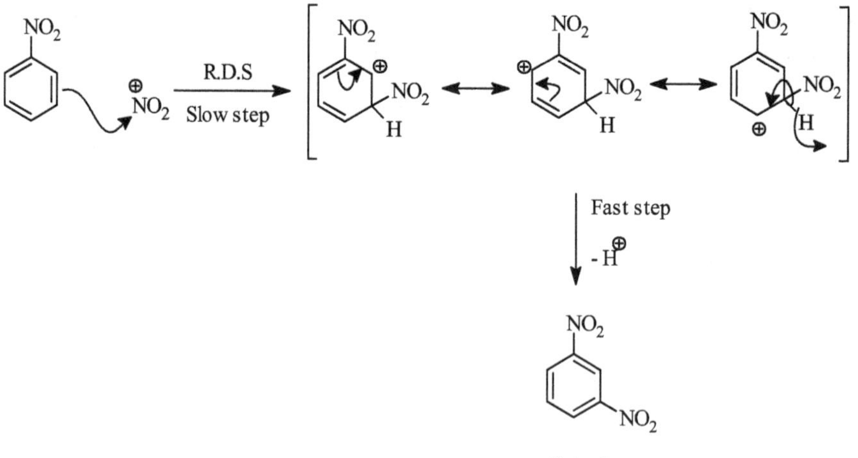

R.D.S
Slow step

Fast step

$-H^{\oplus}$

m-dinitrobenzene

6.10.2 Preparation of *p*-Nitroacetanilide from Acetanilide

Chemicals: Acetanilide, conc. sulphuric acid, conc. nitric acid, acetic acid.

Experimental: Add 5 g of powdered acetanilide to 5 ml of glacial acetic acid contained in a 100 ml beaker and then to the well-stirred mixture add 10 ml of concentrated sulphuric acid. The mixture becomes warm and a clear solution results. The beaker is cooled with a freezing mixture of ice and salt until the temperature of the reaction mixture falls to 0-5°C. Now, while stirring the viscous mixture continuously with the thermometer, add 2 ml of concentrated nitric acid carefully, drop by drop from a dropper, while the temperature is maintained below 10°C. Remove the beaker from the freezing mixture, allow it to stand at room temperature for 30 to 45 minutes. Pour the contents on to 50 g of crushed ice (or 100 ml of cold water), whereby the crude nitroacetanilide is at once precipitated. Allow the mixture to stand for 15 minutes and then filter at the pump, wash thoroughly with cold water until free from acids and drain well. Recrystallise the crude pale yellow product from ethanol or methylated spirit, filter at the pump. wash with little cold spirit and dry in the air upon filter paper. The yellow o-nitro acetanilide remains in the filtrate whilst the *p*-nitro acetanilide is obtained as colourless crystals, m.p. 214°C. The yield is 4 g.

Mechanism

$$HONO_2 + 2H_2SO_4 \longrightarrow \overset{\oplus}{N}O_2 + H_3\overset{\oplus}{O} + 2H\overset{\ominus}{S}O_4$$

R.D.S = Rate determining step

Note: The direct nitration is not possible. The amino group is oxidised during nitration. Hence it is better to protect the amino group by acetylation first followed by nitration to give *p*-nitro acetanilide as the major product. If required, one can isolate the p-nitroaniline by simple hydrolysis of p-nitro acetanilide.

6.10.3 Preparation of *p*-Nitroaniline from *p*-Nitroacetanilide (Hydrolysis)

Chemicals: *p*-Nitroacetanilide, sulphuric acid.

Experimental: Place 2.0 g of *p*-nitroacetanilide in a 50 ml round bottomed flask and add 10 ml of 70 percent w/w sulphuric acid (i). Fit a reflux condenser, add a boiling stone to the solution, and reflux the mixture gently over a small Bunsen flame unitl a clear solution is obtained (20 minutes). Pour the hot, clear solution into 75 ml of cold water in a beaker and make the solution alkaline by the addition of dilute sodium hydroxide solution. Yellow crystals of p-nitroaniline separates. Cool the mixture thoroughly to room temperature and collect the crystalline solid using a Buchner flask and funnel. Wash the product thoroughly with cold water and suck dry.

Recrystallise the material from minimum volume of hot mixture of equal parts of methylated spirit and water. Filter, wash and dry. The yield is 1.5 g m.p. 148°C.

Note (i): A 70% sulphuric acid solution is prepared by adding 60 ml of conc. sulphuric acid cautiously in a thin stream with stirring to 45 ml of water.

Mechanism

6.10.4 Preparation of *o*- and *p*-Nitrophenols

Phenols undergo electrophilic aromatic substitution with great ease. Thus it can be nitrated by dilute nitric acid to give a mixture of o- and p-nitrophenols. These compounds can be readily separated because the former is steam volatile.

Chemicals: Phenol, conc. nitric acid, sodium acetate.

Experimental: Prepare a mixture of concentrated nitric acid (6.5 ml) and water (20 ml) in a 250 ml conical flask. Warm a mixture of phenol (5 g) and water (4 ml) in a small conical flask until the phenol becomes liquid. Shake vigorously to get an emulsion of phenol in water. Cool the nitric acid in ice-water and add the phenol so that the temperature does not rise above 30°C. When the addition of phenol is complete remove the flask from the cooling mixture and allow to stand for at least two hours. During the first hour of standing keep the thermometer in the liquid and do not allow the temperature to exceed 50-55°C.

Add hydrated sodium acetate (17 g) and water (50 ml) and steam distil the mixture. The o-nitrophenol is steam volatile and crystallises out in the cold distillate. If the o-nitrophenol crystallises in the condenser, turn off the cooling water temporarily for a few minutes when the o-nitrophenol will melt and pass into the distillate when the distillation is complete, cool the distillate in ice-water and filter off the o-nitrophenol at the pump. Dry by pressing between sheets of filter paper. m.p. 46°C. The yield of o-nitrophenol 35%.

Mechanism

The p-Nitrophenol can also be isolated from the residue in the steam-distillation flask.

Isolation of *p*-nitrophenol

As soon as possible after the steam distillation, filter the hot residue through a conical funnel in which are two wet fluted papers. Discard the black residue; allow the filtrate to stand overnight. Collect the dark crystals, and suspend them in a 250 ml flask. to gether with cold water (50 ml), conc. hydrochloric acid (2.5 ml) and a small amount of decolourising charcoal. Fit a reflux condenser, bring to the boil (bursen flame), and boil for 10 minutes. Filter hot through a conical funnel, allow the filtrate to cool, and collect the p-nitrophenol as long brownish crystals, m.p. 112°C. Yield is 25%.

Steam Distillation Procedure

For the steam distillation, the two-necked flask should be large enough to hold the sample for steam distillation and any steam that condenses without becoming more than half-full. Under these conditions, neither a spray-trap, nor a special splash-head is needed. For this experiment a 1 litre flask is large enough.

Assemble the apparatus as shown,in the Fig. 6.1 but with the steam supply shut off. The tube delivering the steam should end about 1 cm above the base of the flask. It is convenient to place the flask on a steam bath, to reduce condensation.

Remove the delivery tube from the flask and turn on the steam supply. Wait until any water in the pipe has been emptied and the tube delivers steam. Briefly turn off the steam, insert the delivery tube into the flask, then turn on the steam. Regulate the steam supply so that the dark liquid does not splash into the still-head.

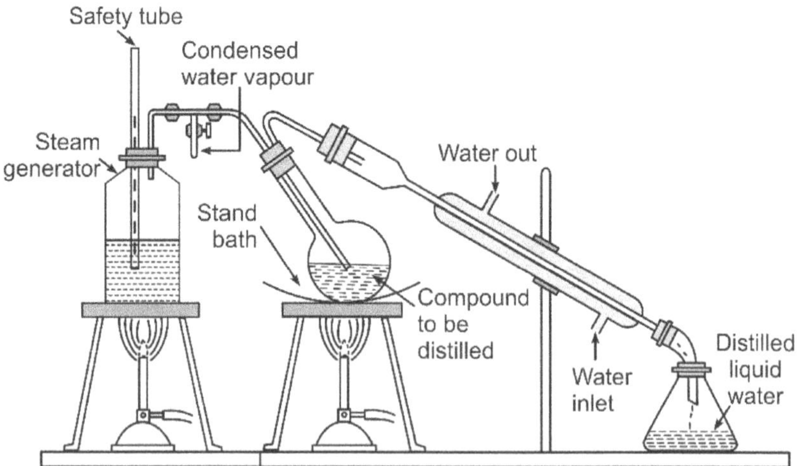

Fig. 6.1 Steam Distillation Procedure

When no more yellow solid collacts in the condenser, turn the steam almost off, raise the end of the delivery tube above the level of the liquid in the flask, then turn off the steam supply completely. If you turn off the steam before the delivery tube is raised, some of the contents of the flask will be sucked back down the steam line.

Questions

1. Why is o-nitrophenol is steam distillable while p-nitrophenol is not?

2. Why is no m-nitrophenol obtained?

6.10.5 Preparation of 1,4-Dichloro-2-Nitrobenzene

In this experiment p-dichloro benzene is reacted with fuming nitric acid in the presence of sulfuric acid. The presence of the two chlorine atoms deactivates the molecule to the extent that a reaction temperature of approximately 100° is required.

Experimental procedure: *p*-Dichlorobenzene (7.0 g) is placed in a 50 ml round-bottomed flask. A mixture of 3 ml fuming nitric acid (density 1.54 g/ml) and 3.5 ml of concentrated sulfuric acid is added in three portions. The flask is swirled vigorously between the additions. A condenser is attached to the flask and refluxed, and the reaction mixture is heated for one hour over a boiling water bath. The reaction mixture is cooled to room temperature. Crystals of the product appear. The slurry is poured into 100 ml ice cold water. The suspension is filtered and the crystals of 1,4-dichloro-2-nitrobenzene are washed thoroughly with water on the filter paper. The product is recrystallized from ethanol. The recrystallization solution is cooled slowly to prevent oiling of the low melting point solid. About 8.0 g of a light yellow solid, melting at about 54°C, will be obtained. (lit., m.p. 56°).

Questions

1. What is the function of sulfuric acid in this experiment?

2. Explain why p-xylene under goes nitration more readily than p-dichlorobenzene?

6.11 Halogenations

Introduction: Halogenation may be defined as a process where in one or more halogen atoms are introduced into an organic compound. The preparation of organic compounds containing chlorine, bromine and iodine.

6.11.1 Bromination of Methyl Cinnamate

Chemicals: Methyl Cinnamate, bromine, dichloromethane

Experimental: Weigh out accurately not more than 1 g of methyl cinnamate and dissolve this in dichloro methane (5 ml) in a 50 ml flask. Stopper lightly with a cork or stopper. Pipette 1 ml of bromine solution in dichloromethane* and add this to the cinnamate solution in one portion. Restopper the flask, swirl gently to mix and leave to stand for 15-25 minutes. Distil off the excess dichloromethane on the steam bath. Don't forget to use boiling stones. Crystallise the solid residue from petroleum ether (60-80°c). Check the melting point and calculate the percentage yield of product obtained.

*A solution of bromine in dichloromethane has to be prepared by the demonstrator. Add 9 ml of bromine to a flask containing 20 ml of methylene chloride by external cooling and thoroughly mixing while adding. Since this is extremely dangerous, great care should be excercised in its use. Rubber gloves should be used and the preparation of bromine solution in dichloromethane should be carried out in a fume cupboard.

General Mechanism

The alkene double bond is an electron-rich system which readily enters into addition reaction with electron seeking (electrophilic) reagents such as bromine.

Bromination of alkenes is considered to involve interaction of the double bond with a polarised bromine molecule in a stepwise process. The mechanism is outlined below using ethane:

[A]

The first intermediate formed is thought to be the cyclic bromonium ion [A]. This intermediate has a three membered ring and the positive charge is on the bromine atom. Since bromine is a large atom this has the effect of blocking one side of the molecule; the only way the second bromine can approach to form a bond to carbon is to attack from the opposite side to form a bond to carbon. This means the bromine end up opposite to each other. This is called trans-addition.

6.11.2 Relative Rates of Bromination of Aromatic Compounds

The rates of electrophilic substitution of benzene and its derivatives are greatly influenced by substituents present. For example, the relative rates of bromination of 4-nitrobenzene and phenol are 1:10.

In this experiment solutions containing equal concentration of four different aromatic compounds in 90% acetic acid will be treated with equal amounts of bromine in the same solvent (90% acetic acid). The compounds to be tested are:

Benzene   Phenol   p-Chlorophenol   p-Nitrophenol

Experimental: In four separate labelled test tubes place 2 ml. of 0.2 M solutions (in 90% acetic acid) of each of the above compounds. Place the tubes in a 400 ml. beaker filled with warm water at 30°c, also place a test-tube containing 10 ml of 0.05 M bromine solution (in 90% acetic acid) in the same water bath and allow the tubes to stand for 10 minutes., until the contents have reached the bath temperature.

Then add 2 ml of bromine solution to each of the labelled test tubes, mix quickly with a glass rod, [Do not place your finger over the top of the test tube and shake] and note the time (in seconds) for the colour of the bromide to disappear. Record the times when the solutions become colourless or a permanent faint yellow. If any of the solutions show no visible change after two hours, simply record the time as greater than 120 minutes.

Questions

(a) From your results with benzene and phenol say whether the OH group is ring-activating or ring de-activating towards bromination in acetic acid. Why?

(a) **Model answer:** Phenols reacts faster than benzene. The –OH group activates the ring towards electrophiles, because of the resonance of phenol. Electrons can be delocalized round the ring in *ortho* and *para* positions.

(b) From your results with the three phenols say whether the Cl and NO_2 groups are ring-activating or ring de-activating towards bromination in acetic acid. Explain the order observed with the three phenols.

(c) Arrange the groups H, NO_2, Cl, and OH in decreasing order of ring activation towards bromination in acetic acid.

(d) Draw the major monobromination product for each of the four compounds brominated.

***Caution:** Phenols and bromine are potentially hazardous chemicals. 0.2M solutions of these compounds and 0.05M solution of bromine in 90% acetic acid are provided by the demonstrator. Wear gloves when preparing these solutions.

6.11.3 *p*-Bromoacetanilide

Chemicals: Acetanilide, bromine, glacial acetic acid

Experimental: Dissolve 2 g of finely powdered acetanilide in 6 ml of glacial acetic acid in a 50 ml. conical flask. In another small flask dissove 0.8 ml of bromine in 5 ml of glacial acetic acid, and add this solution slowly to the acetanilide solution, shaking the latter throughout the addition to ensure thorough mixing; stand the flask in cold water. Allow the final reaction mixture to stand at room temperature for 20 minutes. Pour the reaction mixture in a beaker containing water and if it is coloured, add just sufficient sodium bisulphite solution to remove the orange colour. Filter the crystalline precipitate with suction on a Buchner funnel, wash thoroughly with cold water, drain, and finally recrystallise from rectified spirit. The yield of p-bromoacetanilide is 2.5 g. m.p. 167°C.

6.11.4 Preparation of *p*-Bromoaniline

Chemicals: p-Bromoacetanilide, conc. hydrochloric acid.

Experimental: In a 100 ml round-bottomed flask are placed 2 g of p-bromo acetanilide and 20 ml of 6N hydrochloric acid. The suspension is refluxed for about 15 minutes until a clear solution is obtained. When reaction is completed, the flask is cooled under a cold water tap. White crystals of p-bromoaniline hydrochloride will be deposited because of the limited solubility of the salt in hydrochloric acid solution. A cold solution of 6 g of sodium hydroxide in 10 ml of water is added in portions with cooling. The resulting alkaline solution is cooled. p-Bromoaniline separates as an oil, which soon crystallises. Filter the crystals at the pump, wash with cold water and dry in the air upon pads of filter paper. The yield is 1.6 g m.p. 66°C. Recrystallisation from dilute alcohol is usually unnecessary.

6.11.5 2,4,6-Tribromoaniline

Chemicals: Aniline, bromine, acetic acid

Experimental: Dissolve 1 g of aniline in 4 g of glacial acetic acid in a 200 ml beaker and while stirring add slowly a solution of 1.7 ml of bromine in 4 ml of glacial acetic acid. The beaker should be cooled in ice water during the addition as the reaction is exothermic pour into excess of water, filter at the pump, wash well with water and dry. Recrystallise a small portion from rectified spirit. The yield of tribromoaniline is quantitative m.p. 120°C.

Mechanism

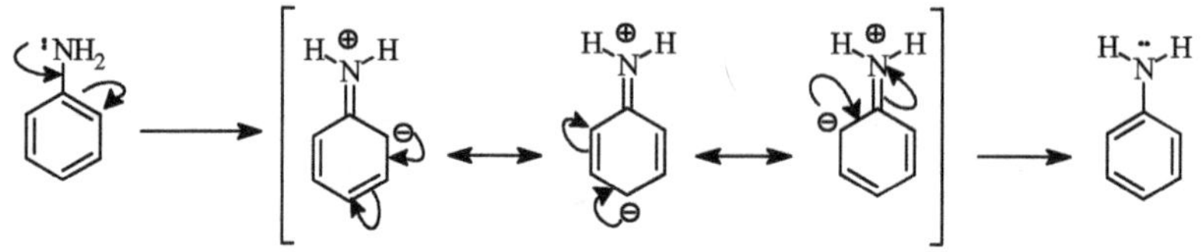

Since the resonance structure of aniline exhibit -ve charge at 2,4,6 position; bromination of aniline take place at 2,4,6--position only.

Mechanism

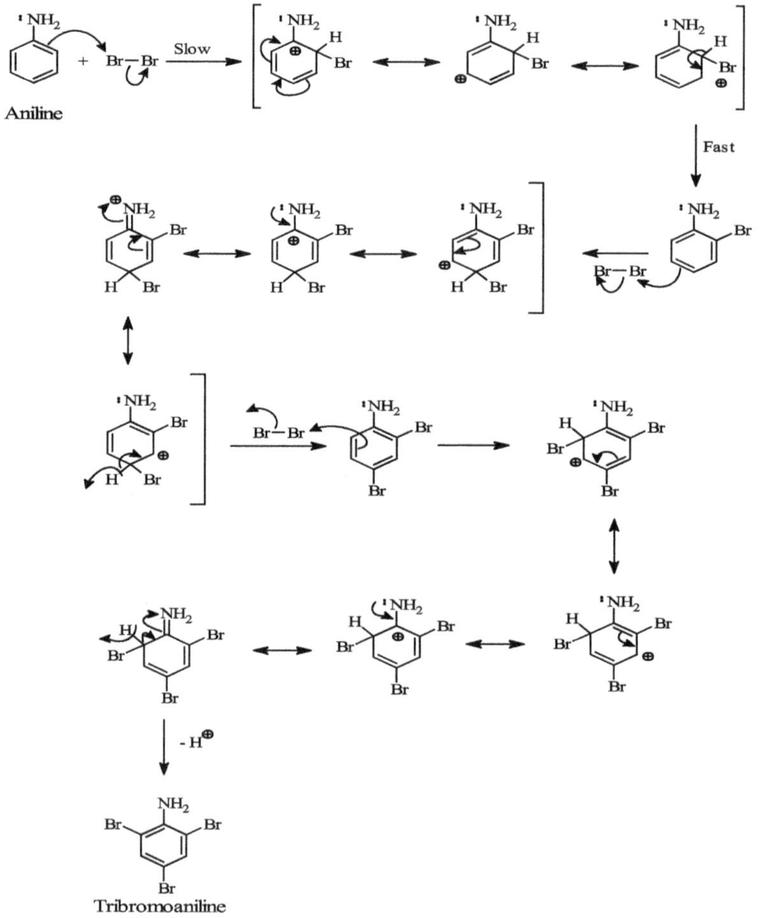

6.11.6 Preparation of 1,3,5-Tribromobenzene

$$2,4,6\text{-}Br_3C_6H_2\text{-}NH_2 \longrightarrow \left[2,4,6\text{-}Br_3C_6H_2\text{-}\overset{\oplus}{N}\equiv N\right] HSO_4^{\ominus} \xrightarrow[\text{- } N_2]{\text{EtOH}} 1,3,5\text{-}C_6H_3Br_3$$

Chemicals: 2,4,6-Tribromoaniline, ethanol, conc. sulphuric acid, sodium nitrite, benzene.

Experimental : Dissolve 2.5 g of 2,4,6-tribromoaniline by heating on a water bath with 15 ml of rectified spirit and 4 ml of benzene in a 100 ml wide-necked flask fitted with a reflux condenser. Now add 1.0 ml of concentrated sulphuric acid (preferably from a small pipette or a dropper) to the hot solution shaking the flask gently and heat the contents on a water bath until the clear solution boils remove the flask from the water bath, add 1.0 g of powdered sodium nitrite in two equal portions, after each addition, replace the stopper and shake the flask vigorously; when the reaction subsides, return the flask with condenser to the water bath and boil the solution for 45 minutes with occasional shaking. Allow the solution to cool for 10 minutes, and then immerse the flask in an ice water. A mixture of tribromobenzene and sodium sulphate crystallises out. Filter at the pump wash with small quantity of methylated spirit and then repeatedly wash with water to remove all sodium sulphate.

Recrystallise from methylated spirit, using animal charcoal and boil the solution. Filter the hot solution through a preheated Buchner funnel and allow the solution to cool. Collect the crystals on a Buchner funnel and dry in the air upon filter paper. The yield is 1.6 g., m.p. 122°C.

Sulfonation

Introduction: Thesulfonic acid group (-SO$_3$H) is one of the more common substituent in dye intermediates. This group is introduced to render intermediates soluble in water, or to provide a route to other substituents, such as hydroxyl group which is obtained by subsequent alkaline fusion.

6.11.7 Preparation of Sulphanilic Acid from Aniline

$$\text{Aniline} \xrightarrow[\text{H}_2\text{SO}_4]{180\text{-}190°C} \left[\overset{\oplus}{NH_3}\,HSO_4^{\ominus}\right] \longrightarrow \underset{SO_3H}{NH_2}$$

Chemicals: Aniline, sulphuric acid, fuming sulphuric acid.

Experimental: Place 5.1 g (5 ml) of aniline in a 100 ml round-bottomed flask and slowly add 10 ml of concentrated sulphuric acid in small portions, shake the mixture gently during the addition and keep it cool by occasionally immersing the flask in cold water, and heat the contents at 180-190°C (fume cupboard) for about 5 hours. Allow the product to cool and pour it carefully with stirring into 100 ml of cold water or of crushed ice. Allow to stand for about 5 minutes and collect the crystalline solid on a Buchner funnel, wash it with water and drain. Purify the crude sulphanilic acid by recrystallising from about 100-125 ml of boiling water; if the resulting solution is coloured, add about 1 g of animal charcoal and boil for 10 minutes. Filter through a hot-water funnel or through preheated Buchner funnel. Dry the crystals between sheets of filter paper. The yield of sulphanilic acid is 5 g. The acid does not melt sharply and it decomposes on being heated. m.p.288° C

Mechanism

$$\text{Aniline} \xrightarrow{H_2SO_4} \text{Anilinium Sulpahte} \xrightarrow{H_2O} \text{Phenyl sulfamic acid} \xrightarrow{H_2O} \text{Sulfanilic acid}$$

6.12 Oxidation

Introduction: Addition of oxygen or removal of hydrogen atom in a organic molecule is known as oxidation. Oxidation constitutes one of the important and powerful tools, for the synthesis of organic compounds. Since a diverse range of oxidizing agents are available to organic chemists, the value of a particular reagent lies in its selectivity. In other words, the reagent of choice for a particular reaction is one which will oxidize one functional group in the molecule while other potentially oxidisable groups present are not affected. For example one of the mildest oxidising agents is manganese dioxide which is prepared in active form by the thermal dehydration of manganese hydroxide. This reagent will selectively oxidize compounds containing activated methylene or hydroxy-methylene groups to ketones.

6.12.1 Synthesis of Benzilic Acid

Aim: To study benzilic acid rearrangement.

Chemical Name: 2, 2-Diphenyl-2-hydroxy ethanoic acid.

Preparation of Benzil from benzoin

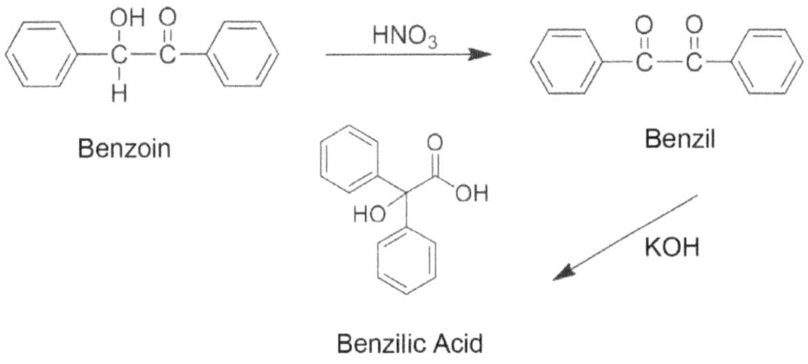

Benzoin

Benzil

Benzilic Acid

Principle : Benzoin undergoes oxidation in presence of conc. nitric acid to give benzil. Benzil in presence of KOH under goes rearrangement to give benzilic acid.

Type of reaction: Benzilic acid rearrangement.

Step I: Benzoin condensation.

Chemicals: Benzoin, conc. HNO_3 and rectified spirit.

Apparatus required: 100 ml R.B. flask, and reflux condensor.

Procedure: Place 2.0 g of powdered benzoin and 10.0 ml conc. HNO_3 in 100 ml R.B. flask fitted with a reflux condensor and heat the flask on a boiling water bath; continue heating for 1.5 hours. When the crystalline benzoin have been completely replaced by oily benzil, pour the mixture into a beaker containing cold water. On vigorous strring the oil will crystallize into a yellow solid. Filter off the latter at the pump and wash thoroughly with water to ensure complete elimination of acid. Recrystallize with rectified spirit. Yield 1.8 g, m.p. 94-6 ^{o}C.

Step - II : Benzilic acid rearrangement

Chemicals: KOH pellets, rectified spirit or ethanol, recrystallized benzil and conc. HCl.

Apparatus required: 250 ml R.B. flask and water bath.

Procedure: In a 250 ml of R.B. flask place a solution of 1.0 g of KOH pellets in a 2.0 ml of water and add 3.0 ml ethanol and 1.0 g of recrystallized benzil. A deep bluish black solution is produced. Boil the mixture on a water bath 10 to 15 min. Pour the contents of the flask in a porcelain dish and cool on a ice water. A potassium salt of benzil acid crystallizes out. Filter the solid at the pump and washed with ice cold ethanol. Dissolve the potassium salt in about 10 ml of water and add slowly with stirring 2 to 3 ml drops of conc. HCl. The precipitate thus produced is reddish brown and sticky, filter it off. The filtrate should be nearly colourless. Continue the addition of HCl with stirring until the solution is acidic to litmus paper. Filter the benzilic acid under suction, wash with cold water. Recrystallise either with hot benzene or hot water using little charcoal. Yeild 0.9 g, m.p. 150 °C.

Reaction & Mechanism:

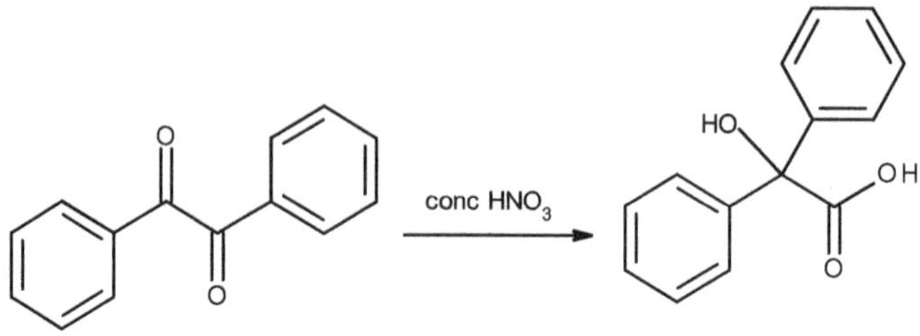

conc HNO$_3$

Mechanism

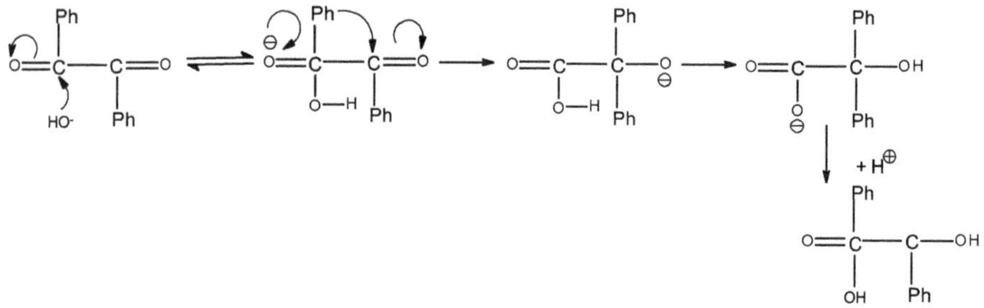

Reference: Elementary practical organic chemister by Arthur

6.12.2 Preparation of Anthraquinone (Oxidation)

Usually benzene is very much resistance to oxidation but anthracene can be easily oxidized to 9,10 –anthraquinone by a mixture of chromium trioxide and conc. Sulphuric acid.

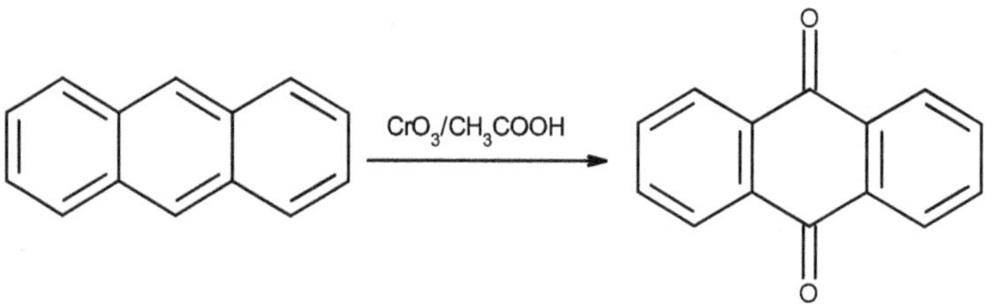

CrO$_3$/CH$_3$COOH

Preparation of organic acid (oxidation)

Chemicals required: Anthracene, chromium trioxide, glacial acetic acid

Experimental: Dissolve 1.0 g of anthracene in 10 ml of glacial acetic acid and place in 50 ml of round-bottomed flask fitted with a reflux condenser. Dissolve 2.0 g of chromium trioxide in 2 ml of water and add 5 ml of glacial acetic acid. Pour the solution down the condenser through a funnel, shake the contents of the flask and boil gently for 30 minutes. Cool and pour the contents of the flask into a beaker containing 20 ml of cold water. Filter off the crude anthraquinone at the pump, wash with water, drain well and dry. It may be further purified by recrystallisation from dilute acetic acid. The yield of anthraquinone is 1.0 g. m.p. 285-286° C.

6.12.3 Benzoic Acid from Toluene

Chemicals: Potassium permaganate, sodium carbonate, toluene, sodium metabisulphate.

The aromatic ring is stable towards most oxidising agents, and it is therefore possible to oxidise a side-chain to a carboxylic acid group without effecting the aromatic nucleus.

Procedure: Place potassium permanganate (4 g) and sodium carbonate (0.5 g) in a small round-bottomed flask. Add water (40 ml) and warm gently to dissolve the solids, and cool the solution. Add toluene (1.25 ml), "Tide" (or a similar detergent, 0.05 g) and a boiling stone. Boil the mixture gently under reflux oven a low Bunsen flame (in the fume-cupboard) for 60-90 minutes. Swirl the flask occasionally to minimise bumping.

Cool the mixture to room temperature, acidify with dilute sulphuric acid (25-30 ml) and check that the solution is acidic using indicator paper. Add, in small portions, sufficient solid sodium metabisulphite (~3.5 g) to destroy the excess of permanganate and manganese dioxide. At this point the colourless solution should contain, in suspension, white crystals of benzoic acid. Cool the mixture thoroughly, collect the crystals by filtration, and crystallise the product from minimum boiling water, collect, dry and weigh the product. Determine the melting point of the acid and place the crystals in a labelled sample tube. Using the above equation calculate the percentage yield of benzoic acid (specific gravity of toluene = 0.87), m.p. 121. The oxidation of toluene with $KMnO_4$ is represented below.

$$C_6H_5CH_3 \; + \; 2KMnO_4 \; \xrightarrow[\text{Alkali}]{\Delta} \; C_6H_5COO^- K^+ + 2MnO_2 + KOH + H_2O$$

6.12.4 Preparation of 4,4′-Dimethylbenzophenone

Chemicals: 4,4'-dimethyl benzhydrol, cyclohexane, manganese dioxide.

Experimental: Dissolve 4,4'-dimethylbenzhydrol (1 g) in cyclohexane (30 ml) and place this solution in a 100 ml round-bottomed flask fitted with a reflux condenser and calcium chloride drying tube. To this solution add manganese dioxide (4 g) and heat the mixture on a steam bath until the solvent refluxes gently (N.B. add two or three pieces of fuming stones to avoid bumping). At intervals of 30 minutes remove an aliquot (Ca 1 ml) of the solution, filter into a test tube and examine the filtrate by thin layer chromatography using a solution of 4,4'-dimethyl benzhydrol in toluene for comparison using toluene as the eluant. Observe the production of 4,4'-dimethylbenzophenone in the reaction by developing the chromatograms in iodine vapour when TLC indicates that the oxidation is essentially complete, allow the reaction mixture to cool and filter off the manganese dioxide. Evaporate the filtrate to dryness and crystallise the residue from methanol. m.p. 90-93 °C

6.13 Reduction

Introduction: Virtually all common inorganic reducing agents have been used in organic chemists, but most useful are hydrogen (with solid catalyst), metal hydrides, active metals, especially, zinc and sodium. The majority of organic reductions involve saturation of unsaturated functional groups, but a number of useful reductive substitution reactions are also known.

Lithium aluminium hydride is the most powerful and it will reduce all types of carbony groups, e.g., in esters, acids, amides, ketones etc.

$$R-CH=CHO+LiAlH_4 \rightarrow R-CH_2-CH_2OH$$

On the other hand sodium borohydride is not so reactive and is also selective towards various carbonyl groups. Thus it will not react with esters or carboxylic acids, but reduces ketones to secondary alcohols and aldehydes to primary alcohols.

$$RCHO+NaBH_4 \rightarrow RCH_2OH$$

6.13.1 Preparation of *p*-Nitrobenzyl Alcohol

Experimental: Dissolve p-nitro-benzaldehyde (0.50 g., 0.0033 mole) in dry methanol (30 ml) (dried over 'Molecular Sieves') contained in a 100 ml R.B. flask fitted with a drying tube. Add to this solution sodium borohydride (0.28 g., 0.008 mole) in two portions, and shake gently to encourage thorough mixing of the reactants. Gas will be liberated initially, but after approximately 15 minutes the reduction should be complete. Add to water (100 ml.) containing dil. HCl (10 ml.). Wash the combined extracts with water, dry over anhydrous sodium sulphate, filter off the desiccant on a Buchner funnel. Examine a small portion of the filtrate by TLC (5% ethyl acetate in toluene as the eluant). Evaporate the filtrate to dryness. Recrystallise the residue from petroleum ether (b.p. 60-80°) and determine the yield and m.p. of the product.

Mechanism: The reduction process occurs in two stages: Nucleophilic addition of H$^-$ (as BH$^-_4$ or AlH$^-_4$) across the $-C=O$ bond is analogous to the reaction of Grignard reagents and yields a salt. This is cleaved in the second stage by the addition of diluteacid.

By products : BH$_3$ (1 equiv)   NaOH (1 equiv)

6.13.2 Preparation of Aniline from Nitrobenzene

Chemicals: Nitrobenzene, granulated tin, conc. HCl.

Experimental: Place 4.5 g of granulated tin and 2.5 g (2.1 ml) of nitrobenzene in a 150 ml round bottomed flask. Place a condenser in the flask arranged for reflux. Pour down the condenser 10 ml of concentrated hydrochloric acid in 5 ml portions. Swirl the contents of the flask well, and cool it after addition under a running tap, if necessary, to keep the temperature from rising above about 60°C. Maintain this temperature for a further 15 minutes after the addition of the acid and then heat the flask on a boiling water bath for a further 30 minutes. During the whole of this time the flask should be shaken well every five or six minutes to ensure complete mixing of the contents. At the end of this time the smell of nitrobenzene should no longer be noticeable. Allow the mixture to cool. During the course of the reduction aniline chlorostannate may separate as a white or yellow crystalline complex. The aniline is obtained from this by treatment with alkali followed by **steam distillation.**

Dissolve 7.5 g of sodium hydroxide in 20 ml of water, cool it to room temperature, and add it gradually to the mixture in the flask prepared above. Keep the contents of the flask quite cool by swirling them round, holding the flask under a running cold-water tap. Now arrange the flask for steam distillation and pass steam into warm mixture until, after the distillate ceased to pass over as a turbid liquid. Collect above 15-20 ml of distillate. This should consist of a lower layer of aniline, above which is a mixture of aniline and water, partly as solution and partly as emulsion. Add to the distillate 5-6 g of sodium chloride, which will dissolve in the water and throw part of the aniline out of solution. Transfer the mixture to a separating funnel, add 10 ml of ether and shake to ensure intimate mixing of the solution and the ether; relieve the pressure by lifting the stopper and run. Allow the two layers to separate; run off the lower aqueous layer into a beaker, and pour the remaining ethereal layer through the mouth of the funnel into a 100 ml flask. Return the aqueous solution to the funnel and extract with a further 10 ml of ether. Proceed as before, and pour the ethereal extract into the flask. Dry with a few grams of anhydrous sodium sulpahte; shake the well-stoppered flask.

When dry, filter the aniline-ether solution into a small flask, add a few porcelain beads and set-up the apparatus for the distillation of ether; observing from all the precautions necessary. When all the ether has been distilled off, take away from the water-bath and heat the flask directly with a low flame. Collect the fraction having b.p. 180-184°C, which will be almost colourless. The yield is about 2 ml.

Mechanism:

$$Ar-NO_2 + 3\,Sn + 12\,H^{\oplus} \longrightarrow 2\,Ar-NH_2 + 3\,Sn^{4\oplus} + 4H_2O$$

The metal (tin) is oxidised to the tin(IV) state. When reduction is complete, a complex amine chlorostannate may separate from which the amine is liberated by treating with alkali.

$$[\,Ar\cdot NH_3]_2^{\ominus}\,[SnCl_6]^{2\ominus} + 8\,\overset{\ominus}{O}H \longrightarrow Ar\cdot NH_2 + SnO_3^{2\ominus} + 6\,Cl^{\ominus}$$

Mechanism

$$H_5C_6-N \xrightarrow{\text{Metal}} H_5C_6-N \xrightarrow{H^\oplus} H_5C_6-N \xrightarrow{\text{OH}} \xrightarrow{\text{Metal}} H_5C_6-N$$

$$H_5C_6-N=O \xrightarrow{\text{Metal}} H_5C_6-N=O \xrightarrow{H^\oplus} H_5C_6-N-OH \xrightarrow{\text{Metal}}$$

$$H_5C_6-N-OH \xrightarrow{H^\oplus} H_5C_6-NH-OH \xrightarrow[H^\oplus]{\text{Metal}} H_5C_6-NH_2$$

6.13.3 Prepartion of *m*-Nitroaniline from *m*-Dinitrobenzene (Partial Reduction)

Chemicals: m-Dinitrobenzene, sulphur, sodium sulphide

Experimental: Dissolve 5.4 g of crystalline sodium sulphide ($Na_2S.9H_2O$) in 15 ml of water in a 100 ml beaker add 0.7 g of finely powdered sodium hydrogen carbonate with stirring. When the carbonate dissolved completely, add 14 ml of methanol and cool below 20°C

Filter off the precipitate sodium carbonate at the pump, using a small Buchner funnel. Wash the precipitate with 5 ml of methanol. Retain the filtrate and washings; these contain about 1.5 g of NaSH in solution and must be used forth with for the reduction.

Dissolve 2.0 g of m-dinitrobenzene in 15 ml of hot methanol in a 100 ml round bottomed flask and add, with shaking, the previously prepared methanolic solution of sodium hydrogen sulphide. Attach a reflux condenser and boil the mixture for 20 minutes. Allow the reaction mixture to cool and fit the condenser for distillation. Distill off most of the methanol (30-35 ml) from a water bath. Pour the liquid residue with stirring into about 50 ml of cold water. Collect the yellow crystals of m-nitro aniline by filtration, wash with water and recrystallise from 75 percent aqueous methanol. The yield is 1.1 g, m.p. 114°C

6.13.4 Preparation of Benzpinacol (Photoreduction)

Hydrogen atom abstraction

Carbonyl compounds are converted into their pinacols. In this type of reductions a hydrogen atom is abstracted from the solvent or from other reactant to give radicals which dimerize. The irradiation of benzophenone in alcoholic solvent gives benzopinacol and oxidation products of the alcohol. Isopropyl alcohol is generally used.

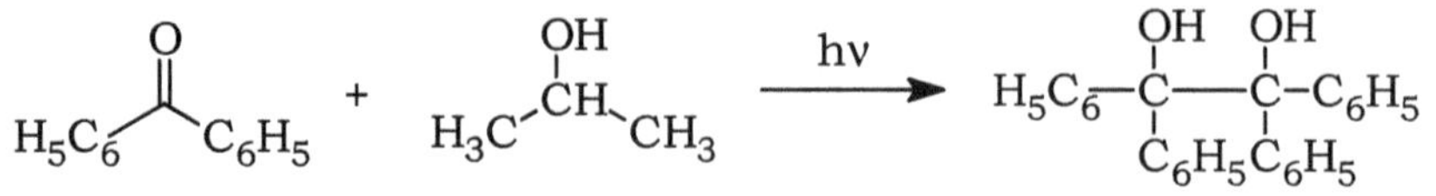

$$H_5C_6\underset{\text{O}}{\overset{\|}{C}}C_6H_5 \; + \; H_3C\overset{\text{OH}}{\underset{\text{CH}}{|}}CH_3 \;\xrightarrow{\;h\nu\;}\; H_5C_6\underset{C_6H_5}{\overset{\text{OH}}{\underset{|}{\overset{|}{C}}}}\underset{C_6H_5}{\overset{\text{OH}}{\underset{|}{\overset{|}{C}}}}C_6H_5$$

Chemicals: Benzophenone, isopropyl alcohol, glacial acetic acid.

Experimental: Dissolve 2.5 g of benzophenone in 20 ml of isopropyl alcohol in a 100 ml round-bottomed flask by warming and add one drop of glacial acetic acid. Fill the flask nearer to the neck with more isopropyl alcohol and stopper the flask tightly. Cover the stopper and joint with aluminium foil, if possible, and place the flask in direct sunlight by inverting the flask in a 100 ml beaker colourless crystals begin to separate within 24-30 hours. Leave the flask to remain in the sunlight, 95% of the reaction is completed in about three to four days. Cool the flask in ice-water and collect the crystals by suction filtration, wash it with 5 ml of ice-cold isopropyl alcohol. The product is almost pure. m.p. 185-186°C, The yield is 2 g.

Mechanism

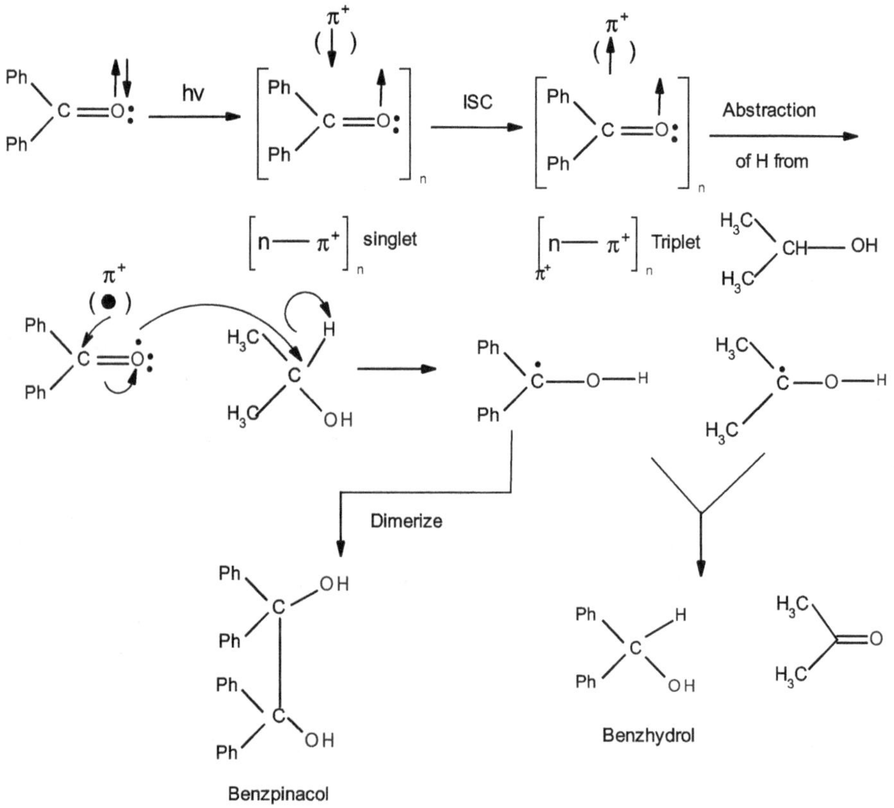

6.14 Benzoylation

Introduction: The insertion of a benzoyl moiety (C_6H_5-CO-) in place of the active hydrogen atom present in hydroxyl group (-OH), primary or secondary amine function (-NH2 or –NH) is usually called the benzoylation reaction. The reagent employed is benzoyl chloride in the presence of a base pyridine or aqueous sodium hydroxide: Schotten-Baumann Reaction.

6.14.1 Preparation of Benzanilide from Aniline (Schotten-Baumann Reaction)

The benzoyl derivative of aniline can be prepared using benzoyl chloride according to the equation.

$$\text{Aniline} + C_6H_5COCl \xrightarrow{\text{Base}} \text{Benzanilide} + HCl$$

Chemicals: Aniline, benzoyl chloride, sodium hydroxide.

Experimental: Add 2.7 ml of benzoyl chloride to 1.5 g of aniline in a small conical flask. Add sufficient dilute sodium hydroxide solution to make the mixture alkaline warm gently for 5 to 10 minutes on the steam bath. Test occationally to make sure that the mixture is still alkaline if necessary add more alkali, cool, pour into water, filter the crystals and wash with water. Recrystallise the benzanilide from aqueous ethyl alcohol. Filter the crystals and dry them. Weigh the dry crystals, determine the melting point, m.p. 163°C. Calculate the percentage yield using the equation given above.

Mechanism

6.14.2 Preparation of Hippuric Acid from Glycine (SN^1 Reaction)

Sodium glycerate reacts as an amine with acid chlorides and with benzoyl chloride the sodium salt of hippuric acid is formed. An excess of strong acid, such as hydro- chloric acid, will yield hippuric acid (benzoyl glycine).

Chemicals: Glycine, NaOH, benzoyl chloride.

$$C_6H_5COCl + H_2NCH_2COONa \longrightarrow C_6H_5CONHCH_2CO_2Na + HCl$$

$$C_6H_5CONHCH_2CO_2Na + HCl \longrightarrow C_6H_5CONHCH_2CO_2H + NaCl$$

Experimental: Dissolve 2.5 g of glycine in 25 ml of 10% sodium hydroxide solution contained in a conical flask. Add 4 ml of benzoyl chloride in 1 ml portions to the solution. Stopper the flask and shake it vigorously after each addition until all the chloride has reacted. Transfer the contents to a beaker and rinse the conical flask with a little water. Place a few grams of crushed ice in the solution and add concentrated hydrochloric acid slowly while stirring until the solution is just acid to litmus paper; this will precipitate the hippuric acid, which is contaminated with a benzoic acid. Filter it using a Buchner flask and then transfer the solid to a small beaker and boil it for a few minutes with about 10 ml of carbon tetrachloride. This dissolves any benzoic acid which may be present. Allow the mixture to cool slightly and filter off the hippuric acid and recrystallise it from hot water. Pure hippuric acid has m.p. 187°C. The yield is 2.5 g.

6.14.3 Preparation of Phenyl Benzoate

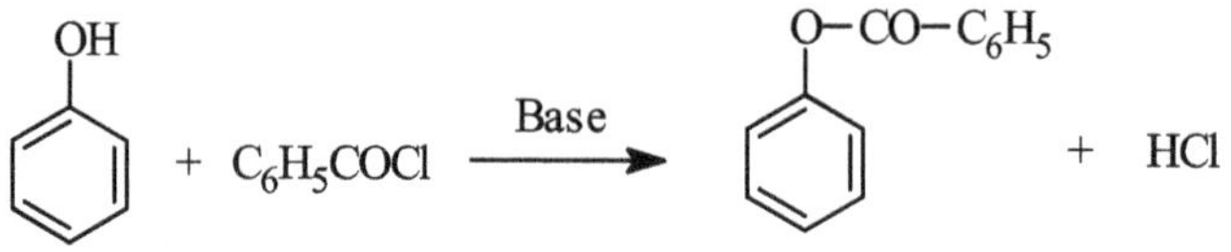

Chemicals: Phenol, benzoyl chloride, sodium hydroxide.

Experimental: Select a wide-mouthed glass stoppered bottle of about 50 ml capacity, and in it place 15 ml of 10% sodium hydroxide solution and 1 g of phenol. Then add 2 ml of benzoyl chloride, cork the bottle securely, shake the mixture vigorously for 15 minutes.

At the end of this time filter the material obtained through a filter pump, for traces of benzoyl chloride may still persist. Break up any lumps on the filter paper with a glass rod (but be careful not to puncture the paper). Wash the solid well with water and discord the filtrate. Recrystallise the solid ester from methylated spirit. The phenyl benzoate is obtained as white crystals, m.p. 60°C. The yield is 1.2-1.5 g.

Mechanism

6.14.4 Preparation of 2,4-Dichlorophenyl Benzoate

$$\text{(2,4-dichlorophenol)} \xrightarrow[\text{NaOH}]{\text{C}_6\text{H}_5\text{COCl}} \text{(2,4-dichlorophenyl benzoate)} + NaCl$$

Chemicals: 2,4-Dichlorophenol, benzoyl chloride

Experimental: Place 1.0 g of 2,4-dichlorophenol in a 50 ml round-bottom flask and add 10 ml of dilute sodium hydroxide to dissolve 2,4-dichlorophenol. Add 1.7 ml of benzoyl chloride to the solution of 2,4-dichlorophenol in sodium hydroxide. Stopper the flask and without heating, shake the flask vigorously for 10 to 15 minutes until a solid precipitate is formed. Test to make sure that the mixture is alkaline. If necessary add more alkali, cool and pour into 100 ml cold water, filter off the crystals and wash with water and purify by recrystallising from methanol.

Filter off the crystals, suck dry, find out the melting point and calculate the percentage of yield using the equation given above and based on the weight of phenol taken.

N.B. This experiment must be carried out in a fume cupboard. Take extra care when handling 2,4-dichlorophenol.

6.15 Acetylation

Introduction: The replacement of active hydrogen of molecules belong to the class phenols or alcohols, in addition to compounds of primary and secondary amines (RNH_2 & R_2NH) may be acetylated directly, whereby the active hydrogen atom is replaced by the acetyl radical (CH_3-CO-). This displacement of an active hydrogen by an acetyl group is known as acetylation.

6.15.1 Preparation of Acetyl Salicylic Acid - Aspirin

In this preparation the phenol, *salicylic acid,* will be acetylated with acetic anhydride using sulfuric acid as a catalyst. This acetylation reaction is also catalysed by certain basic compounds, such as sodium acetate and pyridine.

Chemicals: Salicylic acid (1 g) acetic anhydride (2 ml) conc. H_2SO_4 (2 to 3 drops).

Experimental: Place powdered salicylic acid (1 g) in a 50 ml conical flask and add acetic anhydride (3 ml) using the liquid to wash down any material adhering to the flask walls. Note that there is no visible reaction. Cautiously add two drops of conc. H_2SO_4 and rotate flask for a few minutes to mix the contents and dissolve the salicylic acid. The mixture is heated over a water bath at 50° for 10 minutes. Upon cooling and adding 20 ml of ice-water, crude acetyl salicylic acid is precipitated. The product is filtered and recrystallized from hot water. The yield is about 1 g ; the melting point varies between 130-136°C depending upon the extent of decomposition during the melting point determination.

Mechanism

Acetylation of Aromatic Amines

The common acylating agents are acid chlorides (RCOCl) and acetic anhydrides (R.CO.O.COR). Both of these types of reagents are commonly used for the preparation of acyl derivatives of amines and phenols.

Thus the acetylation of an aromatic amine using acetic anhydride can be represented by the equation:

6.15.2 Preparation of Acetanilide from Aniline

Chemicals: Aniline 2.0 ml, acetic anhydride, 2.5 ml, acetic acid, 2.5 ml.

Experimental: Place 2.5 g of aniline in a clean and dry boiling tube and add carefully 5 ml of a mixture of equal volumes acetic anhydride and glacial acetic acid. Heat the mixture gently on a water bath for 30 minutes. A small piece of boiling stone should be added to the boiling tube to prevent bumping.

Pour the hot solution slowly into 50 ml of cold water in a beaker, stirring the mixture well during the addition. The acetyl derivative will crystallise rapidly. When the mixture is cold, filter the crystals using a Buchner flask and funnel and wash the crude product on the filter with cold water. Transfer the crude product to a 50 ml conical flask and recrystallise from the **minimum** volume of boiling water or aqueous ethanol. Collect the colourless crystals by filtration, suck dry. Weigh the dried crystals, determine the melting point, m.p. 113°C. Calculate the percentage yield of the product as follows:

$$R{-}NH_2 + CH_3CO.O.CO.CH_3 \longrightarrow R.NH.COCH_3 + CH_3COOH$$

Let M = molecular weight of given amine. Then, since acetic anhydride is present in excess; M grams of amine will give (M + 42) grams of acetyl derivative. Thus calculate the theoretical yield of acetyl derivative from 2.5 g of amine and hence the percentage yield.

Mechanism

6.15.3 Preparation of *p*-Acetamol (*p*-Hydroxyacetanilide)

Chemical name: 4-Hydroxyacetanilide

Chemicals: *p*-Aminophenol 1.1 g, acetic anhydride 1.2 ml, acetic acid 1.2 ml.

Experimental: *p*-Aminophenol (2.2 g) is suspended in 6 ml of water containing in a 100 ml beaker or a conical flask and add 2.5 ml of acetic anhydride. Shake the mixture vigorously for 5 min., and warm the solution on a water bath until the solid dissolves. After 15 mins. cool the flask and filter the solid at the pump and wash with cold water. Recrystallised from hot water. The yield is 2.8 g, m.p. 168-169 °C.

Question

1. What is the reason for amino group undergoing acetylation selectively over - OH group?

6.16 Friedel-Crafts Reactions

6.16.1   Preparation of *p-ter*-Butylphenol

Aromatic compounds may be alkylated or acylated in the Friedel-Crafts reaction by alkyl and acyl halides respectively in the presence of a Lewis acid such as aluminium chloride. An example of this is the preparation of p-ter-butylphenol from phenol and t-butyl chloride.

The strong o/p directing effect of the hydroxyl group and the large size of the ter-butyl group results in the almost exclusive formation of the para isomer.

Chemicals: ter-Butyl chloride (1.8 g, 2.2 ml), phenol (1.6 g), anhydrous aluminium chloride (0.2 g).

Experimental: Place ter-butyl chloride (1.86 g, 2.2 ml) and phenol (1.6 g) in a dry boiling tube and stir until the phenol is nearly dissolved. Add aluminium chloride (0.2 g) a few pieces at a time. Hydrogen chloride evolution gas should start immediately. If the reaction mixture becomes warm, cool the boiling tube for a few minutes in a beaker of cold water. Use a glass rod to stir the contents of the flask occasionally to expose new surfaces of the catalyst. After 20-30 minutes, the contents of the boiling tube should solidify.

When the mixture is essentially solid, add 15 ml of methylene chloride and swirl till as much solid as possble has dissolved. Pour off the liquid into a separating funnel, wash with a mixture of water (10 ml) and conc. HCl (5 ml). Finally wash with water. Run off the methylene chloride (lower layer), Dry it with anhydrous sodium sulphate. Distil off methylene chloride to obtain solid.

Transfer the solid to a 50 ml conical flask and add enough petroleum ether (b.p. 60-80 °C) to dissolve the solid (i.e., minimum quantity of boiling solvent). Filter the hot solution rapidly through a preheated funnel. p-ter-butylphenol will crystallise from the filtrate on cooling. Filter the product at the pump and dry. m.p. 99 °C

Mechanism

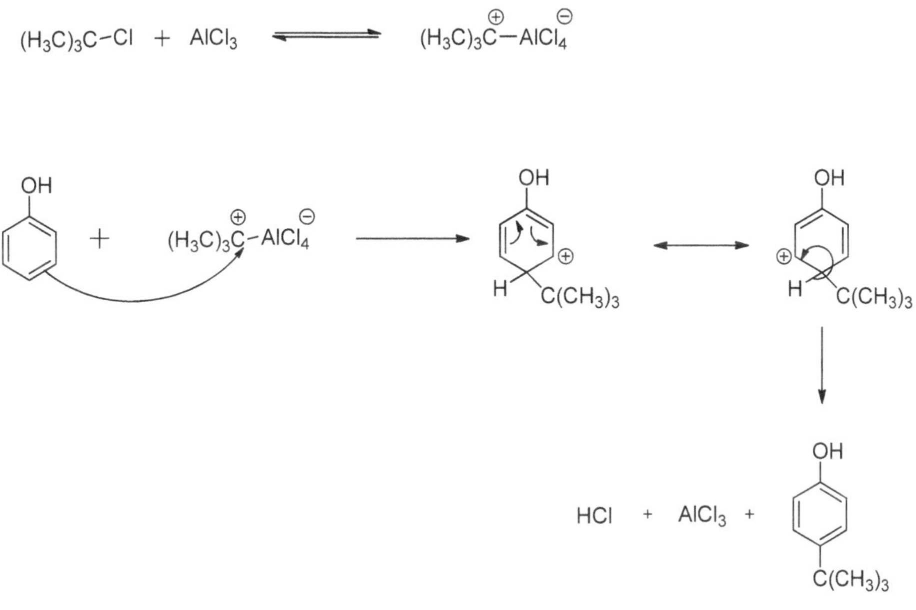

$$(H_3C)_3C\text{-Cl} \; + \; AlCl_3 \; \rightleftharpoons \; (H_3C)_3\overset{\oplus}{C}\text{-}\overset{\ominus}{AlCl_4}$$

6.16.2 Preparation of Benzophenone

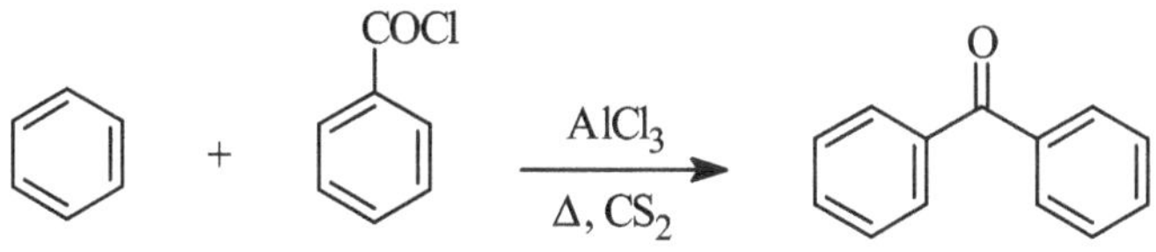

Chemicals: Benzene, finely powdered $AlCl_3$, benzoyl chloride, carbon-disulphide.

Experimental: Weigh out 1 g of finely powdered, anhydrous aluminium chloride into a dry test tube. It is poured with frequent shaking during the course of 10 minutes into a 50 ml dry round bottomed flask containing 3 ml of benzene, 0.75 ml of benzoyl chloride. A reflux condenser with a gas absorption trap attached to the flask, and warmed on a water bath for 2 to 3 hours or until only small amounts of HCl are being evolved. Pour the contents of the flask while still warm into a conical flask containing about 10 ml of water and small pieces of ice. After rinsing out the reaction flask with a little water and adding 4 ml of concentrated hydrochloric acid. The product is taken up in ether (20 ml) and shaken with 10 ml of 5% aqueous sodium hydroxide solution. After the etheral solution has been dried over anhydrous sodium sulphate and the ether has been evaporated the residue is distilled from a flask with a low side tube. A pure product is obtained by vacuum distillation in a round bottomed flask. It solidifies to a white solid on cooling, m.p. 47-48°C.

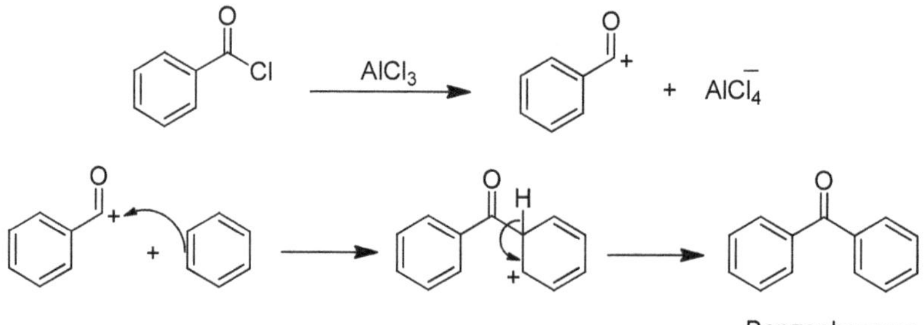

Benzophenone

6.16.3 Preparation of *o*-Benzoylbenzoic Acid

(J. Am. Chem. Soc. 1922, 44, 9, 2055–2060).

The Friedel-Crafts reaction is one of the best methods for the preparation of aromatic ketones and alkyl derivatives of aromatic hydrocarbons. In the preparation of ketones acid anhydrides are frequently used instead of acid chlorides.

In this experiment benzene and phthalic anhydride are the reactants.

Since one mole of aluminium chloride complexes with the carbonyl group and one mole with carboxyl group, a quantity in excess of two moles is required per mole of the anhydride. The benzene must be anhydrous and free of thiophene. The presence of thiophene in even small amounts will bring about tar formation.

Chemicals: Benzene (10 ml), phthalic anhydride (2 g) conc. HCl, 10% solution of sodium carbonate, aluminium chloride (4 g).

Experimental: A 50 ml round-bottom flask and condenser are arranged for refluxing, and a gas trap arranged with a piece of glass tubing extending through the top of the condenser to the surface of water contained in a beaker.

Place 2 g of phthalic anhydride in a 50 ml round-bottomed flask, add 10 ml of anhydrous, thiophene free benzene. The flask is thoroughly cooled in an ice bath until benzene begins to crystallize. An ammount of 4 g of aluminium chloride is weighed in a dry, stoppered test tube and added to the cooled mixture. The flask is connected to the condenser and trap. If reaction does not begin immediatly as evidenced by the evolution of hydrogen chloride, the flask is warmed with water at about body temperature. The beaker containing ice and water should be held in readiness to moderate the reaction should boiling begin. Finally when the vigorous reaction has subsided, the reaction mixture is shaken and is heated over a boiling water bath for 30 minutres.

The flask is removed and cooled in an ice bath, and about 10 g of ice is added. This is followed by 3 ml of concentrated hydrochloric acid. Unreacted benzene is now removed by steam distillation. upon cooling the flask and scratching its sides, o-benzoylbenzoic acid crystallizes. The aqueous solution is carefully decanted through a filter, the solids are washed with 5 ml of water, and the suspension is again decanted. Add 10 ml of 10% solution of sodium carbonate and the resulting mixture is carefully reflxed for 10 minutes to dissolve all solids except aluminium hydroxide. The suspension is cooled a little and the mixture is filtered. The filtrate is placed in a beaker and is carefully made acidic by the addition of small portion of hydrochloric acid. The acid separates as an oil but it soon crystallises on stirring and cooling. Filter when it is cold, and wash with little water. Dry upon a filter paper, the product consists largely of the monohydrate, m.p. 94 $^{\circ}$C.

To prepare pure anhydrous o-benzoylbenzoic acid, the crude wet mono-hydrate is placed in a 100 ml conical flask, 20 ml of benzene is added, and the mixture is heated on a steam bath. After the solids have disolved, the aqueous phase is removed in a separating funnel, and the benzene solution is placed in a dry 100 ml flask. A small amout of charcoal (0.5 g) is added, and is heated on a steam bath for 2 minutes and is then filtered by gravity. The filtrate is concentrated to a volume of about 5 ml. Ligroin is added dropwise to the hot concentrate until a slight turbidity is produced. On cooling in an ice-bath o-benzoyl benzoic acid precipitates and it is filtered. The yield is about 2 g. m.p. 127-128°C. 80% yield.

6.16.4 Carbonium Ion Rearrangement

Ref. G.A. Olah, 'Friefel-Crafts and related reaction, volume II, Part I' Interscience.

Alkylation of benzyne under Friedel-Crafts conditions is the basis for several commercially important syntheses. Alkyl halide, olefins, alcohols, and ethers may be used as alkylating agents together with certain Lewis acids such as HF, BF_3 or anhydrous metal halides (e.g. $AlCl_3$, $SnCl_4$ etc.).

For example ethylbenzene, a precursor in the preparation of styrene, is obtained industrially by the reaction between ethylene and benzene in the presence of $AlCl_3$,, and cumene is obtained from propylene, benzene and $AlCl_3$.

The Friedel-Crafts alkylation process has several disadvantages, including reversibility at higher temperatures and over-alkylation occurring as a consequence of increased reactivity, following the introduction of one substituent. The principal difficulty is associated with a tendency for the alkylating agent to rearrange during reaction. It is straightforward to introduce a mythyl-, ethyl, or isopropyl group, but more difficult to introduce n-propyl-, n-butyl- or isobutyl groups; rearrangement of the latter to isopropyl, or t-butyl groups, respectively, usually occurs. Increase in reaction temperature as well as certain catalysts encourage rearrangement..

Mechanism

The mechanism of the reaction is not certain. For alkyl halides it has been suggested that rearrangement occurs before alkylation, generally to give the product which would result from the most stable carbonium ion. A discrete carbonium ion is probably not involved but

$$\text{e.g.} \quad CH_3\text{-}CH_2\text{-}CH_2Cl + AlCl_3 \rightarrow CH_2\text{-}CH_2\text{-} \overset{+}{C}H_2 \rightleftarrows CH_3\text{-}\overset{+}{C}H\text{-}CH_3$$

primary   secondary

there is evidence for a bimolecular mechanism of this type;

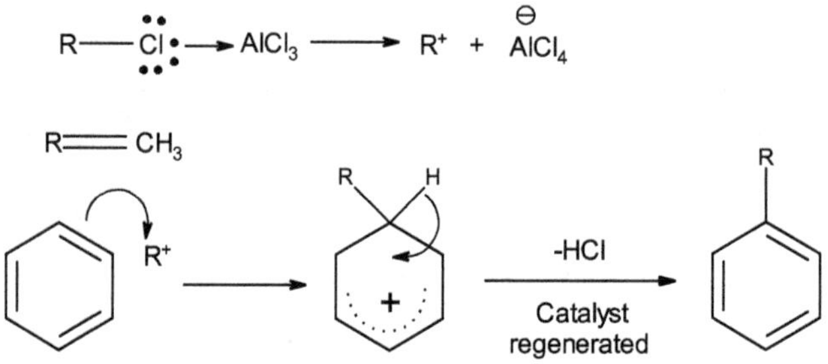

This mechanism involves a transition state best described as the result of nucleophilic attack by the aromatic compound on a polarised $RCl - AlCl_3$ addition compound. Only when the R – Cl bond is strongly polarised, and the resulting carbonium ion is well developed, will appreciable rearrangement occur. It is clear that the nature of R and the Lewis acid will dictate the polarity of the R–Cl bond in the addition compound. (mechanism: example, R = CH_3).

The object of the following experiment is to compare the alkylation of benzene with a primary versus a secondary carbonium ion in the presence of anhydrous $AlCl_3$.

Students should work in pairs, one carrying out an alkylation with n-propyl bromide and the other using isopropyl bromide. The results for each should be collected and compared.

Experimental Procedure

Place sodium-dried benzene (75 ml) and anhydrous* aluminium chloride (7.5 g., 0.056 mole) in a dry 250 ml RB flask fitted with a reflux condenser, dropping funnel, and calcium chloride drying tube. Cool the flask in an ice-bath until the internal temperature is below 10° and add the appropriate propyl bromide (6.6 g, 0.053 mole) from the dropping funnel. Immediately after the addition remove the funnel and replace the drying tube on the condenser. Gas evolution will commence and become more vigorous as the reaction mixture warms to room temperature. Leave the mixture for 24 hours swirling the contents occasionally, and then pour into a separating funnel containing approx. 100 g of ice and conc. hydrochloric acid (50 ml). Separate the benzene layer and wash with dil. hydrochloric acid (2 × 25 ml), water (25 ml), sodium bicarbonate solution (3 × 25 ml) and water (25 ml). Dry the solution (anhydrous $CaCl_2$), filter and distil the product at atmospheric pressure, using a heating mantle. Discard the distillate b.p. $< 130^O$ and transfer the distillation residue into a 25 ml B14 RB flask. Collect the fraction b.p. $145 - 165^O$ for analysis by gas-liquid chromatography. Retain the pot-residue for analysis in case the conversion has been poor. Under the conditions described a yield of at least 3 ml distillate is usual.

n-propylbenzene   b. p. 159 ^{O}C

iso-propylbenzene   b. p. 152-153 ^{O}C

G.L.C Conditions: 6 feet column of silicone rubber SE30 10% on 60-80 mesh firebrick at 140°. Inject samples of pure iso-propyl – and n- propyl benzene as standards before analyzing product.

* Weighed out rapidly in a dry stoppered bottle to prevent hydrolysis.

6.17 Preparation of Dyes

Diazotization/Coupling

Preparation of Dyestuff intermediate, conversion into an azo-dye, and use in dyeing cotton.

6.17.1 Preparation of Azo Dye (Para Red)

(a) Preparation of p-Nitroaniline

This is prepared by acidic hydrolysis of *p*-nitroacetanilide:

$$O_2N\text{—}C_6H_4\text{—}NH\cdot COCH_3 + H_2O \xrightarrow{H^+} O_2N\text{—}C_6H_4\text{—}NH_2 + CH_3\cdot CO_2H$$

The hydrolysis of the anilide is best conducted using strong aqueous sulphuric acid prepared by adding cautiously 20 ml of conc. H_2SO_4 to 15 ml of water. (See page 6.10.2 for experimental procedure)

(b) Preparation of the azo-dye 'Para Red'

p-Nitroaniline is converted into p-nitrobenzene diazonium chloride by treatment with sodium nitrite in aqueous hydrochloric acid solution. This diazonium salt is then couped with b-naphthol in sodium hydroxide solution to give the azo-dye. The equations for these rections are:

a) $O_2N\text{—}C_6H_4\text{—}NH_2 \xrightarrow{HNO_2 + HCl} O_2N\text{—}C_6H_4\text{—}\overset{+}{N}\equiv N\ \overset{-}{Cl} + 2H_2O$

b) $O_2N\text{—}C_6H_4\text{—}\overset{+}{N}\equiv N\ \overset{-}{Cl} \xrightarrow[NaOH]{C_{10}H_7OH} O_2N\text{—}C_6H_4\text{—}N=N\text{—(naphthol)}$

Para Red

In the present experiment the coupling reaction is carried out on a piece of cotton cloth.

Obtain a piece of cotton cloth from the demonstrator and soak it in a solution of b-naphthol prepared as follows: Suspend powdered b-naphthol (0.5 g) in water (100 ml), then with stirring, add sodium hydroxide solution dropwise until the material just dissolves. Soak the cloth thoroughly in this solution and allow it to dry.

Prepare the diazonium salt solution as follows:

Suspend p-nitroaniline (1.4 g) in a mixture of water (30 ml) and 10% hydrochloric acid (6 ml) and heat on the steam bath for 10 minutes with stirring. Cool the suspension to 0-5° in ice (the hydrochloride of the amine will crystallise) and add, all at once, a solution of sodium nitrite (0.7 g) in a little water. Filter at the pump and keep the filtrate ice-cold till required.

The dyeing operation is accomplished by diluting the ice-cold diazonium salt solution with about 300 ml of ice-cold water and dipping the dried, b-napthol-impregnated cloth in the solution. Allow the dyed cloth to dry and keep it for inspection.

6.17.2 Preparation of Methyl Orange and Methyl Red

Chemicals: Anhydrous sodium carbonate, sodium nitrite, sulphanilic acid, hydrochloric acid, dimethyl aniline.

Experimental: In a 150 ml conical flask place 5.25 g of sulphanilic acid dihydrate, 1.3 g of anhydrous sodium carbonate and 50 ml of water and warm the contents until a clear solution is obtained. Cool the solution and add a solution of 1.8 g of sodium nitrite in 5 ml of water. Cool the mixture in ice-water until the temperature has fallen to 5°C. Now add slowly with stirring a solution of 5.2 ml of concentrated hydrochloric acid in 5 ml of water; Do not allow the temperature to rise above 10°C. When all the acid has been added, allow the solution to stand in ice-water for 10-15 minutes. Fine crystals of the diazobenzene sulphonate will soon separate; do not filter these crystals as they will dissolve during the next stage of the preparation. Dissolve 3.1 ml dimethylaniline in 1.5 ml of glacial acetic acid, and add it with vigorous stirring to the suspension of diazotised sulphanilic acid. Allow the mixture to cool for 10 minutes; a pale red colouration is developed. Then add slowly with stirring 18 ml of 20 percent sodium hydroxide solution until the mixture attains a uniform orange colour.

The sodium salt of methyl-orange separates as very fine particles, direct filtration may be very slow. Hence warm the mixture to 55-60°, stirring carefully with a glass rod; when all the methyl orange dissolved. Add about 5 g of powdered sodium chloride, and warm at 80-90°C until the salt is dissolved. Allow the mixture to cool for 25 minutes in ice-water. Filter off the methyl orange at the pump, but apply only gentle suction so as to avoid clogging the pores of the filter paper. Drain well, and recrystallise from hot water (about 75 ml are required); filter the hot solution, if necessary, through a pre heated Buchner funnel. The deep reddish orange crystals of methyl orange separates as the solution cools. The yield is 6.5 g.

Methyl Red

Methyl orange is a weak base, and in the presence of acids takes up a proton at the basic nitrogen atom, to give an ion which has a colour (red) different from that of the original molecule (orange).

Preparation of a Dyestuff and its use in dying cotton

6.17.3   Preparation of 1-Phenylazo-2-Naphthol

Primary amines of the aromatic series will yield diazonium salts when reacted with sodium nitrite in aqueous hydrochloric acid. These relatively unstable diazonium salts are very important intermediates in the synthesis of many different types of aromatic compounds. By the elimination of nitrogen from the diazonium group it is possible to introduce such function as halogens, cyano, hydroxy and nitro.

The second important application of diazonium salt is in coupling reactions. In this experiment 1-phenylazo-2-naphthol is prepared. The dye is prepared by diazotizing aniline hydrochloride and coupling the diazonium salt formed with beta-naphthol. The equations for the reaction are :

Experimental: Obtain a piece of cotton cloth from the demonstrator and soak it in a solution of 2-napthol prepared as follows.

Dissolve 2-naphthol (0.5 g) in dilute sodium hydroxide solution (2 ml) and add water (100 ml) in a 250 ml beaker. Soak the cloth thoroughly in this solution and allow it to dry.

Dissolve 1.85 ml aniline (measure this using a graduate pipette, do not pipette by mouth) in a mixture of 25 ml of water and 6 ml of concentrated hydrochloric acid in a 100 ml conical flask. Cool the solution to 0-5°C by keeping in crushed ice and add in four portions a solution of sodium nitrite (1.4 g) in water (5 ml) during a ten minute period. Keep the solution in ice cold throughout the addition.

Take half of the diazonium salt solution prepared as above into another beaker and dilute this with 25 ml ice-cold water. Dip a piece of cotton cloth previously impregenated with 2-naphthol (see above) into the solution. Allow the dyed cloth to dry and keep it for inspection.

Add the remainder of the diazonium salt solution very slowly to another beaker containing a stirred solution of 2-naphthol (1.5 g) in 10 ml dilute sodium hydroxide solution. After the addition is complete, allow the reaction mixture to stand in an ice-bath for 20 minutes. Filter off the precipitate on a Buchner funnel, wash well with water and suck dry. Recrystallise the red crystals of p-phenylazo-2-naphthol from methanol. Pure 1-phenylazo-2-naphthol has m.p. 131°C.

1-Phenylazo-2-napthol

N.B. : 1-Phenyl azo-2-naphthol is an excellent orange dye. Take care not to get this on yourself or your cloths as the dye is very difficult to remove.

6.18 The Beckmann Rearrangement

6.18.1 Preparation of Benzanilide

It involves two steps (i) Preparation of benzophenone oxime from benzophenone

Step (i) : Preparation of benzophenone oxime

$$H_5C_6-\overset{\overset{O}{\|}}{C}-C_6H_5 \quad \xrightarrow{NH_2OH \cdot HCl} \quad H_5C_6-\overset{\overset{N-OH}{\|}}{C}-C_6H_5$$

Chemicals: Benzophenone, hydroxylamine hydrochloride, sodium hydroxide.

Experimental: Place a mixture of 2.5 g pure benzophenone, 1.5 g of hydroxylamine hydrochloride, 5.0 ml of rectified spirit and 1.0 ml of water in a 100 ml round bottomed flask. To this add 2.8 g of sodium hydroxide (pellet form) in portions with shaking; If the reaction becomes vigorous, cool the flask under running tap water. When the addition of sodium hydroxide is complete, attach a reflux condenser to the flask, heat to boiling for 5 minutes. Cool, and pour the contents of the flask into a solution of 7.0 ml of concentrated hydrochloric acid in 50 ml of water contained in a 200 ml beaker. Filter the precipitate at the pump, wash with cold water and dry. The yield of the benzophenone oxime is 2.6 g., m.p. 142°C. If may be crystallised from methanol. The oxime is gradually decomposed on exposure to oxygen. It should be kept in a vacuum desiccator until further use.

Step (ii): Preparation of benzanilide from benzophenone oxime

$$Ph_2C{=}N{-}OH \ (\text{Oxime}) \ + \ SOCl_2 \ \longrightarrow \ Ph{-}\underset{\underset{\displaystyle}{\|}}{\overset{\overset{\displaystyle O}{\|}}{C}}{-}NH{-}Ph \ (\text{Benzanilide})$$

Chemicals: Benzophenone oxime, anhydrous ether thionyl chloride.

Experimental: Dissolve 2.0 g of benzophenone oxime in 20 ml of dry ether in 100 ml conical flask and add 3 ml of thionyl chloride; leave the contents for 10 minutes at room temperature. Distil off the solvent and other volatile products on a water bath (Caution ether) and add 25 ml of water. Boil the mixture for 10 minutes and break-up any lumps which may be formed. Decant the supernatant liquid, and recrystallise in the same flask, from the boiling ethyl alcohol. The yield is 1.6 g, m.p. 163°C.

Mechanism: Step (i)

Step (ii)

6.18.2 Preparation of Caprolactam

Principle: Cyclohexanone condense with hydroxylamine hydrochloride to give cyclohexane oxime. The oxime undergoes Beckmann rearrangement to give cyclic amide (lactam).

Step (i) Preparation of cyclohexanone oxime

Chemicals: Cyclohexanone, hydroxylamine hydrochloride conc. sulphuric acid.

Experimental: Place a mixture of 2.0 g (2.1 ml) of pure cyclohexanone, 1.7 g of hydroxylamine hydrochloride and 5 ml of water in a small conical flask; cool the mixture in ice-water. To this add a solution of 1.3 g of anhydrous sodium carbonate in 4 ml of water while stirring the contents, and heat the mixture on a water bath at 50°C for 20-25 minutes cool, the oxime rapidly separates on cooling. Filter the precipitate at the pump, drain thoroughly and dry it in a (vacuum) desiccator. Recrystallise from petroleum ether. The yield of the oxime. is 1.6 g, m.p. 88°C.

Step (ii) Beckmann rearrangement: Prepare 85% sulphuric acid by adding 50 ml of the conc. sulphuric acid cautiously to 10 ml of water and then cool the dilute acid in ice cold water. Place 5 ml of conc. sulphuric acid in a beaker or conical flask, add 2.5 g. of the pure oxime prepared above, and warm the mixture cautiously until effervescence begins and at once remove the heat. A vigorous reaction occurs and is soon complete. Cool the contents in ice water and stir it mechanically with the help of glass rod, whilst, slowly adding 25% aqueous potassium or sodium hydroxide (about 20 to 30 ml) until the mixture is faintly alkaline to litmus; ensure that the temperature does not rise above 20^0 C during operation. A considerable amount of potassium sulphate crystallizes from the mixture. Filter the inorganic salt at the pump. Finally extract the aqueous layer two times with chloroform, using 30 ml each occasion. Dry the organic layer with sodium sulphate, filter and distill the chloroform. The caprolactum separates on cooling. Take small amount and recrystallise from petroleum ether. m.p.70-72° C, Yield 2.0 g.

Mechanism

e-Caprolactam

6.19 Fischer Indole Synthesis

6.19.1 Preparation of 2-Phenylindole

Preparation is carried out in two steps

Step (i): Formation of Schiff's base by the reaction between acetophenone and phenyl hydrazine.

Step (ii): Cyclisation of the phenyl hydrazone derivative in presence of polyphosphoric acid to form 2-phenylindole.

Chemicals: Acetophenone, phenyl hydrazine, glacial acetic acid, phosphoric acid, phosphorus pentoxide.

Step (i) : Preparation of acetophenone phenylhydrazone

Experimental: In a 100 ml conical flask place a mixture of 2.5 ml of acetophenone, 2.3 ml phenylhydrazine in 15 ml of ethanol. Add a few drops of glacial acetic acid and warm the contents on a steam bath for 30 minutes. Cool and filter the product, wash with dilute hydrochloric acid followed by about 5 ml of cold rectified spirit. Recrystallise a small portion from hot ethanol.,m.p. 106°C.

Step(ii): Preparation of 2-phenylindole

Place 1.5 g of the crude phenylhydrazone, prepared in step (i), in a 100 ml conical flask or a beaker containing polyphosphoric acid. Heat on a boiling water bath, stir with a glass rod cautiously and maintain the temperature at 100°C for 1 hour. To this add 25 ml of cold water and stir well to dissolve unreacted polyphosphoric acid. Collect the solid on a Buchner funnel, wash well with cold water. Recrystallise the crude solid from rectified spirit. The yield of 2-phenylindole is 1.0 g, m.p. 187-188°C.

Mechanism

Step (i)

Step (ii)

6.20 Preparation of 1,2,3,4-Tetrahydrocarbazole (Method-1)

Aim: To prepare 1,2,3,4-tetrahydrocarbazole from cyclohexanone and phenyl hydrazine following Fischer-indole synthesis.

Chemicals required: Phenyl hydrazine (1 ml), cyclohexanone (1 ml), acetic acid (5 ml).

Experimental: Dissolve 1 ml of cyclohexanone in 5 ml of glacial acetic acid taken in a clean and dry boiling tube. Add 1 ml of phenyl hydrazine and one porseline chip. Heat the reaction mixture gently with a low flame using wire gauge for 5 minutes. Cool the boiling tube to room temperature and then pour the reaction mixture into 30 ml of ice cold water with stirring.

Filter the flesh coloured crude precipitate through a Buchner funnel. Wash the solid on the filter with cold water; suck almost dry. Spread the crude solid upon absorbent paper. Recrystallise the sample from methanol; add a little decolourising carbon and filter through a hot-water funnel. The yield is 1.3 g, m.p. 116-117°C.

Mechanism

1,2,3,4-Tetrahydrocarbazol

(b) Preparation of 1,2,3,4-tetrahydrocarbazole (Method-2): Indole synthesis by Pd-catalysed annulation of ketones with o-iodoaniline (1H) carbazole, 2,3,4,9-tetrahydro-). The indole nucleus is a common and important feature of a variety of natural products and medicinal agents. The traditional approach for preparing the indole nucleus is the Fischer indole reaction. As the reaction has shortcoming, the palladium-catalysed coupling of ortho-haloanilines is becoming an excellent alternative (Chen, C.Y et al., J. Org. Chem. (1997), 62, 2676).

Chemicals: Cyclohexanone, o-iodoaniline, 1,4-diazabicyclo [2,2,2] octane (DABCO), N,N-dimethyl formamide, palladium acetate.

Experimental: To a 100 ml two necked flask, is added a mixture of cyclohexanone (1.5 g, 15 mmol), o-iodoaniline (1.1 g, 5 mmol), and

1,4-diazabicyclo[2[2[]octane   (DABCO)   1.7   g   (15   mmol)   in   N,N-dimethylformamide (DMF) (15 ml). The mixture is degassed three times via nitrogen/vacuum, followed by the addition of palladium acetate 0.56 mg, 0.025 mmol). The mixture is degassed twice and heated at 105°C for 3 hr or until completion of the reaction (note). The reaction mixture is cooled to room temperature and partitioned between isopropyl acetate (45 ml) and water (15 ml). The organic layer is separated, washed with brine (15 ml), and concentrated under vacuum to dryness. The residue is chromatographed on 20 g of silica gel using 175 ml of ethyl acetate-heptane (1:6) as the eluent to give 0.55 g of 1,2,3,4-tetrahydrocarbazole (65%) as pale yellow brown solid. m.p. 116-118°C.

Note: The reaction generally takes 3-5 hrs to complete and is monitored by TLC (Rf = 0.50, SiO_2, eluted with ethyl acetate-heptane, 1:4).

6.21 Hoffman Rearrangement

Amides which do not have a substituent on the nitrogen display a molecular rearrangement on treatment with solution of bromine or chlorine to give primary amines. In this rearrangement the carbonyl carbon atom of the amide is lost and the R group of amide gets attached to the nitrogen of the amine.

Aim: To study multistep synthesis of heterocyclic system.

6.21.1 Synthesis of Acridone

It involves five steps.

Step-(i)   :   Preparation of phthalimide from phthalic anhydride

Step-(ii)   :   Preparation of anthranilic acid from phthalimide (Hoffman rearrangement).

Step-(iii) :   Preparation of orthochlorobenzoic acid from anthranilic acid.

Step-(iv) :   Preparation of N-phenyl anthranilic acid

Step-(v)   :   Preparation of acridone from N-phenyl anthranilic acid by cyclisation with acid catalyst.

Step (i) : Preparation of phthalimide

Aim: To prepare phthalimide from phthalic anhydride by amidation using urea.

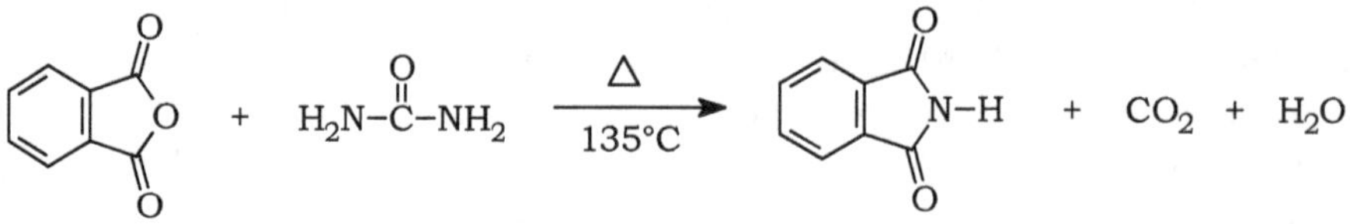

Chemicals required: Phthalic anhydride (5 g), urea (1.5 g)

Experimental: Intimately powder 5 g of phthalic anhydride, 1.5 g of urea and place this mixture in a clean and dry conical flask. Heat the flask directly with a Bunsen burner using sand bath at 135°C. After few minutes the contents of the flask melt completely and effervesance starts slowly and gradually increases in vigour. After heating for another 5 minuts, the reaction mixture suddenly froths upto three times of its original volume and then becomes a solid. Remove the flame and allow the contents to cool down to room temperature. At this stage add 5 ml of water and disintegrate the solid. Filter the colourless solid of phthalimide at the pump and dry. The phthalimide (90%) is practically pure and melts at 233-234°C. If desired, the phthalimide may be recrystallised from industrial spirit.

Mechanism

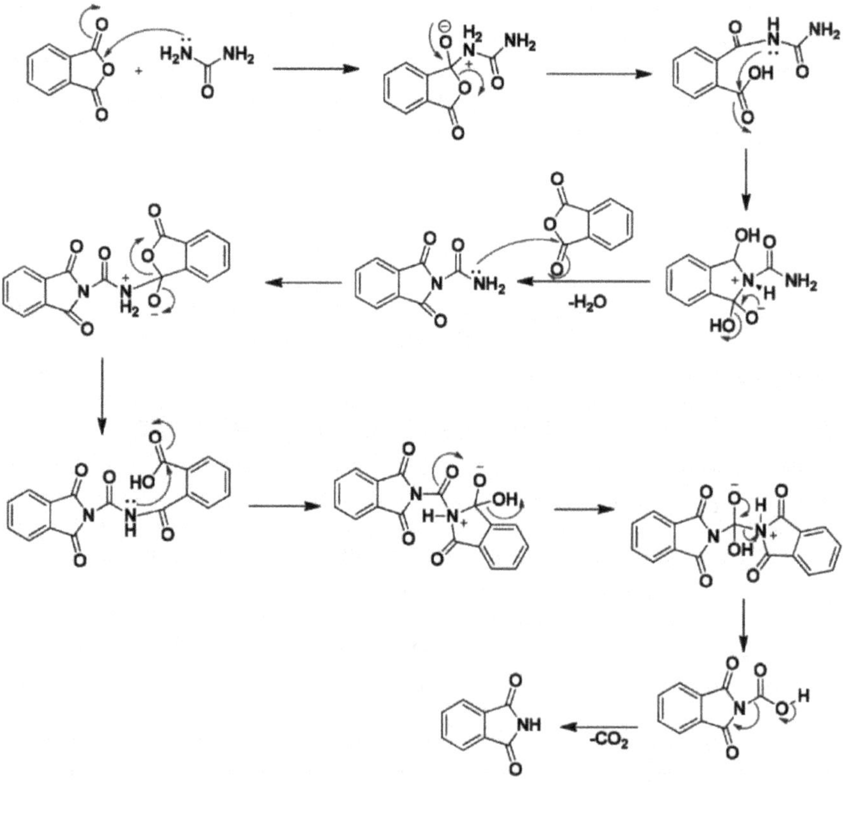

Step (ii): Preparation of anthranilic acid from phthalimide

Principle: Phthalimide undergoes Hoffman rearrangement in the presence of bromine and sodium hydroxide to give anthranilic acid.

Chemicals: Phthalimide, sodium hydroxide, bromine.

Experimental: A solution containing 6 g of sodium hydroxide in 25 ml of water is prepared in a 250 ml conical flask while cooling the flask in an ice bath. To this solution is added 1.7 ml of bromine in one portion and shake until all bromine reacts and cool the mixture again to 0° or below, Meanwhile, prepare a solution of 4 g of sodium hydroxide in 15 ml of water. Add 5 g of finely powdered phthalimide in one portion to the cold solution of sodium hypobromite and stir vigorously while swirling the contents of the flask. Remove the flask from the cooling bath and add conc. hydrochloric acid slowly with stirring until the solution is just neutral. Precipitate the anthranilic acid by the gradual addition of glacial acetic acid. It is advisable to transfer the mixture to a 500 ml beaker as some foaming occurs. Filter the solid at the pump and wash with a little cold water. Recrystallise from hot water with the addition of a little charcoal; collect the acid on a Buchner funnel and dry at 100°C. The yield is 60%. m.p.145°C.

Mechanism:

(ii) Hydrolysis

Formation of sodium hypobromide

The mechanism involves on the following lines:

Ø It is a base promoted bromination of an amide

Ø The base abstracts a proton from the nitrogen to give a bromo-amide anion.

Ø Separation of the halide ion gives an electron deficient nitrogen atom, a nitrene.

Ø This is followed by actual migration of R group to the electron deficient nitrogen.

Ø Lastly the isocyanate undergoes hydrolysis to give an amine.

$$NaOH \ + \ Br_2 \longrightarrow NaOBr \ + \ HBr$$

Step (iii): Preparation of *ortho* chlorobenzoic acid from anthranilic acid

Chemicals: Anthranilic acid (3 g), conc. HCl, sodium nitrite (1.7 g), copper sulphate (5.5 g), sodium chloride (2.5 g), copper turnings (3 g).

Experimental: In a 150 ml conical flask 3 g of anthranilic acid, 5 ml conc. HCl is dissolved in 20 ml of water by gentle heating. The flask is then placed in an ice-salt bath and cooled to 0° and diazotise it by adding dropwise an ice-cold solution of 1.7 g of sodium nitrite in 20 ml of water. Keep the diazonium solution below 5°C.

Preparation of reducing agent: Place 5.5 g of copper(II) sulphate, 2.5 g of sodium chloride, 2.5 ml of water in a 100 ml R.B. flask equipped with a reflux condenser. Heat the solution to boil and add 15 ml of conc. HCl, 3.0 g of copper turnings into it and continue boiling until the solution becomes practically colourless. Cool the solution to 0-5 degrees in an ice bath and add to it in portions with constant stirring cold diazonium salt solution; the reaction proceeds rapidly with frothing. After the addition is over, allow the reaction mixture to stand at room temperature for an half an hour with occasional shaking. Filter the product with the Buchner funnel and wash with cold water. Recrystallise from water containing a little alcohol. m.p. 138°

Mechanism

Step (iv): Preparation of N-phenyl anthranilic acid from o-chlorobenzoic acid

Chemicals: Ortho chlorobenzoic acid (3 g), aniline (11 ml), anhydrous potassium carbonate (4 g) and CuO (1/4 g).

Experimental: Place a mixture of ortho chlorobenzoic acid (3 g), aniline (11 ml), anhydrous K_2CO_3 (4 g) and CuO (1/4 g) in a round-bottomed flask (50 ml) equipped with an air condenser. Reflux the mixture for about 2 hours in an oil bath. Allow it to cool, remove excess of aniline by steam distillation; boil with 3 g of decolourising carbon for 10 minutes and filter at the pump. Add the filtrate with stirring to a mixture of 4.5 ml of conc. hydrochloric acid and 9 ml of water and allow the mixture to cool. Filter the crude precipitate at the pump and recrystallise from a solvent mixture of acetic acid and water.

Mechanism

N-Phenyl anthranilic acid

Step (v): Preparation of acridone from N-phenyl anthranilic acid

Principle: Acridone can be prepared by cyclisation of N-phenyl anthranilic acid with acid.

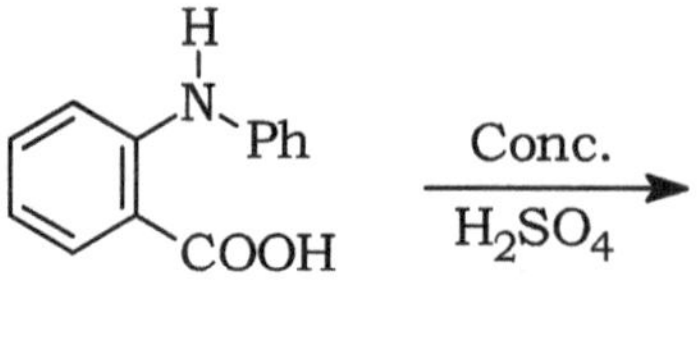

Chemicals required: N-Phenyl anthranilic acid (2 g), Conc. H_2SO_4.

Experimental: Prepare a mixture of N-phenyl anthranilic acid (2 g) and conc. sulphuric acid (5 ml) in a 100 ml conical flask and heat the contents for 1½ hrs. on a steam bath. Then pour the hot dark solution slowly and continuously into 100 ml boiling water in 400 ml beaker (allowing the acid to run down the side of the beaker to prevent spurting). The mixture is boiled for a further period of 5 minutes and filter while hot through Buchner funnel. The precipitate is separated on cooling. Filter at the pump and wash with water and dried. Recrystallise from acetic acid using animal charcoal. m.p. 352°C.

$$H_2SO_4 \rightleftharpoons H^+ + HSO_4^-$$

Mechanism:

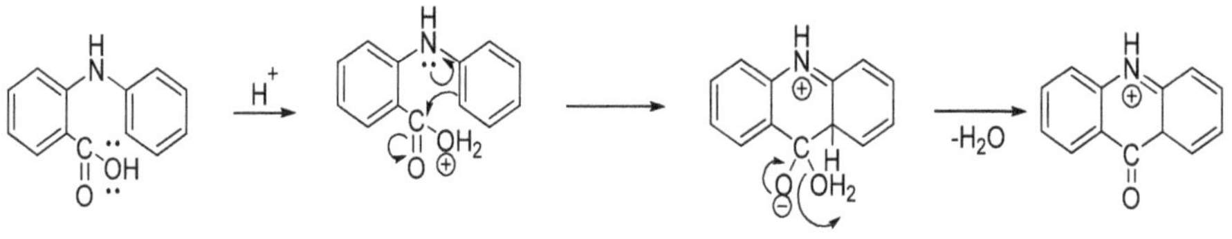

Acridone

***

6.22 Claisen-Schmidt Reaction

6.22.1 Preparation of a Chalcone (Base-catalysed Condensation)

Ref: C D Gutsche, the chemistry of carbonyl compounds (1967).

In presence of a base, aldehydes condense with compounds containing activated (acidic) methylene or methyl groups to give β-hydroxy carbonyl compounds.

$$R-CHO \ + \ R^1CH_2COR^2 \longrightarrow R-\overset{\overset{\displaystyle OH}{|}}{CH}-\overset{\overset{\displaystyle R^1}{|}}{CH}COR^2$$

An example of this reaction is the aldol condensation in which two molecules of an aliphatic aldehyde react in presence of base to give β-hydroxy aldehyde.

$$\text{e.g.,} \quad 2\ RCH_2CHO \longrightarrow RCH_2-\overset{\overset{\displaystyle OH}{|}}{CH}-\overset{\overset{\displaystyle R}{|}}{CH}CHO$$

These β-hydroxy carbonyl compounds undergo ready dehydration to give, α, β-unsaturated aldehydes and ketones which constitutes the best method of preparation of these derivatives.

For example, aromatic aldehydes and methyl ketones react in presence of base to give 1,3-diphenylprop-2-en-1-ones ('Chalcones').

$$Ar-CHO \ + \ Ar^1COCH_3 \longrightarrow ArCH=CH-COAr^1$$

Experimental: A solution of potassium hydroxide (0.14 g, 0.0025 mole) in water (2 ml) is added with stirring to a solution of an acetophenone (3 g, 0.025 mole) in ethanol (25 ml). To this mixture is added slowly with stirring a solution of a benzaldehyde (2.65 g, 0.025 mole) in ethanol 50 ml and the mixture is allowed to stand until precipitation of the product is complete (during overnight). The crude product is filtered off, dried. Recrystallise the chalcone with ethanol. Record its melting point and calculate the percentage yield of the product.

Questions

1. What is the function of the base in this reaction?

2. Reaction of propionaldehyde with aqueous sodium hydroxide solution affords the aldol (I). Why is the alternative aldol (II) is not formed?

$$CH_3\,CH_2\,CH\,(OH)\,CH\,(CH_3)\,CHO \rightarrow CH_3\,CH_2CH\,(OH)\,CH_2CH_2CHO$$
$$\qquad\qquad\quad \text{I} \qquad\qquad\qquad\qquad\qquad\qquad\qquad \text{II}$$

3. How the following transformations are carried out?

   (a) $R-Br \rightarrow R\,CH_2C(=O)\,CH_3$

   (b) $R-Br \rightarrow R\,CH_2COOH$

6.23 Sandmeyer Reaction

6.23.1 Preparation of 3-Chlorotoluene

Aromatic diazonium salts, prepared by diazotization of aromatic amines, are reactive compounds, serving as useful intermediates for the introduction of a variety of substituents into the aromatic ring. One of the most important applications of Sand- Meyer reaction, is the treatment of a diazonium salt with CuCl, CuBr, or CuCN resulting in replacement of the diazonium group by Cl, Br, or CN respectively.

The preparation of m-chlorotoluene from m-toluidine (3-methylaniline) is a good example, since direct chlorination of toluene would give o- and p-chlorination.

At low temperatures, treatment of a diazonium salt with copper(I) chloride yields a complex, which at higher temperatures, decomposes by means of an electron-transfer reduction, with resultant generation of an aryl radical. The radical then undergoes a ligand-transfer reaction with copper(II) chloride, generated in the first stage of the reaction; this yields the aryl chloride, while copper(I) chlorde is regenerated.

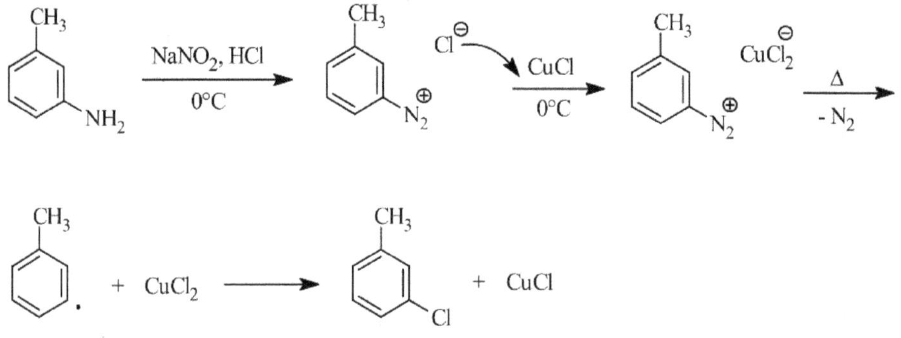

Chemicals: m-Toluidine, cuprous chloride, hydrochloric acid, sodium nitrite.

Experimental: Dissolve 2.3 g cuprous chloride in 10 ml concentrated hydrochloric acid contained in a conical flask, and cool the solution in an ice-salt bath while the diazotization is being carried out.

Dissolve 2.0 g m-toluidine in 5 ml concentrated hydrochloric acid and 4 ml of water, contained in a 100 ml conical flask or beaker cool the solution to 0° in an ice-salt bath with vigorous stirring or shaking, and the addition of a little crushed ice. The salt, m-toludine hydrochloride, will separate as a finely crystalline precipitate. Add dropwise during 10-15 min a solution of 1.3 g sodium nitrite in 4 ml of water; shake or stir the mixture well, keeping it at a temperature of 0-5°C by addition of a little crushed ice from time to time.

Maintaining the correct temperature at this stage is crucial. The hydrochloride will dissolve as the soluble diazonium salt is formed.

Pour the cold diazonium chloride solution slowly and with shaking into the cold cuprous chloride solution. The mixture should thicken. Allow the mixture to warm to room temperature with occasional shaking, but without external heating. When the temperature reaches around 15°C, the solid complex starts to breakdown, with liberation of nitrogen and formation of an oily layer of m-chlorotoluene. Warm the mixture on a water bath to about 60°C to complete decomposition, shake occasionally.

When the evolution of nitrogen ceases, steam distil the mixture until no more oily drops are present in the distillate. Transfer the distillate to a separatory funnel and remove the layer of m-chlorotoluene. Extract the aqueous layer once with 10 ml methylene chloride and combine the two organic layers. Wash the organic layers with 10 ml of 10% sodium hydroxide solution and 10 ml water, and dry over about 1.5 g anhydrous calcium chloride. After filtration, distil off the methylene chloride on a steam bath. Transfer the residual m-chlorotoluene to a small distillation flask, and distil over a small Bunsen flame in a fuming cupboard. Collect the m-chlorotoluene, b.p. 160-165° C, Record the yield.

6.24 7-Hydroxy-4-Methyl Coumarin (Pechmann Synthesis)

It is a simple synthesis of coumarins involving the condensation of phenols with malic acid or b-ketoesters in presence of conc. sulphuric acid or other condensing agents.

$$\text{Resorcinol} + \text{ethyl acetoacetate} \xrightarrow{H_2SO_4} \text{7-Hydroxy-4-methyl coumarin} + C_2H_5OH + H_2O$$

Chemicals: Resorcinol 1 g; ethyl acetoacetate 1.3 ml conc. sulphuric acid 10 ml.

Place 10 ml of concentrated sulphuric acid in a 150 ml conical flask with external ice-water cooling until the temperature of the acid is below 10°C. When the temperature falls below 10°C, add a solution of 1 g of resorcinol in 1.3 ml of ethyl acetoacetate, prepared in another test tube, dropwise, so that the temperature of the mixture does not rise above 10°C. Keep the reaction mixture at room temperature with occasional stirring. Pour the contents into 100 ml ice-water contained in a beaker. Collect the precipitate by suction filtration and wash it with cold water. For purification, dissolve the solid in 20 ml of 5% aqueous sodium hydroxide solution, filter and reprecipitate it by addition of dilute hydrochloric acid with stirring until the solution is acid to litmus. Collect the crude product by filtration at the pump, wash with cold water and dry. Recrystallise from ethanol or methylated spirit, using charcoal if necessary. The yield of 7-hydroxy-4-methyl coumarin is 1.5 g, m.p. 185°.

Mechanism: The reaction proceeds by the formation of b-hydroxy ester intermediate, which then cyclises and dehydrates to yield the coumarins.

7-Hydroxy-4-methylcoumarin

6.25 7- Methoxy-4-Methyl Coumarin

Caution: Like all alkylating agents, dimethyl sulphate is a toxic reagent. It should be used in fume cupboard and never allowed to come into contact with the skin.

This preparation is carried out on a small scale. The use of dropping pipettes for the transfer of all solutions is strongly recommended.

Experimental: Dissolve 7-hydroxy-4-methyl coumarin (0.250 g) in acetone (8 ml) in a 25 ml round bottomed flask and add anhydrous potassium carbonate (0.5 g) and dimethyl sulphate (0.4 ml) using a graduated α pipette. Heat the mixture under reflux for 2 hours. Filter the cooled solution through a small funnel and wash the residue with acetone (3 × 5 ml). Distil the filtrate and washings to dryness. If the mixture is not solid, triturate to induce crystallization. Recrystallise the product from dichloromethane or 60-80 petroleum ether. Record the yield. m.p. 158-160 °C.

6.26 7-Hydroxy Coumarin (Pechmann Synthesis)

Chemicals: Resorcinol 2 g, malic acid 2.5 g, conc. sulphuric acid 5 ml.

Experimental: Place a mixture of resorcinol (2 g), malic acid (2.5 g) and concentrated sulphuric acid in a 150 ml conical flask. The flask is held with a clamp, and is gently swirled by hand over the yellow flame of a Bunsen burner in fume cupboard. The heating is continued until the mixture begins to foam. By cautious heating, maintain the foaming for several minutes. Let it cool at room temperature for about 5 minutes and with good swirling pour the contents into 50 ml of ice-water contained in a beaker. Use a little more water as a rinse. The cold suspension is suction filtered. Dissolve the crude material in ethanol, add a pinch of decolourising carbon and boil for 2 minutes. Filter the hot solution through a Buchner funnel and cool the filtrate in ice-water collect the crystals by filtration. The 7-hydroxy coumarin is obtained as pale pink prisms. m.p. 227-228°C, yield 2 g.

Mechanism

6.27 Preparation of N-Phenylsuccinimide

Discussion: Amides are formed in the presence of ammonia or primary or secondary amines. In this experiment aniline reacts with succinic anhydride to form N-Phenylsuccinamide acid. The latter then dehydrated with acetyl chloride to form the cyclic compound, N-phenylsuccinimide.

Experimental: The succinic anhydride (2.0g) prepared in earlier experiment is placed in a 50 ml round-bottom flask provided with a reflux condenser and is dissolved in a minimum volume of hot benzene. The theoretical amount of aniline, dissolved in 10 ml of benzene, is added in one portion. The flask is heated in a boiling water bath for 30 minutes. The mixture, is cooled and filtered, and the crystals are washed with a small volume of benzene. A small portion is reserved for a melting point determination. The melting point of N-Phenylsuccinamic acid is 147^0 C.

The remainder of this product is placed in a 50 ml round-bottomed flask provided with a reflux condenser protected by a calcium chloride drying tube, and 15 ml of acetyl chloride added. The reaction mixture is heated on a water bath $(60 - 65^0)$ for 15 minutes. Remove the complete assembly from the water bath, allow it to cool in ice water. The crystals obtained are filtered and wash with small portions of cold water. The yield is 4 g. m.p. 156^0 C.

Ref: Laboratory Exercises Organic Chemistry by James M Sugihara 1969, Burgess publishing company.

6.28 Preparation of 4-Chloro-2-Nitroanisole

In this experiment, 2-nitro 1-chloronitrobenzene is converted into 4-chloro-2-nitroanisole by reaction with methoxide ion. The Latter is generated in the reaction of sodium hydroxide and methanol.

Experimental procedure: Sodium hydroxide (2.0 g) and methanol (25 ml) are placed in a 50 ml round bottomed flask. A condenser is attached and the mixture heated over a hot water bath until solution of the base has been effected. A few white particles of undissolved sodium hydroxide will remain. The flask is disconnected from the condenser and add 1,4-dichloro-2-nitrobenzene (8.0 g). The condenser is reattached and the reaction mixture is heated over a hot water bath, maintained at a temperature such that gentle refluxing occurs, for a period of one and one- half hours. The reaction mixture is cooled slightly and poured into about 100 ml of cold water. The solids obtained are filtered and thoroughly washed with cold water on the filter. The product is recrystallized from a solvent mixture of ligroin and benzene (3:1). About 5 to 5.5 g of a cream coloured solid is obtained melting at 94-96°. (lit., m.p. 98°)

6.29 Preparation of 4-Chloro-2-Aminoanisole

In this preparation 4-chloro-2-nitroanisole is reduced using stannous chloride and hydrochloric acid.

$$\text{(4-chloro-2-nitroanisole)} + 3SnCl_2 + 6HCl \longrightarrow \text{(4-chloro-2-aminoanisole)}$$

Experimental procedure: A solution of 25 g of stannous chloride dehydrate in 25 ml concentrated hydrochloric acid is prepared in a 250 ml conical flask, and 5 g of 4-chloro-2-nitroanisole is added. Cold water is place in a pan or a 600 ml beaker in order that the subsequent exothermic reaction may be moderated when necessary. The reaction mixture is heated cautiously while swirling the contents. The flask is removed from the burner when reaction is initiated as evidenced by the disappearance of some of the solid. If the reaction becomes vigorous, the reaction flask is immersed momentarily in the coldwater. When solution of the solid nitro compound has been effected, the reduction is complete. The solution is cooled thoroughly in an ice bath. The salt of the amine, which precipitates, is filtered using a Buchner funnel. The salt cake is returned to the conical flask and 10 ml water is added followed by 20 ml of 40% sodium hydroxide solution. All larger particles are broken with a stirring rod. The suspension is filtered, and the solid amine on the filter is washed thoroughly with water. The product is dissolved in ethanol, warm it, and resulting mixture is filtered hot to remove inorganic impurities (a fluted filter paper placed in a short stem funnel should be used in this operation). The alcohol solution is concentrated to about 20 ml (or to a proportionately smaller volume) and allowed to cool in an ice water bath and filtered. About 2 to 2.5 g of nearly colorless needles of 5-chloro-2-methoxyaniline is obtained melting at 81-82° C (lit., 84°C).

Questions

1. 5-Chloro-2-methoxyaniline and 4-chloro-2-nitroanisole differ markedly in their comparative solubility in hydrochloric acid. Explain why?

2. Why is an excess of concentrated sodium hydroxide added to the salt obtained in the reduction reaction?

Compare the relative stability toward oxidative degradation of o-aminophenol and m-aminophenol.

6.30 Methyl Carbamate Formation via Modified Hofmann Rearrangement Reaction: Methyl *N*-(*p*-Methoxyphenyl) Carbamate

Experimental Procedure: To a 1 litre round bottomed flask equipped with a stirring bar are added p-methoxybenzamide (10g. 66mmol), N-bromosuccini mide (NBS) (11.9 g. 66 mmol), 1,8-diazabicyclo[5,4,0]undec-7-one (DBU) (22 ml., 150 mmol) and methanol (300 ml) The solution is heated at reflux on an oil bath for 15 min., at which point an additional aliquot of NBS (11.9 g. 66 mmol) is added slowly. The reaction is allowed to continue for another 30 min. (progress can be followed by TLC, eluted with EtOAc/hexane (1:1). Methanol is removed by rotary evaporation and the residue is dissolved in 500 ml of ethyl acetate. The ethyl acetate solution is washed with 6N hydrochloric acid (2 x 100 ml), 1 N sodium hydroxide (2 × 100 ml) and saturated sodium chloride, and then dried over magnesium sulfate. The solvent is removed by rotary evaporator and the product, methyl N-(p-methoxyphenyl)carbamate, is purified by flash column chromatography [50 g of silica gel, EtOAc/hexane (1:1) togive a paleyellow solid (11.1g. 93%), which is further purified by recrystallization from 500ml of hexane. m.p. 88-89°C. FTIR Spectrum (CHCl$_3$)cm^{-1}: 3437, 3080, 2963, 1734, 1600, 1511, 1464, 1298, 1226, 1181, 1076.

Ref: Organic Synthesis by William R. Roush, Volume 78.

6.31 1,3-Cyclopentadiene: A Retro-Diels-Alder Reaction

Cyclopentadiene is a low-boiling (b.p. 41°C) unsaturated hydrocarbon which finds use as a precursor in a large number of organic synthesis. Since the molecule contains a conjugated diene group, it can be used in Diels-Alder reactions.

e.g.,

A further example of this behaviour is the self-condensation of cyclopentadiene by a Diels-Alder reaction in which the compound acts both as diene and dienophile. The product of this reaction is dicyclopentadiene which is slowly formed from cyclopentadiene on standing.

Dicyclopentadiene is a colourless liquid which breaks down at its boiling point (170 °C) to give cyclopentadiene. This is an example of a retro-Diels-Alder reaction.

Experimental Procedure: The apparatus for this experiment has been assembled in a fume cupboard, and is for geenral use.

The distilling flask is one-third filled with dicyclopentadiene and placed in the heating mantle. The heating rate is adjusted in such a way that the liquid refluxes gently and foaming, which is unavoidable, is limited. Liquid should not be allowed to choke the fractionating column (12" packed with glass beads). Volatile, monomeric cyclopentadiene which is produced by thermal 'cracking' of dicyclopentadiene passes through the column. The hydrocarbon is collected in a flask fitted with a $CaCl_2$ drying-tube, and immersed in solid CO_2 to prevent re-polymerisation. The temperature at the top of the fractionating column should not be allowed to exceed 41°C, other-wise entrained dicyclopentadiene will contaminate the product. This material should be used immediately for the preparation of thallium cyclopentadienide.

Questions

1. Cyclopentadiene losses a proton very readily where as   cyclopropene and cycloheptatriene are non-acidic. Suggest an explanation for this behavior

2. How could cyclopentadiene be converted into t-butylcyclopentadiene?

3. On standing for prolonged periods, cyclopentadiene is converted into a dimer (dicyclopentadiene). In addition, small amounts of trimeric and higher polymeric compounds are formed. Suggest possible structures for the trimer of cyclopentadiene.

6.32 A Stable Carbanion:
Thallium Cyclopentadienide

References Meister, *Angew. Chem.*, **69**, 533 (1957); Hunt and Doyle, *Inorg. Nucl. Chem. Lett.*, **2**, 283 (1966).

Carbanion species are important intermediates in a wide range of synthetically important organic reactions. They are normally generated *in situ* and used without isolation. Removal of a proton from the methylene group in cyclopentadiene occurs in presence of bases or alkali metals. The cyclopentadienyl carbanion produced is isoelectronic with benzene and possesses aromatic stabilisation.

Because of its high lattice energy, thallium cyclopentadienide is a particularly stable salt of the cyclopentadienyl carbanion which may be conveniently isolated from the reaction between cyclopentadiene and thallous salts in aqueous solution. The compound is obtained as a white solid which is stable in air but slightly photosensitive.

Experimental Procedure: Dissolve thallous nitrate (4.0 g) in a solution of potassium hydroxide (10 g) in water (100 ml). To this solution, add *freshly distilled* cyclopentadiene (2 ml) and shake the mixture in a stoppered R/B flask (250 ml capacity) for 1minute (prolonged agitation gives a fine suspension of product difficult to filter). Filter off the white product (thallium cylopentadienide) and wash with absolute ethanol (2 x 10 ml). Dry the crude product in a desiccator over anhydrous calcium sulphate, or chloride, weigh, and determine the yield.

Thallium cyclopentadienide can be purified by high vacuum sublimation but this is usually unnecessary. It should be stored in the absence of light. Convert the product prepared into ferrocene *after* it has been inspected by a demonstrator.

6.33 Preparation of Ferrocene

References: The Chemistry of the Iron Group Metallocenes'. M. Rosenblum, Angewandte Chemie Vol 5 (1966)

Ferrocene and related metallocenes (i.e., dicyclopentadienyl-metal complexes) are most usually prepared by the reaction between cyclopentadienide salts or cyclopentadienyl-magnesium halides and the appropriate metal halide.

$$2 \; \text{Cp}^- \;+\; MCl_2 \;\longrightarrow\; M(\text{Cp})_2 \;+\; 2Cl^-$$

M = transition metal

Ferrocene (dicyclopentadienyliron) is an extremely stable orange-yellow compound which possesses high aromatic reactivity. The structure of the compound has been described as a 'molecular sandwich' since the iron atom lies symmetrically between the cyclopentadienyl rings which occupy parallel planes. In solution or in the vapour phase, the rings may rotate freely about the central axis of the molecule passing through the iron atom.

Ferrocene is formed in almost quantititative yield from ferric chloride on reaction with an excess of thallium cyclopentatadienide in tetrahydrofuran solution.

$$Tl^{\oplus} \; \text{Cp}^- \;\xrightarrow{\;FeCl_3\;}\; Fe(\text{Cp})_2$$

Experimental Procedure: Use the thallium cyclopentadienide prepared in the previous experiment and adjust accordingly the weights of the reactants given below. Add freshly prepared thallium cyclopentadienide (6 g) and *anhydrous**ferric chloride (1.3 g) to ice-cold tetrahydrofuran (40 ml) contained in an R/B flask (100 ml capacity) fitted with a reflux condenser and $CaCl_2$ tube. When the colour of the ferric chloride has been discharged (<u>Ca</u>. 2 minutes), heat the mixture on a steam bath under reflux for 2 hr. and allow to stand overnight.[#] Filter the solution and wash the residue with tetrahydrofuran (2 x 10 ml). Evaporate the filtrate and washings to dryness and extract the residue with benzene. Evaporate the benzene in a 250 ml. B19 R/B flask obtaining crude ferrocene. Purify the product by vacuum sublimation. Determine the yield of the product based on ferric chloride and record its melting point in a <u>sealed</u> capillary tube.

Note:*: Anhydrous ferric chloride which appears to be a black crystalline solid is *very* hygroscopic. It should be weigh in a small beaker and exposed to the atmosphere for minimum time. : If excess ferric chloride has been used, then the solution will be blue-green or green-brown at this stage, due to the presence of ferricinium ion, the one-electron oxidation product of ferrocene. Proceed thus:

Evaporate the THF and treat the residue with titanous chloride solution (25 ml) to effect the reduction of ferricinium ion. Transfer the mixture to a separating funnel with methylene chloride (2 × 25 ml) and shake thoroughly. Run off the lower orange phase and re-extracts are colourless. Combine all the extracts, wash with sodium bicarbonate solution (25 ml) and then water (25 ml) and dry over anhydrous $CaCl_2$. Filter the solution and evaporate to dryness in a 250 ml. B19 R/B flask. Purify the residue by vacuum sublimation.

Questions

Explain the following:

(a) Ferrocene has a zero dipole moment

(b) The proton magnetic resonance spectrum of ferrocene shows only a sharp singlet resonance at *delta* 4

(c) The infrared spectrum of ferrocene contains only one stretching frequency in the C-H region.

(d) Why is an excess of thallium cyclopentadienide used in the preparation of ferrocene from ferric chloride

6.34 Friedel-Crafts Acylation of Ferrocene

This experiment provides an additional example of electrophilic substitution of an aromatic compound. It is a convenient method of adding a carbon chain to an aromatic compound.

The cyclopentadienyl rings in ferrocene may be acylated under Friedel – Crafts conditions to give both mono- and disubstituted derivatives. In practice it is found that the principal disubstitution product is the 1,1'- isomer together with a trace of the 1,2 – isomer. The 1,3 – disubstituted isomer has not been detected.

The products of this acylation are readily separated from each other, and from unchanged ferrocene, by column (adsorption) chromatography on alumina. Since each of the components of the mixture is coloured the progress of the separation is easily followed.

Experimental procedure: This preparation should be carried out under anhydrous conditions and all reagents and apparatus must be dry.

Fit a 100 ml two-neck or three-neck flask with a mechanical stirrer and a dropping funnel fitted with a fresh calcium chloride drying tube, and stand the flask in a bowl or beaker which can be used as a cooling bath. Place in the flask ferrocene (1g., 0.0054 mole) and freshly crushed anhydrous aluminium chloride (1.44 g., 0.011 mole) (note 1) and add dry methylene chloride (30 ml). Cool the flask externally with ice for approx. 15 minutes and stir the mixture gently. Add the acyl chloride (0.011 mole) in methylene chloride (note 2) drop wise over 5 minutes and leave the mixture stirring for one and half hours at 0^0 C. Cautiously add the contents of the flask to crushed ice (approx. 100 g) in a beaker, using a little water and chloroform to rinse the apparatus. Separate the lower organic layer, wash this successively with water (50 ml) , sodium bicarbonate solution (2 × 50 ml),and water (50 ml), and dry the methylene chloride solution with anhydrous calcium chloride. Filter or carefully decant this solution into a small flask and distil off the solvent until remaining volume is about 10 ml. Using a dropper, carefully transfer this solution into prepacked alumina chromatography column (35 x 3 cm). Elute unreacted ferrocene with petroleum ether, the mono-acyl ferrocene with 30% ether-petroleum ether and the di-acyl ferrocene with pure ether. # Evaporate the total eluate containing each band separately, using conventional distillation equipment.

Calculate the recovery of ferrocene and the yields of the products before and after recrystallisation. Recrystallise each product from a mixture of benzene and petroleum ether. It is not necessary to recrystallise the ferrocene.

Procedure for packing the column: Place a plug of cotton wool at the bottom of the chromatography column which should be held vertically by two clamps on a retort stand. Fill the column to within 5 cm. of the joint with petroleum ether (b.p. 40-60^0) and allow this to drip out slowly. At the same time tip alumina into the column through a funnel, a little at a time, with sufficient agitation to enable the solid to settle without trapping air-bubbles. Fill the column with alumina, within 5 cm. of the joint, replacing the solvent as this becomes necessary. Run approx. 100 ml solvent through the column before adding the solution to be chromatographed, and ensure that the flat surface of the alumina is not disturbed by this latter procedure.

Note 1. Trust the aluminium chloride rapidly in a dry motar and pestle and weigh out required quantity using a stoppered weighing bottle.

Note 2. Measure out 10 ml of the solution which will contain 0.011 mole acyl chloride from the respective burette

If it is necessary to leave the column overnight, then it should be wrapped in black paper to prevent photochemical decomposition of ferrocenes.

6.35 Preparation of Triphenyl Carbinol (Triphenylmethanol)

The Grignard reaction is applicable with equal facility in the aromatic series. In this experiment phenyl magnesium bromide will be prepared in an ether solution and then reacted with benzophenone to form triphenylcarbinol.

In a 250 ml three necked flask equipped with a mechanical stirrer, a water condenser, a drying calcium chloride tube and a dropping funnel, place 0.525 g of magnesium turnings in 10 ml sodium dry ether and add a small crystal of iodine. From the dropping funnel add slowly a solution of 3.3 ml of bromobenzene in 10 ml of dry ether with constant stirring. If the reaction does not start spontaneously, a warm water bath is helpful to initiate the reaction. When the addition is of bromobenzene is complete reflux the mixture for 5 minutes to complete the formation of Grignard reagent.

Cool the flask in cold water, and add to it a solution of 3.8 g of benzophenone in 15 ml of dry ether taken in the dropping funnel. The reaction mixture is refluxed for 30 minutes on a water-bath. Cool the flask and pour the contents in a beaker containing 20 g of ice and 3 ml of concentrated hydrochloric acid. Stir the mixture with a glass rod to decompose the magnesium compound. Transfer the contents into a separating funnel, separate the aqueous layer and wash the organic layer with 5 ml of 10% H_2SO_4 followed by water. Steam distil the mixture till no more oil passes over. The residue is cooled, filtered and recrystallise the product from carbon tetrachloride (4 ml per gram of solid) gave 3.6 g of white solid. M.p. 160-162 ^{o}C.

6.36 Preparation of Carbonium Ions

Reference: Gould, Mechanism and structure in Organic Chemistry, p.303et seq.

To account for a wide variety of observations that has been made concerning the chemistry of alkenes, alcohols, alkyl halides, and many other classes of organic compounds, the existence of reactive intermediates called carbonium ions have been postulated. These carbonium ions usually have a short life, and only in exceptional circumstances can they be observed.

Alkyl carbonium ions can be generated and observed spectroscopically in fluorosulphuric acid solution. Some carbonium ions may be generated and isolated as salts from aprotic media. Typical of the latter type are the tripheny –carbonium ion (I) [trityl], cycloheptatrienyl cation (II) [tropylium], and triphenycyclopropenium cation.

Triphenylcarbonium ion (I) can be prepared in acetic anhydride solution by the reaction of fluoroboric acid, or other strong acids, on triphenylcarbinol. This saltmay be isolated, or generated and used in situ for the removal of hydride ion from cycloheptatriene. Trityl salts are excellent reagents for removing hydride ion from a variety of substances.

Prepare either triphenylmethyl tetrafluoroborate, or tropylium tetrafluoroborate by the following procedures.

Experimental Procedure

6.36.1 Triphenylmethyl Tetrafluoroborate

Dissolve triphenyl carbinol (1.0 g., 0.0038 mole) in acetic anhydride (10 ml.) by warming gently in a 50 ml round-bottomed flask fitted with a calcium chloride tube. Cool the solution for 15 minutes in an ice – bath and add fluoroboric acid (1. 2 ml., 48% solution) in small portions (caution*). Agitate the solution gently from time to time, and leave in an ice-bath for approximately 30 minutes, after the addition of the fluoroboric acid. cautiously decant off the acetic anhydride into a beaker of ice, and repeatedly wash the residual orange-yellow salt with anhydrous ether [5 x 20 ml) until the odour of acetic anhydride is no longer apparent. Place the flask and contents in a vacuum desiccator and evacuate on a water bath pump to dry the salt. This is very hygroscopic and should be handled in the air for the minimum time. Yield about 1 g.

6.36.2 Tropylium Tetrafluoroborate

Dissolve triphenyl carbinol (2 g., 0.0076 mole) in acetic anhydride (15 ml) by warming gently in a 50 ml round-bottomed flask fitted with a calcium chloride tube. Cool the solution for 15 minutes in an ice-bath and add fluoroboric acid ((1.2 ml., 48% solution) in small portions (caution*). Agitate the solution gently and add cycloheptatriene (1.5 ml). Leave the mixture at room temperature until the characteristic yellow colour of the trityl carbonium ion has disappeared and precipitation of tropylium tetrafluoroborate has commenced. Add dry ether (30 ml) to precipitate the remainder of the white slat which should be removed by filtration, washed with ether and dried. Yield about 1 g.

Caution: Wear safety goggles: Fluoroboric acid is an exceptionally corrosive acid. Acetic anhydride containing filtrates should be destroyed by pouring into ice in a beaker and disposed of in the fume cupboard only after hydrolysis is complete.

Questions

1. Which is the stronger acid of triphenylmethane and cycloheptatriene?

2. Why are tropylium and trityl cations stable ?

3. How would the stability of the trityl cation be affected by (a) p-nitro substituents (b) p-methoxy substituents?

6.37 Arynes

Reference: 'Carbenes, Nitrenes and Arynes', T. L. Gilchrist and C. W. Rees (Nelson, 1969)

The commonest aryne is benzyne, or 1,2 – dehydrobenzene. Arynes are highly reactive species of very short lifetime which intervene in a number of organic reactions (cf. carbenes). Evidence for their existence is therefore usually indirect, e.g., the reaction of o-chlorotoluene with potassium amide in liquid ammonia gives a mixture of *ortho*- and *meta*–toluidine through formation of a 2-methylbenzyne intermediate.

Arynes are very strong dienophiles and enter into a wide variety of Diels-Alder reactions e.g. benzyne reacts with anthracene to give triptycene and with furan to give 2,3 – benzo-7-oxa-bicyclo[2 : 2 : 1] hept-5-ene.

Benzyne may be generated in a variety of ways, the simplest involving the thermal decomposition of the diazonium salt from anthranilic acid.

6.37.1 Preparation of Triptycene

Experimental: Anthracene (2 g), isoamyl nitrite (2 ml) (note 1) and 1,2-dimethoxyethane (20 ml) are placed in a 250 ml three necked RB flask fitted with a dropping funnel in the centre neck, and a reflux condenser. The mixture is heated to gentle reflux, using a heating mantle, until most of the anthracene has dissolved. A solution of anthranilic acid (5.2 g) in 1,2 –dimethoxyethane (20 ml) is prepared and one half of this solution is slowly added drop-wise from the funnel, during 20 minutes. A further volume of isoamyl nitrite (2 ml) is then added to the flask and the remainder of the anthranilic acid solution added during a further 20 minutes. The mixture is refluxed for 10 minutes after the addition of the anthranilic acid has been completed, ethyl alcohol (10 ml) and dilute sodium hydroxide solution (40 ml) are added, and the mixture is allowed to cool in ice. After 30 minutes the solid is filtered off, washed with ice-cold methanol:water (4:1) and dried for at least for 24 hours in a vacuum-desiccator which is re-evacuated at least three times during that period. Examination of the dried product by t.l.c. will reveal a mixture of triptycene and anthracene.

Removal of the latter is accomplished by heating the mixture for 5 minutes with maleic anhydride (1 g) and triglyme (20 ml) [note 2]. Thr mixture is cooled to approx. 100^0 and ethanol (10 ml) then dilute sodium hydroxide solution (40 ml) is added. The mixture is cooled in ice and crude triptycene crystallizes out. This is washed with ice-cold methanol:water (4:1), dried in a vacuum desiccator, and recrystallised from 60-80^0 petroleum ether. The product melts at 253-254 ^{0}C

Note 1: Isoamyl nitrite is a dangerous reagent. It is extremely inflammable and can decompose spontaneously and explosively. It should be stored in the refrigerator. It is also a heart stimulant and inhalation of the vapour should be avoided.

Note 2: Triglyme is the trivial name for triethylene glycol dimethyl ether. It boils at approx. 222^0 and a reflux air-condenser should be used.

Questions

1. What product(s) are formed if benzyne is generated in the absence of a dienophile in an inert solvent?

2. Compare the electronic structures of benzyne with an alkyne and account forthe much greater reactivity of the former.

3. Indicate the stereochemistry expected of the product(s) formed in the reactionof benzyne with *cis*-2-butene.

6.38 Wittig Reaction

Preparation of 4-methyl–trans-stilbene and its photochemical oxidation to 3-methylphenanthrene

This preparation illustrates (i) the Wittig reaction, an important method for the synthesis of olefins from aldehydes and ketones; and (ii) a chemical reaction induced by irradiation with ultraviolet light.

In one version of the Wittig reaction a phosphorylide (II), prepared by the action of a base on a phosphonium salt (I), is reacted with an aldehyde or ketone. The products are a phosphine oxide and an olefin. The latter is usually obtained as a mixture of *cis*- and *trans*-forms, although in the experiment below only the trans-form is isolated.

A photochemical reaction is brought about by the absorption of ultraviolet light (or less often visible) light. In the example below, irradiation induces

(a) isomerisation of 4-methyl-*trans*-stilbene into the less stable (i.e. higher energy) *cis*–isomer, and (b) coupling of two benzenoid rings forming a dihydrophe nanthrene system. Both reactions are reversible, and the equilibrium is displaced by oxidation (with iodine and air) of the dihydrophenanthrene to the phenanthrene.

Benzyltriphenyphosphonium chloride. This quaternary salt is prepared by nucleophilic substitution of benzyl chloride with triphenylphosphine.

$$(C_6H_5)_3P + C_6H_5 - CH_2Cl \rightarrow (C_6H_5)_3P^+ - CH_2 - C_6H_5Cl^-$$

Experimental: A mixture of benzyl chloride (3.5 g., 3.2 ml), triphenylphosphine (6.5 g), and nitromethane (5 ml) is heated on the steam-bath for 1.5 hours. Cool the almost solid reaction mixture and add ether (50 ml). Stir well for 2 minutes, filter and wash the residue thoroughly with more ether. Record the yield of benzyltriphenylphosphonium chloride which is sufficiently pure for use in the next stage without recrystallisation.

6.38.1 4-Methyl *trans*-Stilbene

The phosphorylide, prepared from benzyltriphenylphosphonium chloride with sodium methoxide, is reacted with p-tolualdehyde to give 4-Methyl-trans-Stilbene.

$$(C_6H_5)_3P^+ - CH_2 - C_6H_5Cl^- \xrightarrow{\text{NaOCH}_3} (C_6H_5)_3P^+ - CH^- - C_6H_5$$

$$(C_6H_5)_3P^+ - CH^- - C_6H_5 + p - CH_3 - C_6H_4 - CHO \rightarrow$$

$$C_6H_5 - CH = CH - C_6H_4 - CH_3 - p + (C_6H_5)_3P^+ - O^-$$

Experimental: To a solution of sodium methoxide (0.46 g of clean sodium dissolved in 50 ml of dry methanol) in a flask fitted with drying tube, add benzyltriphenylphosphonium chloride (8 g) and swirl the flask gently until all the phosphonium salt has dissolved (usually about 2 to 3 minutes; a fine precipitate of sodium chloride appears simultaneously). After a further 5 minutes, add freshly distilled p-tolualdehyde (2.5 g, 2.6 ml) to the solution of the phosphorylide, stopper the flask and set it aside at room temperature for 2 days. Pour the contents of the flask into a mixture of water (35 ml) and concentrated hydrochloric acid (15 ml), stir for a few minutes, cool in ice for 30 minutes, filter and wash the crude stilbene with water. Recrystallise the 4-methyl-trans-stilbene from hot ethanol, wash the crystals with a small volume cold ethanol to remove traces of colour, and record the yield and m.p. (lit. m.p. 119 -120^{0}).

Check the purity of your product by t.l.c. on silica plates using toluene as the mobile phase. Examine the plates under ultraviolet light, or develop by spraying with ceric sulphate solution and heating in the oven at 100-120^{0} for 10-15 minutes.

Characterise the 4-methyl-trans-stilbene by preparation of its dibromide. In the fume cupboard dissolve the stilbene (0.4 g) in hot glacial acetic acid (7 ml) in a flask fitted with a reflux condenser. To the solution add slowly a solution of bromine in acetic acid (molar, 2 ml). (CAUTION, on no account allow the bromine solution to touch the skin. If it does, wash it off immediately with soap and water or mild alkali). Boil the solution gently for one minute, cool and collect the dibromide by filtration. Recrystallise from a suitable solvent and record the yield and m.p.

6.38.2 Irradiation of 4-Methyl-*trans*-Stilbene

Experimental: Dissolve 4-Methyl–*trans*-Stilbene (0.05 g) in aerated light petroleum ether (b.p. 40-60^{0}, 20 ml) and add a small crystal of iodine. Place the solution in the clear quartz tube provided and insert the tube into the U.V. irradiator. (N.B. Do not look directly at the U.V. lamp as it is harmful to the eyes). Leave the tube in the irradiator for 24 hours, filter, transfer the filtrate to a flask and evaporate to dryness on the rotary evaporator. To the residue add 1,3,5–trinitrobenzene (0.02 g) and dissolve the mixture in hot ethanol (0. 5–1 ml). Transfer the solution in a small test-tube and rinse out the flask with hot ethanol (0.5 ml). Transfer the washings to the test-tube, heat on the steam-bath to obtain a clear yellow solution and allow to cool. Collect the yellow crystals of the 3-methylphenanthrene–trinitrobenzene complex and record the yield and m.p.

In this experiment it is not necessary to isolate the 3-methylphenanthrene as such, but this could be done easily since trinitrobenzene (and picrate) complexes dissociate readily, even on attempted chromatography. Show this dissociation by running a silica t.l.c of the complex using toluene as the mobile phase.

6.39 Microwave–Assisted Laboratory Experiments

Microwave Synthesis

There have been several articles published describing the possible use of microwave synthesis in the under graduate laboratory. The most important reason reactions proceed faster in the microwave than with conventional heating is that energy is transferred directly to the reactants. Microwave energy is an inherently efficient way to transfer energy to a reaction, as infra-red transfers kinetically rather than thermally. One can use water, ethanol or other environmentally benign solvents. There have been many reports of remarkable decreases in reaction time for reactions carried out under microwave irradiation in domestic microwave ovens.

The Diels-Alder reaction is selected to test the suitability of microwave acceleration in an under graduate teaching laboratory setting.

6.39.1 Preparation of *cis*-1,2,3,6-Tetrahydro-4,5-Dimethyl-Phthalic Anhydride

Procedure: Add 0.5 g of freshly distilled 2,3 dimethylbuta-1,3-diene to 0.59 g of finely powered maleic anhydride contained in a small beaker. The beaker was covered with a watch glass and place in the microwave oven. The irradiation is carried out for 4 minutes at a medium power setting (level 5). The beaker was removed from the oven and allowed to cool until the mixture attains the room temperature. White crystals of the product was collected upon a filter paper in the air, and then recrystallise from light petroleum ether (40-60 $^{\circ}$C). The yield of the tetrahydrophthalic anhydride 97%, m.p. 78-79 $^{\circ}$C.

The reaction of anthracene with maleic anhydride would be a suitable example except that it requires a reflux period of 90 minutes as carried out conventionally.

6.39.2 Reaction of Anthracene with Maleic Anhydride

A mixture of 1.8g (0.01 mol) of anthracene and 0.98g (0.01 mol) of maleic anhydride was ground thoroughly in a mortar and then transferred to an 250 ml beaker. After the addition of diglyme (solvent), the mixture was shaken gently. The beaker was covered with a watch glass and place in the microwave oven. The irradiation was carried out for 90s at a medium power level (level 5). After beaker was removed from the oven and allowed to cool to room temperature, the adduct crystallized out and was collected by suction filtration. The product, after washing with methanol (2 × 5 ml) and drying, had mp 258-260 C. Yield 80%.

Anthracene   Maleic anhydride   9,10-dihydroanthracene-9,10-succinic anhydride

6.39.3 Preparation of Phthaloylglycin

A mixture of 1.48g (0.01 mol) of phthalic anhydride and 0.75g (0.01 mol) of glycine was ground thoroughly in a mortar. The mixture was transferred to a 250 ml beaker and 5 ml of N,N-dimethylformamide (solvent), followed by 0.25 ml of N-methyl-morpholine was added. The beaker was covered with a watch glass and place in microwave oven. The irradiation was carried out for 60s at a medium power setting (level 5). The beaker was removed from the oven and, after the mixture had cooled to room temperature, 10 ml of water was added. The precipitated phthaloylglycine was filtered and recrystallized from 95% ethanol, mp. 192-195 C. Yield 75%.

Structure *contd...*

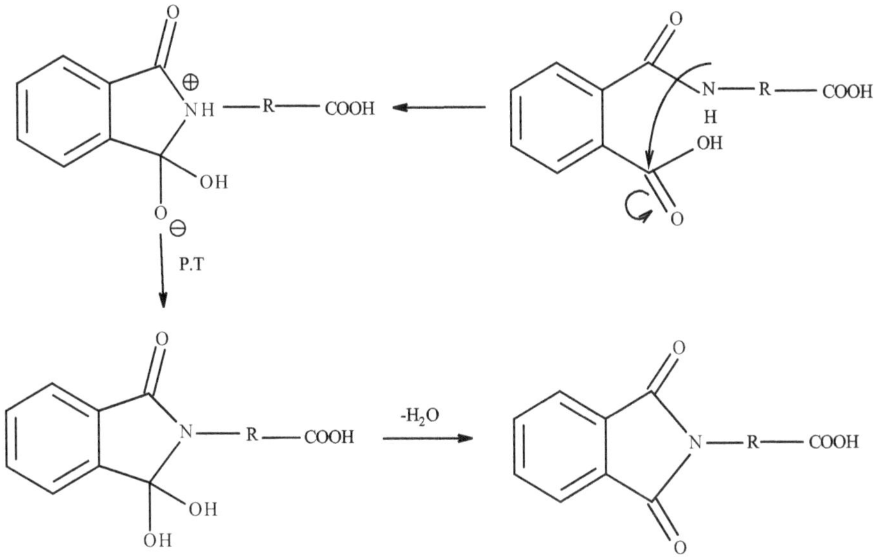

Ref: Shamsher S.Bari et al (Department of Chemistry and chemical Engineering, Stevens Institute of Technology, Hoboken, NJ 07030.

Microwave-assisted Reactions in Water

Water-based microwave reactions have been relatively well studied for hydrolysis and hydrogen peroxide oxidation reactions, the natural solvent of choice would be water.

6.39.4 Hoffmann Elimination Reaction

In this example a poorly mixed two phase water-chloroform system was used. Being polar the starting quaternary ammonium compound was water-soluble.

Microwave irradiation quickly heated the water phase to over 100 ^{O}C, causing rapid elimination (reaction time 1 minute). The less polar product rapidly partitioned into the chloroform phase, which being less polar had only reached a temperature of 48 ^{O}C. This low temperature enabled the product to be isolated in 97% yield, twice that using conventional heating (Scheme 1).

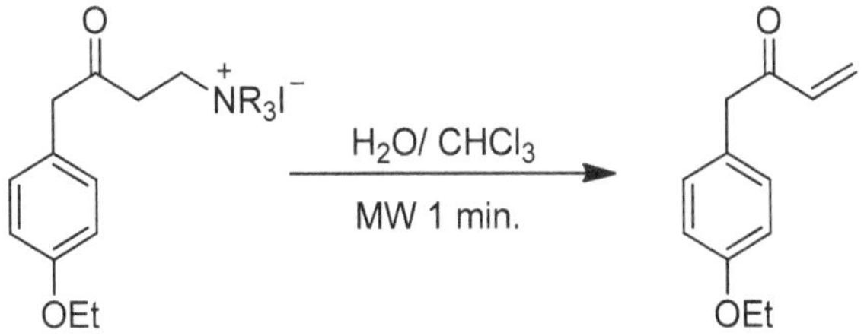

(b) In the hydrolysis of benzamide with sulfuric acid quantitative conversion was achieved in 7 minutes under microwave irradiation at 140 ^{O}C compared to a 90% conversion after 1-hour reflux using a conventional heating source.

R-CH$_2$-OH with Benzamide → Benzoic acid (COOH) + NH$_3$ (Ammonia), reagents H$_2$O$_2$, Δ over H$^+$ or OH$^-$

(c) A range of primary alcohols have been oxidized to the corresponding carboxylic acid using sodium tungstate as catalyst in 30% aqueous hydrogen peroxide. Yields, although variable, of up to 85% have been achieved in a rapid, clean, safe and atom efficient reaction.

$$R\text{-}CH_2\text{-}OH \xrightarrow[\;Na_2WO_4\;]{\;30\%\;H_2O_2\;MW\;} R\text{-}C(=O)\text{-}OH$$

6.39.5   Solvent-Free Reactions

Solvent-free reactions are especially suitable for microwave heating. Since energy conduction is not required, unlike when using conventional heat sources. In the absence of solvent, the radiation is directly absorbed by the reactants, giving enhanced energy efficiency.

There are several examples of N-alkylation and acylation being successfully carried out, sometimes using a solid catalyst.

$$R\text{-piperazine(NH)} + R'\text{-CO-Cl} \xrightarrow{\;SiO_2\;} R\text{-piperazine-CO-}R'$$

The rapidity of microwave-assisted reactions is well exemplified in the field of oxidation chemistry. By simply mixing the solid oxidizing agent (Clayfen and MnO$_2$ impregnated on silica have been widely used) with a range of secondary alcohols and irradiating for periods of less than 1 minute yields of ketones in excess of 90% are often obtained.

$$R_1R_2CH\text{-}OH \xrightarrow[\;MW,\;15\text{ - }60\text{ sec}\;]{\;clayfen\;} R_1R_2C\text{=}O$$

$$R_1 = alkyl,\ aryl\ ;\ R_2 = H,\ alkyl,\ aryl$$

An interesting selective oxidation of thiols has been achieved using sodium periodate on silica. Depending on the amount of oxidant used either the sulfoxide or sulfone can be obtained in very high selectivity in less than 3 minutes.

A final example of the simplicity of microwave reactions is the simple irradiation of aromatic aldehydes in air to give benzoic acid. Although yields are only moderate it is a useful indication of the potential simplicity and convenience of what can be achieved using this technology

Reference: Green Chemistry an Introductory Text by Mike Lancaster, (2010), published by the RSC.

6.40 Enzymatic Reduction: A Chiral Alcohol from Ketone

Reduction of an achiral ketone with the usual laboratory reducing agents such as sodium borohydride or lithium aluminum hydride will not give a chiral alcohol because the changes for attack on two sides of the planer carbonyl group are equal. However if the reducing agent is chiral, there is the possibility of obtaining a chiral alcohol. In this experiment we will use the enzymes found in baker's yeast to reduce ethyl acetoacetate to S (+) –ethyl 3- hydroxy butanoate. This compound is very useful synthetic building block.

6.40.1 Yeast Reduction of Ethyl Acetoacetate: (*S*)-(+)-Ethyl 3-Hydroxybutanoate

Aim: Optically active ethyl 3-hydroxybutanoate is a very useful chiral building block for natural product synthesis.

Procedure

A 4-L, three-necked, round-bottomed flask equipped with mechanical stirrer, bubble counter, and a stopper is charged with 1.6 L of tap water, 300 g of sucrose, and 200 g of baker's yeast, which are added with stirring in this order.

The mixture is stirred for 1 hr at about 30°C, 20.0 g (0.154 mol) of ethyl acetoacetate is added, and the fermenting suspension is stirred for another 24 hr at room temperature. A warm (ca. 40°C) solution of 200 g of sucrose in 1 L of tap water is then added, followed 1 hr later by an additional 20.0 g (0.154 mol) of ethyl acetoacetate. Stirring is continued for 50–60 hr at room temperature. When the reaction is complete by gas chromatographic analysis, the mixture is worked up by first adding 80 g of celite and filtering through a sintered-glass funnel (porosity 4, 17-cm diam). After the filtrate is washed with 200 mL of water, it is saturated with sodium chloride and extracted with five 500-mL portions of ethyl ether (Note 1). The combined ether extracts are dried over magnesium sulfate, filtered, and concentrated with a rotary evaporator at 35°C bath temperature to a volume of 50–80 mL. This residue is fractionally distilled at a pressure of 12 mm through a 10-cm Vigreux column, and the fraction boiling at 71–73°C (12 mm) is collected to give 24–31 g (59–76%) of (*S*)-(+)-ethyl 3-hydroxybutanoate and (Note 2); the specific rotation $[á]^{25}D$ + 37.2° (chloroform, *c* 1.3) corresponds to an enantiomeric excess of 85% (Note 3).

The enantiomeric excess may be enhanced by several crystallizations of the 3,5-dinitrobenzoate derivative or else by using "starved" yeast (Note 4).

1. In the case of emulsions, addition of methanol may be helpful.

2. This ester should be stored in a refrigerator as there has been some indication that it may undergo a trans esterification–oligomerization upon standing at room temperature.

3. The specific rotation $[\alpha]^{25}D$ varies from +35.5° to +38° (82–87% enantiomeric excess).

4. Activation of the enzyme(s) producing the *S*-enantiomer of ethyl-3-hydroxybutanoate. The procedure is as follows. A suspension of 125 g of baker's yeast in 1000 mL of H_2O/EtOH (95 : 5) was shaken (120 rpm) at 30°C in a 2-L Erlenmeyer flask with indentations for 4 days. After the addition of 5 g (38 mmol) of ethyl acetoacetate the reaction was followed by GLC. When the reaction had reached completion (2–3 days), the mixture was centrifuged and the supernatant was extracted continuously with ether (4 days). The organic layer was dried over magnesium sulfate, filtered, and concentrated with a rotary evaporator at 35°C bath temperature. The crude product was purified by bulb to bulb distillation to give ethyl 3-hydroxybutanoate, 3.54 g (70%), as a colorless liquid with an optical purity of 94% e.e. (enantiomeric excess).

Reference: Dieter Seebach, Marius A. Sutter, Roland H. Weber, and Max F. Züger[1]. *Org. Synth.* **1985**, *63*, 1

6.41 Methyl 2-Naphthyl Ether (Methylation)

Chemicals required: 2-Naphthol (2 g), dimethyl sulphate (1.4 ml), 10% sodium hydroxide solution (6 ml).

OH   dimethyl sulphate   OCH3

NaOH

2-napthol   Methyl, 2-Naphthyl ether

Experimental: In a 100 ml flask place 2-naphthol (2 g) 10% sodium hydroxide solution (6 ml) and shake the mixture. To this add dimethyl sulphate (1.4 ml) dropwise and warm the flask in a water bath at 70-80° for 1 hour to complete the reaction. Allow it to cool and the methyl naphthyl ether rapidly separates as a greyish-white powder. Filter the methyl ether at the pump, wash with water (20 ml) and drain.

Recrystallise from methylated spirit, from which the methyl 2-naphthyl ether separates readily as colourless crystals, m.p. 72°C. Yield is 1.8 g.

Note: Vapours of dimethyl sulphate highly poisonous. If the sulphate is splashed on the the hands, wash immediatly with plenty of ammonia solution followed by water.

6.42 Preparation of Soap

Fats and oils belongs to the family of esters particularly glycerol and long chain carboxylic acids. Preparation of soaps are carboxylate salts with very long hydrocarbon chains. Soap can be made from the base hydrolysis of a fat or an oil. This hydrolysis is called saponification.

Fats and oils belongs to the family of esters particularly glycerol and long chain carboxylic acids. Soaps are carboxylate salts with very long hydrocarbon chains. Soap can be made from the base hydrolysis of a fat or an oil. This hydrolysis is called saponification, and the reaction has been known for centuries. Traditionally, soaps were made from animal fat and base (NaOH). An example of a saponification reaction is shown below.

H₂C-O—CO—(CH₂)₁₄—CH₃
|
HC-O—CO—(CH₂)₁₄—CH₃   +   3 NaOH $\longrightarrow$ H₂C-OH / HC-OH / H₂C-OH   +   3 NaO—CO—(CH₂)₁₄—CH₃
|
H₂C-O—CO—(CH₂)₁₄—CH₃

Triglyceride   Glycerol   Sodium palmitate (a soap)

As you may remember, fats and oils are triesters of glycerol and three fatty acids. Esters can be hydrolyzed to their alcohol and carboxylic acid components in the presence of acid or base. Fats, oils, and fatty acids are insoluble in water because their hydrophobic tails are so long. If a base is used for hydrolysis, the fatty acids produced are deprotonated and are present as the corresponding carboxylate salts. Because these product carboxylate salts are charged, they are much more soluble in water than the corresponding uncharged fatty acids. Since the carboxylate salts also each have a long nonpolar tail, they are also compatible with nonpolar greases and oils.

Experimental Procedure

Saponification (preparation of soap): In a 250 ml conical flask or a beaker weigh 20 g of cotton seed oil or any other fat, and add 12 ml of ethanol and 20 ml of 20 % NaOH to the beaker. Heat the mixture of the beaker on a water-bath (with constant stirring) maintaining the bath temperature between 80-90° C. and continue heating for 1 hour or so, until the solution no longer has two separate layers. The solution should be transparent at this point. When saponification is complete, carefully remove the beaker from the water-bath. After this period add 150 ml of saturated sodium chloride solution and cool the mixture. This process is called "salting out". It increases the density of the solution and causes the soap to precipitate. Filter it through double the thickness of cheese cloth. Wash the soap on the cloth with 50 ml cold water and mould it into cake in a china dish or other suitable moulds.

To recover glycerol (a by-product) slightly acidify the filtrate obtained from above with HCl and evaporate it to dryness. Extract glycerol with 20 ml of absolute alcohol. Decant the alcohol solution from the salt and evaporate on a hot water-bath.

6.43 Synthesis of 2,4-Dihydroxy Ethylbenzene by Clemmensen Reduction

Chemical required: 2,4- Dihydroxy acetophenone, Zinc amalgam

Preparation of Zinc amalgam: A mixture of 100 g of mossy Zinc, 5 to 10 g of mercury chloride, 5 ml concentrated hydrochloric acid and 100-150 ml of water is stirred for 5 minutes. The aqueous solution is decanted and amalgamated zinc is converted with 75 ml of water and 100 ml of concentrated of hydrochloric acid. The material to be reduced, usually 40 to 50 g, is then added directly and the reaction is started

Experimental: A mixture of amalgamated zinc (prepared from 20 g of mossy zinc and 1 g of mercury chloride as described above) 15 ml of water and 15 ml of concentrated hydrochloric acid, and 5 g of 2,4,-dihydroxyacetophenone (resacetophenone) is refluxed in a 100 ml round bottom flask until a drop of the liquid in ethanol gives no color with aqueous ferric chloride. A portion of the above 1-2 ml of concentrated hydrochloric acid is added hourly. When the colour test indicated that the reaction is completed (3-4 hours), the mixture is cooled and the solution is decanted from any unchanged zinc amalgam. The solution is saturated with sodium chloride and extracted with ether to remove the reaction product. Removal of the solvent yields a light yellow solid which crystallizes from benzene or chloroform as thick white prisms, m.p. 97 °C. The yield is 4.4g.

ISOLATION OF NATURAL PRODUCTS

7.1 Isolation of Piperine from Black Pepper

Aim: To isolate piperine from black pepper.

Chemical Name: 5-(3,4-methylene dioxy) phenyl-2,4-pentadienyl piperidamide.

Structure:

CH=CH–CH=CH–C–N

Uses: Flavoring agent and insecticide.

Principle: Piperine is an alkaloid, weakly basic substance, present in black pepper. Piperine constitutes about 10% weight of black pepper. Piperine gives punguent taste to the black pepper. It can be extracted from the black pepper with 95% ethyl alcohol. In an ideal case the extraction should be carried out in a soxhlet apparatus.

Chemicals: requried: Black pepper, 95% ethanol, 10% of KOH.

Apparatus required: Beaker, water bath, RB flask (500 ml) mortar and pestle, funnel.

Weigh 30 g of black pepper grind it finely and place the powder into the soxhlet apparatus and fix it to the 500 ml R.B. flask containing 250 ml of 95% ethanol. The flask heated on a boiling water bath for 3-4 hour. Concentrate the extract to the minimum solvent and add 125 ml of 2N ethanolic KOH solution. Stir the warm soluion and filter any insoluble material. Warm the solution on a steam bath and add 15-20 ml of water. At this stage turbidity appears and yellow needles of piperine may separate. Keep this solution over night and filter the crude piperine. Recrystallize from acetone to give fine yellow needles. m.p. 129-131°C yield 1.5 g.

7.2 Caffeine from Tea Leaves

Caffeine, one of the most important naturally occurring methyl derivatives of xanthine, is present in a variety of teas to the extent of 2-4.6%. It acts as a stimulant of the central nervous system, relaxes the smooth muscle of the bronchial, it is a mild diuretic

Aim: To isolate caffeine from tea leaves.

Chemical name: 1,3,7-trimethyl xanthine

Structure

Uses: It has a stimulating effect on the CNS, heart, blood vessels and kidney and also diuretic.

Principle: Caffeine is a xanthine derivative and constitutes 4% of tea leaf. It is soluble in hot water and organic solvents like chloroform and benzene. It is bitter in taste. On addition of lead acetate solution tannins get precipitated. Tannins are naturally occurring polyphenols and largely responsible for the aroma, colour and quality of tea.

Requirements: Tea powder (4 teabags), lead acetate, beaker, muslin cloth, glass rod, filter paper, distilled water and chloroform.

Procedure: Boil 20 gram (4 tea bags) tea leaves in 500 ml beaker containing 200 ml water for 20 minutes. Filter the contents through a Buchner funnel without using the filter paper at the pump to remove tea leaves. To the clear filtrate add 60 ml of 10% lead acetate solution to precipitate tannins. Leave the mixture for 2 days. The precipitate is filtered. The filtrate is concentrated on a sand bath for about 30 minutes or to a minimum solution. On cooling the solution is extracted two to three times with 20 ml portions of chloroform. Combine the extracts and remove the solvent by distillation. The residue obtained is cooled and add 25 ml of petroleum ether and stir for 5 minutes. Filter the crude caffeine. It can be recrystallised from hot water m.p. 235-237°C. Record the yield obtained.

Obtain the infrared (KCl disc) and ultraviolet spectra of the caffeine and compare these with the data recorded below:

Infrared Spectrum of Caffeine

v_{max} at 3134 cm^{-1} (C – H stretching), 2850 (N – CH$_3$), 1705 (C = O), 1660 = C), 1604, 1548, 1440 (pyrimidine ring), 1470, 1358 (CH$_3$ bending), 1230, 1197, 1020 (C – N) stretching, 740 (C – H deformation).

Ultraviolet Spectrum

λ max = 278 nm. (E, 11, 000) in water.

7.3 Cineole from Eucalyptus Leaves

Aim: Isolation of cineole from eucalyptus oil by steam distillation.

Cineole occurs aa a mixture of with camphene and phellandrene.

Structures

Cineole   Camphene   Phellandrene

Uses: Local antiseptic, It is used as a pain balm.

Principle: Steam distillation is a means of separating organic compounds having boiling point below 100°C passing steam into a flask containing natural product to separate the volatile substances.

Apparatus required: R.B. Flask (1 lit.,) distillation apparatus, steam head.

Procedure: Cut fresh leaves into small pieces and transfer them into a round bottom flask pass steam through the steam head kept into the eucalyptus leaves present in the R.B. flask, which is connected to the condenser and receiver. Collect the first few ml that is the pure eucalyptus oil, extracted from eucalyptus leaves. Eucalyptus oil obtained as colourless liquid with pleasant smell. Collect the oil and report with a label.

7.4 Nicotine from Tobacco

Aim: Extraction of nicotine from tobacco leaves. Fresh tobacco leaves contain roughly 3 to 3.5% soluble nitrogenous compounds.

Nicotine is the major component of cigarette smoke which causes many respiratory problems. Nicotine is a colourless, volatile oily liquid and is soluble in water. Pure nicotine is highly poisonous and has unpleasant smell. Boiling point 246-247°C.

Structure

Chemical Name: 1-Methyl-2-(3-pyridyl)pyrrolidine.

Apparatus required: Conical flask, beaker, china dish.

Procedure: Cut fresh tobacco leaves (25 g) into small pieces and transfer them into a conical flash and add 30 ml of concentrated (30%) sodium hydroxide solution. Warm the contents on low flame with continuous shaking till a strong smell of tobacco starts evolving from the flask. Cool the contents and transfer to a separating funnel (100 to 200 ml). Add about 40 to 50 ml ether and shake gently. Allow the two layers separate and collect the upper layer (ether layer) in a separate beaker or china dish. Allow the ether to evaporate on water bath, preferably in the fume-cupboard (make sure keep it away from flame as ether is highly inflammable). Collect the residue left in the china dish and weight it and submit the sample with a label.

Result: Compare the nicotine content in different samples of tobacco. Amount of amino acids and nicotine content increases when the nitrogen supply is more in soil.

Note: Take a few mg of the compound and add few drops of conc. HCl, a light brown to dark brown coloured precipitate appears, shows the presence of nicotine.

7.5 Isolation of Citric Acid from Lemon Juice

Plant acids are defined as simple organic compounds containing not more than six carbon atoms and two or three carboxyl groups. Many of them function as intermediates in cellular respiration. Several, such as citric acid and succinic acid, occur widely in plant tissues in variable concentrations. Others, less well known, that occur in smaller quantities, include á-ketoglutaric acid and malic acid.

Citric acid is one of the most widely distributed plant acids, occurring in cranberries, red currants, strawberries, raspberries, beets, citrus fruit, and animal tissues. It may be isolated from citrus fruits as the comparatively soluble calcium salts. This compound displays unusual solubility properties in that it becomes less soluble at increased temperature. This information kept in mind when carrying out the following experiment. The formula for citric acid is:

$$\underset{\displaystyle HO}{O}\!\!\diagup\!\!C-\underset{H_2}{C}-\underset{\underset{\displaystyle C}{|}}{\overset{\overset{\displaystyle OH}{|}}{C}}-\underset{H_2}{C}-C\diagup\!\!\underset{OH}{O}$$

Procedure: Measure 90 ml thawed frozen lemon juice concentrate into a 250 ml beaker and carefully add a 10% NaOH solution with stirring until the mixture is slightly alkaline. A distinct colour change occurs at this point, the solution passing from a clear yellow to a brownish colour. Strain the solution through muslin cloth to remove large particles of pulp and then filter through a paper in a Buchner funnel. The pores of the filter paper may tend to become clogged by the extract in spite of the previous straining. Should this occur, change the paper in the funnel once or twice as required to complete the filtration. Measure the filtrate and record.

Result :________ml.

Now place the filtrate in a beaker and add 5 ml of calcium chloride, stirring constantly for each 10 ml of filtrate. Boil and filter off the copious precipitate of calcium citrate ($Ca_3C_{12}H_{10}O_{14}$) from the hot solution using a Buchner funnel. Wash the precipitate with a small quantity of boiling water. Now resuspend it in a minimum quantity of cold water, heat to boiling, and once more collect the insoluble calcium citrate by filtration. Allow the salt to air dry. Wash and calculate the yield.

Calculations: Citric acid may be prepared from the calcium citrate by hydrolysis using 1N H_2SO_4. The air dried salt is placed in a beaker. Add sufficient 1N sulphuric acid required to convert the salt to the acid.

The equation for the reaction:

$$Ca_3C_{12}H_{10}O_{14} + 3H_2SO_4 \rightarrow 2(C_6H_8O_7)_2 + 3CaSO_4$$

Allow the mixture to stand for a few minutes, filter off the insoluble calcium sulphate, and concentrate the filtrate to a small volume on a steam bath. Citric acid crystallizes out. Filter, dry and weigh the acid, forming anhydrous acid m.p.153 C. Calculate the percentage of citric acid in the lemon juice sample used.

7.6 Isolation of Eugenol from Cloves by Steam Distillation and it's Identification by Infra-red Spectroscopy

Introduction: Essential oils are the volatile components associated with the aromas of many plants. In this experiment, the essential oil eugenol (the main component of oil of cloves will be isolated from ground cloves using the technique of steam distillation, which is often used to isolate liquid natural products from plants.

Ground cloves and water will be charged into the distillation flask. The mixture will then be heated to boiling on a hot plate with an aluminum heating block and the distillate (eugenol/water mixture) will be collected. The eugenol will then be separated from the water by extraction with methylene chloride. The methylene chloride extract will then be dried, decanted and evaporated to afford the liquid eugenol.

Experimental: The apparatus was assembled using a 25 ml round bottom flask as the distillation pot. The distillation pot was charged with 10 gms of ground cloves and 15 ml distilled water. The cloves were allowed to soak in the water until thoroughly wetted (about 15 min.), then the mixture was distilled, the distillate being collected at the rate of about one drop every 2-3 seconds. After about 6 ml of distillate were collected, the distillate was extracted with 2.0 ml. of CH_2Cl_2 DCM, then again with (2 x 1.0 ml) of DCM. The combined extracts were dried over Na_2SO_4, and evaporated to give the product eugenol as a pale yellow oil.

X g of euginol was recovered from 10 g of cloves:

Percentage recovery   =   amount of euginol isolated/amount of clove used

=   X g/10g.

The percent recovery from cloves will be determined, and is expected to be about 10%, based on literature data.

The product will be analyzed by recording infra red spectra neat sample using NaCl plates to confirm its structure. The major IR absorption are expected to be 3200 – 3500 cm^{-1} (OH stretch), 3000 – 3150 cm^{-1} (sp^2 C – H stretch), 1600 – 1680 cm^{-1} (alkene C = C), and 1400 – 1500 cm^{-1} (aromatic C = C).

7.7 Extraction of Cinnamaldehyde from Cinnamomum Zeylanicum

The most important components which contribute to the fragrance of cinnamon are cinnamyl acetate and *trans*-cinnamyldehyde.

Cinnamon essential oil has high antimicrobial properties which formed clear zone when tested with gram positive bacteria *Bacillus subtilis s.p* and a gram negative bacterium *Escherichia coli.*

(a) **Raw materials:** 25.5 g of Cinnamon sticks, 300 mL of dichloromethane (DCM),

Cinnamon sticks were cut into smaller pieces and placed into a mortar, where they were crushed with the help of a pestle. In order to avoid foaming during distillation, the sticks were not completely crushed into powder form.

(b) **Steam distillation method:** About 30 g of freshly crushed Cinnamon sticks were added into a 250 ml three necked distilling flask. 200 ml of distilled water was added to it. Later, a 100 ml of water was added to an addition funnel attached to one of the necks. The flask is heated with the help of a heater. A 200 ml round-bottomed flask was used to collect the distillate. During the distillation process, a cloudy distillate travelling down the condenser column was noticed. 100 ml of the cloudy distillate was collected. The cloudiness of the distillate is due to the presence of an insoluble suspension of Cinnamaldehyde. 100ml of the distillate collected was transferred into an Erlenmeyer flask. To collect further amount of distillate, 100 ml of water was added to the distilling flask. The next batch of distillate obtained was clear indicating low amount of Cinnamaldehyde.

(c) **Extraction of oil:** The combined distillate transferred into a 250 ml separating funnel and extract with three 50 ml portions of dichloromethane (DCM). The combined DCM washings washed with sodium chloride solution and then separate organic layer (lower layer, DCM being a higher density than water) from the aqueous layer. Dry the DCM layer with anhydrous magnesium sulfate and filter the solution. Remove the dichloromethane using a rotary evaporator and collect the colorless cinnamon essential oil. Record your yield. Calculate the yield of *trans*-cinnamaldehyde from the dry mass of cinnamon sticks.

B.p. 252^0C ;UV absorption of cinnamaldehyde at 260 nm.

IDENTIFICATION OF ORGANIC MOLECULES BY SPECTROSCOPY

8.1 Energy - The Electromagnetic Spectrum and the Absorption Spectrum

Energy increases going to the left. The electromagnetic radiation interacts with the electromagnetic fields of the electrons to raise their energy levels from one state to the next.

The nature of that interaction depends on the energy available. Ultraviolet and visible have sufficient energy to effect electronic transitions.

Infrared has sufficient energy only to effect transitions between vibrational energy states.

Microwave has only enough energy to effect transitions between rotationally energy states. Thus the radiation absorbed tells us different information.

Radio waves have insufficient energy to effect molecules but affect nuclear spin energy states found in magnetic fields. This latter interaction is most important because it is used in Nuclear Magnetic Resonance spectroscopy

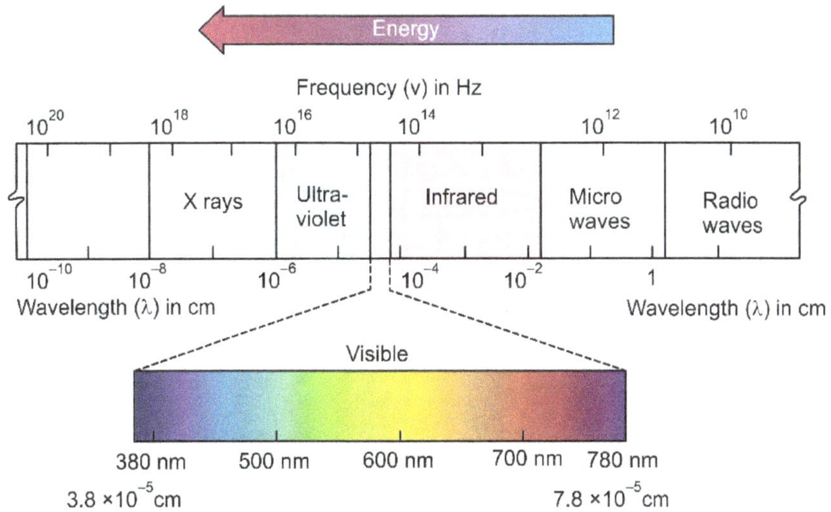

8.2 Important Spectral Data for Determining the Structure of Organic Molecules

8.2.1 Ultraviolet Absorption Spectroscopy

The ultraviolet region falls in the range between 190-380 nm, the visible region falls between 380-750 nm.

Terminology for absorption shifts

(i) **Chromophore:** A covalently bonded unsaturated group is responsible for absorption of UV – Visible radiations.

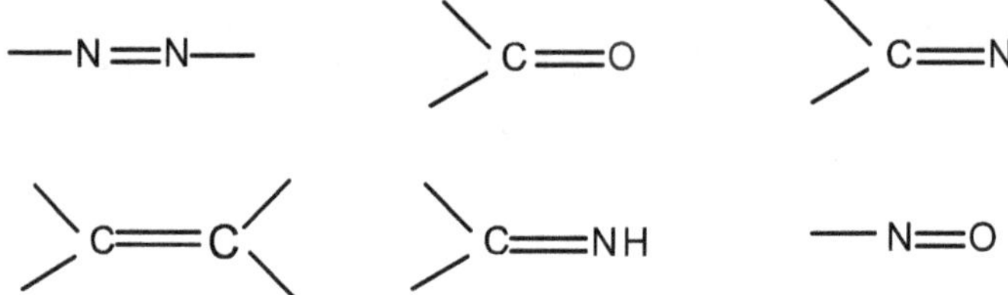

(ii) **Auxochrome:** A saturated group with 'n' electrons, which, when bonded to a chromophore, alters the wavelength & intensity of the absorption.

$$— OH \qquad — NHR \qquad — Cl \qquad — Br$$

$$— CH_2 \qquad — NO_2 \qquad — NR_2 \qquad — COOH$$

$$— NH_2 \qquad\qquad — SO_2H$$

(iii) **Bathochromic Shift:** The shift of absorption to a longer wavelength due to substitution or solvent effect (Red shift).

$$H_3C – CH_2 – CH_2 – CH = CH_2 \qquad\qquad H_2C = C\,(CH_3) – CH = CH_2$$

$$178\ nm \qquad\qquad\qquad\qquad 222\ nm$$

(iv) **Hypsochromic Shift:** The shift of absorption maxima, to shorter wavelength due to substitution or solvent effect (Blue shift)

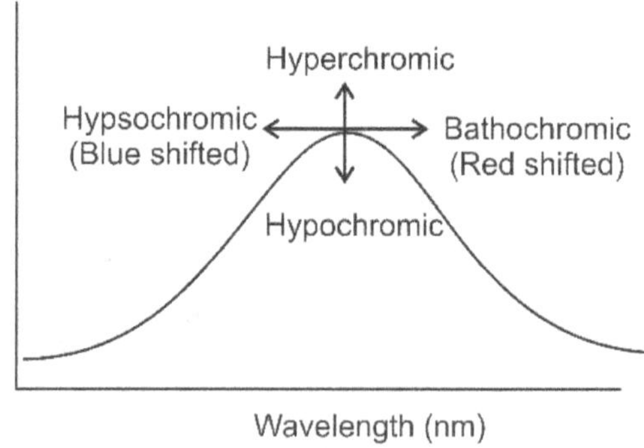

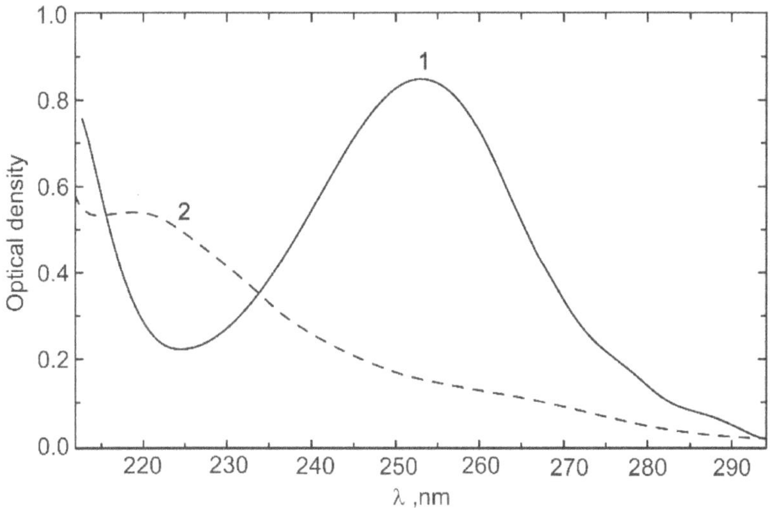

Fig. 8.1 Absorption spectra of benzaldehyde
(Dotted line shows continuing spectrum beyond 295 nm)

Characteristic absorption of Organic Compounds

The ability of an organic compound to absorb UV – Visible radiations is dependent on its electronic structure.

8.2.2 Compounds containing only σ Electrons

For example saturated hydrocarbons (alkanes & cycloalkanes)therefore, exhibits σ to σ^* transitions; it requires around 185 K.cal/mole energy which is available in radiations with wavelength 155 nm. These hydrocarbons are transparent in U.V. region.

Saturated compounds with '*n*' electrons, saturated compounds containing hetero atoms such as oxygen, sulphur, nitrogen &halogens possess '*n*' electrons in addition to electrons. Hence, σ to σ^* and n to σ^* (both transitions require high energy).

8.2.3 Compounds containing π Electrons

(a) Alkenes – non-conjugated dienes display no absorption above 200 nm.

$$\text{Ethylene } CH_2 = CH_2 \ \lambda_{max} \ 165 \,\&\, 199 \, nm$$

(b) Alkyl substituted on alkenes moves the λ_{max} to longer wave lengths.

$$H_3C - CH = CH_2$$

(c) Attachment of heteroatom to the olefinic bond brings a bathochromic shift.

$$H_2C - S - C \ H = CH_2 \ \lambda_{max} \ 228 \,\&\, 199 \, nm$$

(d) Alkynes, alcohols, ethers, esters do not show absorption at 200-400 nm; i.e., they are transparent in UV region.

8.2.4 Benzene and its Derivatives

(i) Benzene displays absorption bands at 184 nm ($\varepsilon = 60{,}000$) — ε_1 band

204 nm ($= 7{,}400$) ——— ε_2 or K band)

254 nm ($= 200$)   ——— B band

(ii) All these bands originates from $\pi - \pi^*$ transitions

ε_1 band is intense and is due to allowed transitions.

ε_2 **B** bands are weaker due to symmetry forbidden transitions.

Intensity order $= \varepsilon_1 > \varepsilon_2 >$ B

(iii) Electronic excitation of benzene and its derivatives is influenced by

(a) the nature of chromophore;   (b) the nature of auxochrome and

(c) hyperconjugation

8.2.5 Applications of Ultraviolet Spectroscopy

Alkanes, alkenes (non-conjugated), alkynes, alcohols, ethers, aliphatic carboxylic acids, aliphatic esters, aliphatic amides, nitriles: no absorption in U.V., no peak in $200 - 400$ nm (i.e., transparent in U. V).

(a) Aliphatic aldehydes and ketones only n to π^* transition

A weak peak in the region $275 - 295$ nm ($\log \varepsilon = 1 - 2$)

(b) Compounds having sulfur and bromine n to σ^* transition

A weak peak in the region $210 - 250$ nm ($\log = 1 - 2$)

(c) Aromatic compounds: π to π^*(B - band)

Medium intensity peak/s in the region $250 - 285$ nm ($\log = 2$)

(d) Conjugated dienes π to π^* $220 - 230$ nm ($\log = 4$)

(e) Enones: 2 peaks in the region 220 -250 nm, intense π to π^*

300 to 350 nm weak n to π^* transition

8.3 Infrared Absorption Spectroscopy

Infrared spectroscopy is the study of how molecules absorb IR radiation and ultimately convert it to heat.

What changes occur in the Infrared absorption frequencies of C, H, O, and N compounds in the conventional NaCl region broadly classified into frequency regions?

All frequencies (strictly speaking wave numbers) are given in reciprocal centimeters (abbreviation v, units cm^{-1}). The relationship to wavelength (λ) is given by

$$\lambda = 1000/v,$$

when λ is in microns (μ)

The following changes occur when a molecule absorbs IR radiation.

1. IR radiation absorbedcauses increases in vibrational and rotational movements to occur within the molecule.

2. At room or ambient temperature majority of the molecules have ground state vibrational levels

3. When a sample is irradiated with IR radiation, absorption of IR energy takes place by the molecules

4. As a result, they gain more energy and they undergo upward transitions in vibrational levels i.e., $V_0 \rightarrow V_1$ or $V_0 \rightarrow V_2$ etc.

5. During the upward transition in vibrational levels changes in rotational levels also take place because changes in rotational movements need less energy than what is required for vibrational changes.

   Therefore, IR spectroscopy is also referred to as vibrational – rotational spectroscopy.

6. In other words absorption of IR radiation causes atoms and groups of atoms of organic compounds to vibrate with increased amplitude about the covalent bonds that connect them.

   Tables of Infrared absorption frequencies of C, H, O, N compounds in the region $4000 - 625$ cm^{-1}

   The region $4000 - 2000$ cm^{-1} (2.5 to 5μ).

   The region $1500 - 650 cm^{-1}$(7.5 to 15 μ) contains relatively large numbers of bands, the assignments of which are often not certain. It includes the so-called **'finger print region'**.

8.3.1 Alkanes and any Alkyl Groups

$Sp^3C - H$ Stretching vibrations	$3000 - 2800$ cm^{-1}
$C - H$ bending vibrations	$1370 - 1470$ cm^{-1}
Alkenes $= C-H$ (Sp^2) stretching	$3050 - 3000$ cm^{-1}
$C = C$ stretching	$1650 - 1600$ cm^{-1}

cis-alkenes stretching		700 cm^{-1}
trans-alkenes stretching		950 cm^{-1}
Alkynes	stretching(Sp)	$2150 - 2250 \text{ cm}^{-1}$ (weak)
	stretching around	3300 cm^{-1}
Aromatic:	(C = C) ring stretching	$1600 - 1450 \text{ cm}^{-1}$ (2-3 peaks).
	= C–H Stretching vibrations	$3050 - 3000 \text{ cm}^{-1}$
	C–H bending vibrations (overtones)	$2000 - 1600 \text{ cm}^{-1}$
O – H	stretching vibrations	
Alcohols:	not hydrogen bonded	3625 cm^{-1}
	Intramolecular H bonded	$3450 - 3500 \text{ cm}^{-1}$
	Intermolecular H bonded	$3200 - 3400 \text{ cm}^{-1}$
Phenols:	not H-bonded	3605 cm^{-1}
	Intramolecular H-bonded	$3300 - 3450 \text{ cm}^{-1}$
	Intermolecular H-bonded	$3150 - 3350 \text{ cm}^{-1}$
Carboxylic acids:	not H-bonded	3530 cm^{-1}
	O – H hydrogen bonded	$3500\text{-}3300 \text{cm}^{-1}$
	C = O stretching	1700 cm^{-1}

Acid derivatives	
Carboxylic acids (monomers)	1720cm^{-1}
α, β-Unsaturated acids	1700 cm^{-1}
Aryl acids (monomers)	1700 cm^{-1}
Esters and 6-membered ring lactones	1740 cm^{-1}
5-Membered ring lactones	1770 cm^{-1}
Acid halides	$1790 - 1800 \text{ cm}^{-1}$
Enol esters	1775 cm^{-1}

In most cases hydrogen bonded OH groups are broad; exceptions are salicylate ester OH's and other chelated OH groups.

C – O stretching vibrations in ethers

Dialkyl ethers: $1100 \text{ cm}^{-1} \pm 50$

Aryl and vinyl ethers: 1250 cm^{-1}±20

1,2– Epoxides: 1250 and 890 cm^{-1}(*trans*), 1250 and 830 cm^{-1} (*cis*)

C – O stretching vibrations of esters

Enol esters: 1775 cm^{-1}

Formates: 1190 cm^{-1}±10

Acetates: 1240 cm^{-1}±10

Phenolic acetates: 1205 cm^{-1}

α, β -unsaturated esters: 1250 and 1125 cm^{-1}

Nitriles: stretching 2150 - 2250 cm^{-1} (intense)

Nitro: N=O stretching 1370 - 1300 cm^{-1} and 1575 - 1500 cm^{-1}

Amides: C=O stretching 1640 - 1620 cm^{-1}

    NH$_2$ stretching 3500 -3300cm^{-1}

Amines:

Primary   NH$_2$ stretching 3500 - 3300 cm^{-1} (2 peaks)

Secondary N-H stretching 3600 - 3500 cm^{-1} (1 peak)

Aromatic compounds (strong peaks only)

Five adjacent free H,   monosubstitutedbenzene (2 bands) 750 and 700 cm^{-1}

Four adjacent free H,   1,2 - disubstituted(1 band) 750 ± 15 cm^{-1}

Three adjacent free H, 1,3 – disubstituted and

        1,2,3 – trisubstitutedbenzene 770 cm^{-1}± 20

Two adjacent free H, 1,4 - disubstituted benzenes 830 cm^{-1}± 30

In addition to these bands, absorption occurs in the 1000 region, which is similar to that of the corresponding alcohol. A summary of the absorption of an acetate, as an example of a typical ester is:

$C = O$ absorption in the region 1730 cm^{-1}

$C - O$ absorption in the region 1250 cm^{-1}

$O - R$ absorption in the region 1050 cm^{-1}

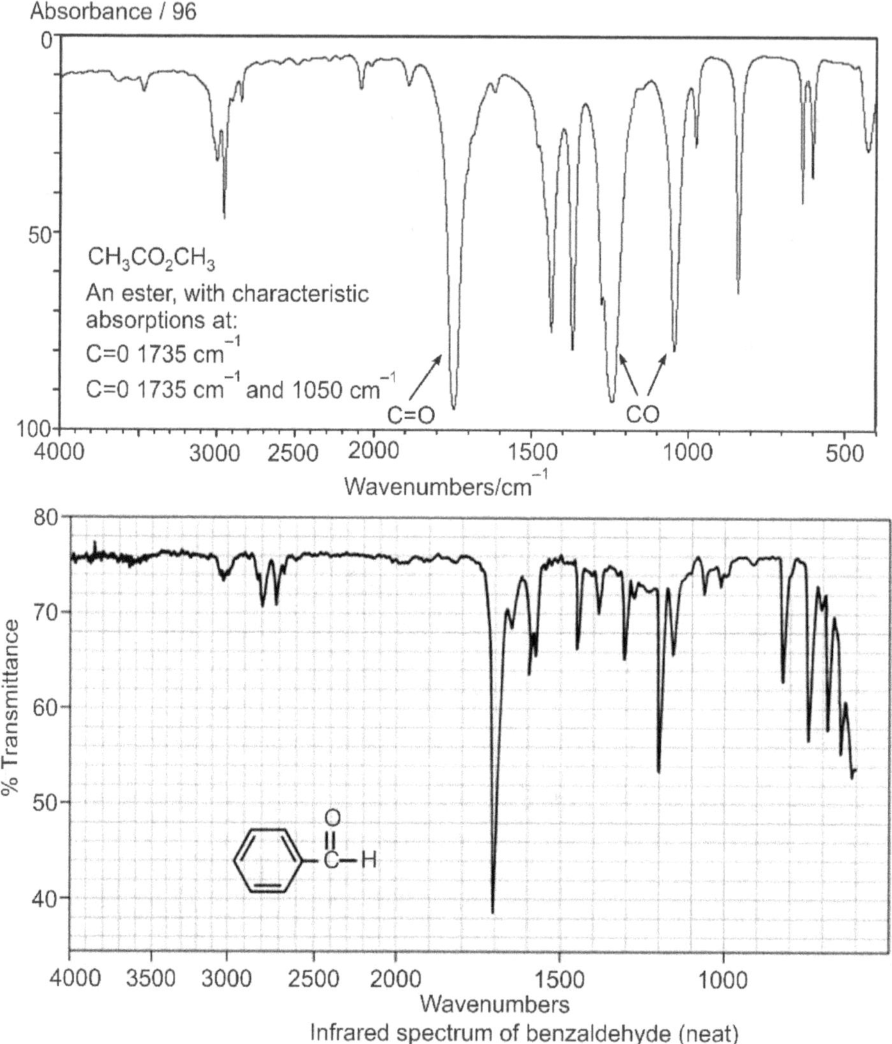

Infrared spectrum of benzaldehyde (neat)

Fig. 8.2

8.4 Nuclear Magnetic Resonance Spectroscopy

Nuclear magnetic resonance spectroscopy has become the preeminent technique for determining the structure of organic compounds. Of all the spectroscopic methods, it is the only one for which a complete analysis and interpretation of the entire spectrum is normally expected.

Although larger amount of sample is needed than for mass spectra, N. M. R. is non-destructive and with modern instruments good data may be obtained from samples weighing less than a milligram.

To be successful in using N. M. R. as an analytical tool, it is necessary to understand the physical principles on which the method is based.

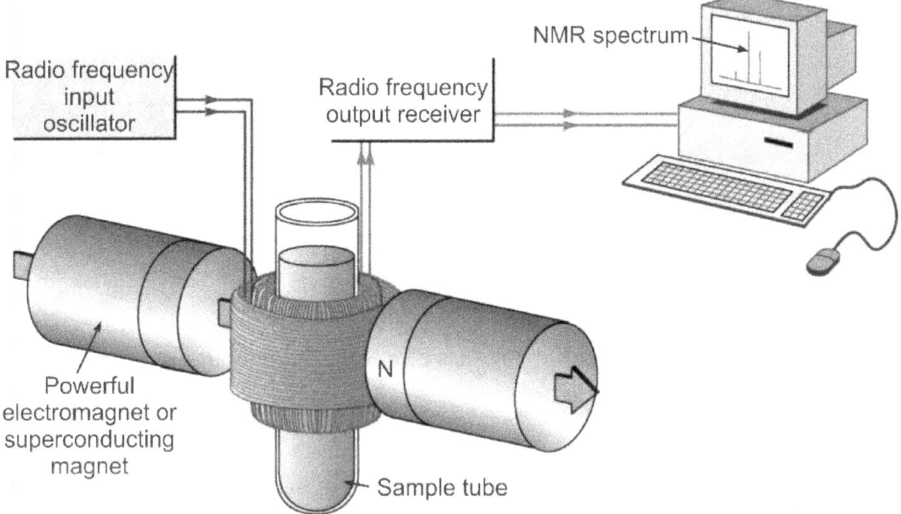

Fig. 8.3 Schematic diagram of a Nuclear Magnetic Resonance Spectrometer

As is implied in the name, Nuclear Magnetic Resonance (NMR), is concerned with the magnetic properties of certain atomic nuclei-notably the nucleus of hydrogen atom. ** Nuclei with I = ½ have a magnetic moment and align themselves with the field of the spectrometer. Radiofrequency radiation can then be absorbed causing the aligned nuclei to orientate against the applied magnetic field. The nuclei of many elemental isotopes have a characteristic spin (**I**) and in organic chemistry the most useful atoms concerned are those with I=1/2, namely ^1H, ^{13}C, ^{31}P and ^{19}F. The main isotope of carbon ^{12}C has I=0 and does not show NMR effects.

Under appropriate conditions, in an external magnetic field, a sample of an organic compound can absorb electromagnetic radiation in the radio frequency (r.f) region at frequencies governed by the characteristics of the sample.

A plot of the frequencies of r.f. radiations absorbed by the sample versus intensities of the r.f. absorption constitutes the "NMR spectrum".

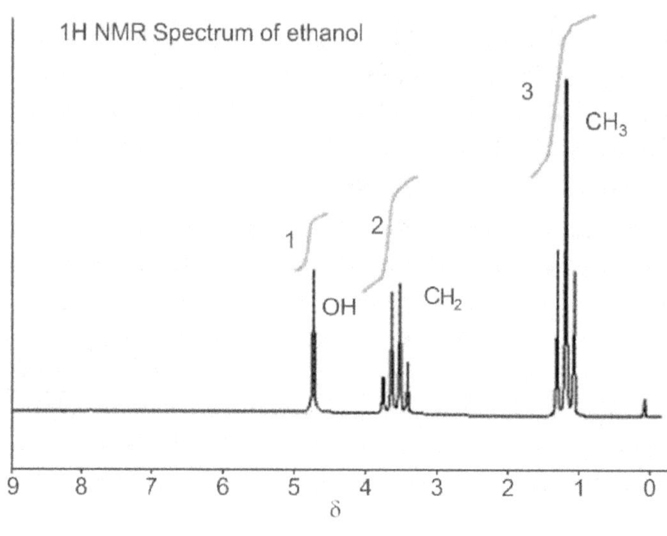

Fig. 8.4

Peak area and Proton Counting

We are not only interested in the position of a NMR signal but also interested in the relative intensities of each peak.

Integration Curves

NMR instruments draw stepped curves on the top of each signal. These stepped curves are known as "integration curves'. The heights of integration curves are proportional to peak areas, which are proportional the number of protons.

By comparing the heights of integration curves, we can establish the ratios of different protons in the molecule.

8.4.1 General Notes on Presentation of Spectra

By convention all field-sweep NMR spectra are plotted with the direction of increasing magnetic field going from left to right. In the case of frequency-sweep spectra the equivalent presentation is obtained with direction of decreasing frequency going from left to right.

The direction from left to right is referred to (in the case of both field- and frequency – sweep) as the **upfield** direction, the opposite direction is referred to as the **downfield** direction.

A nucleus giving rise to a signal further upfield than another nucleus is said to be more **shielded**. A nucleus giving rise to a signal further downfield than another is said to be **de-shielded**.

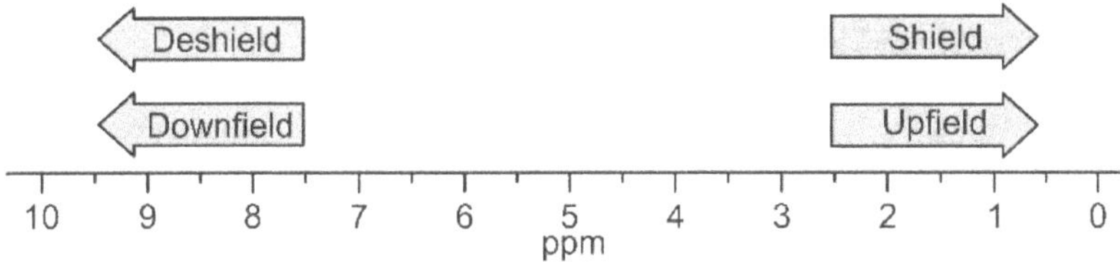

Fig. 8.5

Movement of a resonance signal upfield is often referred to as a **diamagnetic** shift, while movement in the downfield direction is known as a **paramagnetic** shift.

8.4.2 Chemical Shift Measurements

Chemical shifts are measured with respect to the resonance signal of a reference substance. Corrections for the magnetic susceptibility of the solvent are avoided if an internal reference. (i.e., a reference substance added to the sample is used). For proton resonance in non-aqueous solutions the single resonance line of the internal standard tetramethysilanes. TMS is almost universally used. Since most instruments (both field and frequency sweep) have their scales calibrated in frequency units (Hz), the initial measurements are made in these units. e. g. 247 Hz downfield from TMS at 60 MHz (note that the instrument operating frequency must be stated) but it is preferable to employ units which are independent of the operating frequency and field of the instrument. Two of these frequency independent scales are used. The delta (5)scale is defined:-

$$\delta = \frac{v(sample - v(TMS)}{v(TMS)} \times 10^6 \, (ppm)$$

In practice the value of a resonance line (or multiplet) is found by dividing the separation of the resonance line from the TMS line measured in Hz by the operating frequency of the instrument in MHz. Resonances downfield from TMS gave positive values.

The second Tau scale (τ) can be calculated by:-

$$\tau = 10 - \delta$$

Tetramethylsilane (TMS) is a good reference substance for non-aqueous solutions since it is inert, readily volatile and gives rise to a single resonance line which is at a field higher than those of most organic substances. It is insoluble in water and for aqueous solutions the sodium salt of 3 - (trymethylsilyl) propanesulphonic acid (TMSPSA) is used (Tiers and Cook, J. Org. Chem., 1961, 26, 2097).

8.5 Aromatic Compounds

In aromatic compounds π - electrons are cyclically delocalized over the aromatic ring. When these compounds are kept in a external magnetic field, H_O –these closed loop of - electrons are induced to rotate further.

Shielding: If the induced magnetic field opposes the applied magnetic field at the proton, then the proton is said to be shielded and the phenomenon is known as shielding.

De-shielding: If the induced magnetic field aligns with the applied magnetic field at the present, then the proton is said to be deshielded and the phenomenon is known as de-shielding.

In aromatic rings the circulation of electrons produce a substantial electric current called the ring current and of course an associated magnetic field. This field in the centre of the ring opposes the applied external field (diamagnetic effect) but the field around the outside of the ring enhances the instrument field and the aromatic protons are deshielded (see below). Aromatic protons therefore appear in the spectrum at around (see chemical shift tables).

Tables of chemical shifts of various types of protons

Type of proton	Chemical shifts (δ values)
CH_3 Resonances	
CH_4	0.24
$CH_3 - CR_3$	0.88 to 0.94
$CH_2 F$ 4.26	
$CH_3 Cl$	3.05
$CH_3 Br$	2.68
$CH_3 I$	2.16
$CH_3. CHO$	2.07
$CH_3. CO. R$	2.10
$CH_3. CO. Ar2. 62$	
$CH_3. CO. OH$	2.07
$CH_3. CO. OR$	2.00
$CH_3. CO NH_2$	2.02
$CH_3. CR' = CR'_2$	1.63 to 1.70
$CH_3. CR' = CR' CO. R''$ (*cis*)	2.14
$CH_3. CR' = CR'. CO. R''$ (*trans)*	1.88

$R'_2 C = C (CH_3) . COR''$	1.90
$CH_3 .Ar$	2.34
$CH_3. OR$	3.30
$CH_3 . OH$	3.45
$CH_3. O. CO. R$	3.60
$CH_3. O. CO Ar$	3.90
$CH_3. O. Ar$	3.80
$CH_3. NR_2$	2.15
$CH_3. S. R$	2.10
$CH_3. S. Ar$	2.47

CH_2 Re*sonnance*

(a) Methylene groups carrying 1 alkyl

    group and 1 simple

(or complex) functional group	(δ values)
$R_3C . CH_2. CR_3$	1.25
$R. CH_2 F$	4.35
$R. CH_2 Cl$	3.40
$R. CH_2 Br$	3.31
$R. CH_2 I$	3.14
$R CH_2. CHO$	2.20
$R. CH_2. CO. R$	2.40
$R. CH_2.CO. OR$	2.34
$R. CH_2. CONH_2$	2.23
$R. CH_3. CR' = CR'_2$	2.00
$R. CH_2. CR' = CR' . CO. R''$ (*cis*)	2.65
$R. CH_2. CR' = CR' . CO. R''$ (*trans*)	2.22
$R'_2 C = C (CH_2R). CO. R''$	2.00
$R. CH_2Ar$	2.62
$R. CH_2. OR$	3.40
$R. CH_2. OH$	3.56
$R. CH_2. O. CO. R$	4.10
$R. CH_2. O. CO. Ar.$	4.23

R. CH_2. OAr3.90	
R. CH_2. NR_2	2.30-2.70
(b) Methylene groups carrying 2	
Functional groups	(δ values)
CH_2 Cl_2	5.30
CH_2 $(CO_2R)_2$	3.37
CH_2 $(OR)_2$	4.75
CH_2 Ar_2	3.02
(c) Methylene groups in rings	(δ values)
Cyclopropane	0.22
Cyclopentane	1.51
Cyclohexane	1.43
Cycloheptane	1.53
CH_2. CO - (6 membered) -	2.30
CO. CH_2. CO - (6 membered)	3.30
-CH_2. O - (6 membered)	3.70
-CH_2 – O - (5 membered)	3.75
-CH_2- O- (epoxides)	2.29
-O-CH_2-O	4.79
CH Resonance	(δ values)
CH R_3	0.89
R_2 CH-Cl	4.02
R_2 CH –Br	4.10
R_2 CH-I	4.20
R_2 CH.CHO	2.39
R_2 CH. CO.R	2.48
R_2 CH. CO. Ar	3.27
R_2 CH. CO_2H	2.57
R_2.CH.CR ' = CR'$_2$	2.55
R_2CH.Ar	2.87
R_2CH.OR	4.00
R_2 CHOH	3.85
R_2 CH.O.COR	5.01

$R_2CH.O.CO.Ar$	5.12
$CH (OH) Ar_2$	5.60
$R. CH (OR)_2$	4.60
$CH Ar_3$	5.30
$CH (OR)_3$	5.50

Formyl Proton Resonances (aldehydes and formic acid derivatives etc).	(δ values)
$R - CHO$	9.2 - 10.2
$Ar - CHO$	9.6 - 10.5
$H.CO_2R$	8.0
$H.CO.NR'_2$	7.8 - 8.7
$R.CH = NOH$	6.8 - 7.9
$Ar. CH = NOH$	7.2 - 8.6
$R.CH = N.CO.NH_2$	6.5 - 7.5
$R. CH = N.NH.Ar$	6.1 - 7.7

Ethinyl Proton Resonances	(δ values)
$R - C \equiv CH$	1.80
$Ar - C \equiv CH$	2.7 - 3.4
$R.CO. C \equiv CH$	2.2 - 3.3
$R. (C \equiv C)_n.H$	1.8 - 2.4

OH Protons (δ values)	
Alcohols	0.5 - 5.5
Phenols - normal	4.0 - 5.5
intra molecularly H-bonded	10.5 - 15
$O -$ alkoxy	6.5
Enols (intramolecularly H-bonded)	15 - 19
Carboxylic acids (COOH)	10 - 13
Oximes	7.0 - 10.0

Water, HDO		4.8
$R.SO_3H$		11 - 12

NH Protons (δ values)

$R.NH_2$	$R_2 NH$	0.5 - 3.5
$Ar.NH_2$	$Ar_2 NHAr NHR$	3.0 - 5.0
$R.CO.NH_2Ar CO. NH_2$		3.0 - 5.0
R.CO.NH.R Ar. CO.NH.R		6.0 - 8.0
R.CO.NH.ArAr.CO.NH.Ar		8.0 - 10.0
R.SH		0.9 - 2.5
Ar SH		3.0 - 4.0
Inside NH in porphyrins		4.0
Aromatic (Ar – H) Benzylic Ar – CH_2 -		6.0 - 8.5
Allylic ($H_2 C = CH – CH_3$)		2.2 - 3.0

With reference to the spectrum of ethanol shown above note that the methyl signal is split into three peaks (a triplet) and the methylene appears as a quartet. Since there is free rotation about the single bonds in this molecule the methyl protons are equivalent but are affected by the magnetic fields generated by the methylene protons adjacent. The two methylene protons can be in the following states (-1/2,-1/2), (-1/2,+1/2) (+1/2,-1/2), (+1/2,+1/2), or for clarity (—) (-+) (+-)(++).The two central states are the same and are thus twice as probable as the others giving the 1:2:1 ratio of the methyl signals. Similarly the methyl protons have states (—) : (—+) (-+-) (+—) : (++-) (+-+) (-++) : (+++) this gives a ratio of 1:3:3:1 and thus the methylene signals appear as a quartet. The separation of the splits are equal, denoted as J value, or coupling constant, and are usually given in Hertz. (In the case of ethanol J=7 Hz approximately). These splitting patterns are very helpful in assigning molecular structure.

Since the ^{12}C isotope has no magnetic moment and is 99% of all terrestrial carbon there is no complication of the proton spectra. ^{13}Chas a magnetic moment, is 1% of all terrestrial carbon, and its NMR can be useful in chemistry. The ^{13}C NMR spectrum can be made to appear as a single line for each carbon position in the molecule. This is done by irradiating the proton bandwidth such that all the protons resonate rapidly and the average field causes no splitting of the ^{13}C signals. Using a lower power level of the proton irradiation produces partial splitting of the ^{13}C signals which can be helpful in determining the number of protons attached to each carbon atom. The probability of two ^{13}C atoms occurring next to each other and therefore causing splitting can be ignored due to the low abundance.

Questions for Students

1   What does an isopropyl group give in ^{1}H NMR? (see problem 14 example later).

2   If a resonance is split by two non-equivalent protons what is observed.

(Hint it is not a quartet)

8.6 Mass Spectrometry

Mass spectrometry is not an absorption spectroscopy unlike UV, IR and NMR spectroscopy.

It is a destructive method. Therefore the compound cannot be recovered.

Fortunately, we require only very small quantity of the compound (often less than 1 pg)

Mass spectrometry is useful for

(a)   Determining the molecular weight of the compound

(b)   Determining the molecular formula

(c)   Detection of S, N, Cl, Br, I, in a molecule

(d)   Detection of certain groups in organic compounds like

$-CH_3$; $-CH_2-CH_3$; $-CH_2 - CH_2 - CH_3$; $-OH$, $-NH_2$, $-CN$, $-Ph$, $Ph - CO$

etc.

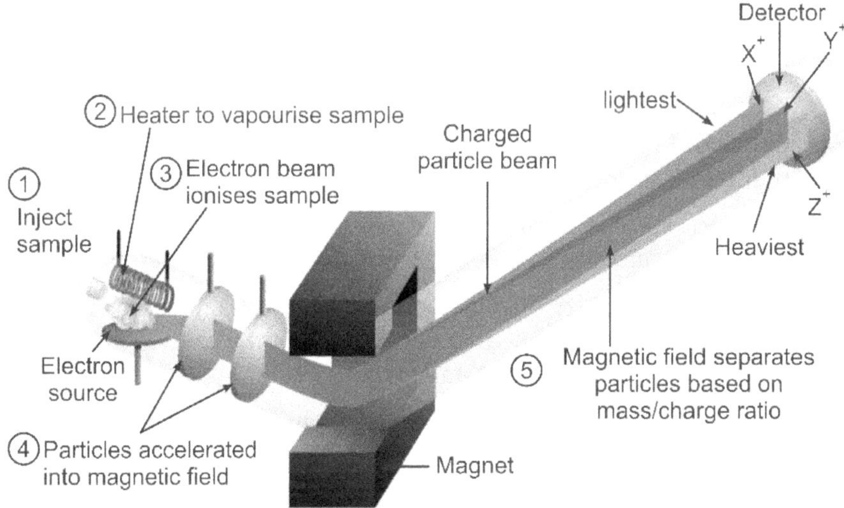

Fig. 8.6 Schematic drawing of a simple mass spectrometer

8.6.1 Basic Principle

In the mass spectrometer, organic molecules are bombarded with a high energy electron beam. This result in the loss of one loosely bound electron from the molecule and forms a highly energetic positively charged ions called molecular ions or parent ions.

These molecular ions, further break & produce smaller ions called daughter ions which give a characteristic ''fingerprint'' of the molecule.

8.6.2 Origin of Molecular Ion

All the organic molecules contain even number of electrons.

High energy e- beam (kinetic energy of e⁻ is enough to displace an electron from the molecule to produce a positive ion M+).

Sample must be in gaseous state, because in vapor state interaction with e⁻ beam is perfect

Generally, the ionization potential of organic molecules is 7–15 eV.

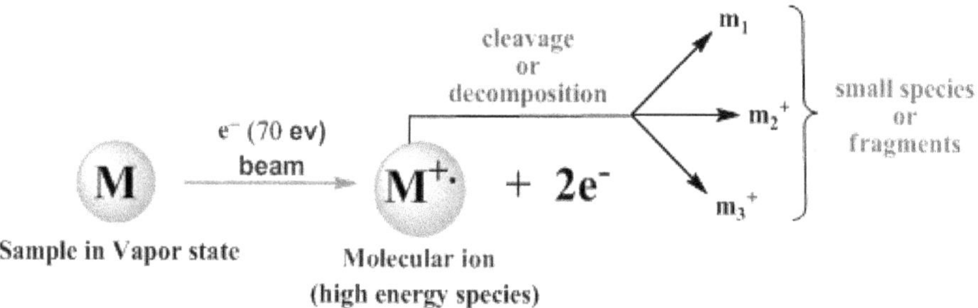

Fig. 8.7

Primary process: Formation of molecular ion by the loss of one electron from the neutral molecule is known as theprimary process.

Secondary process: This occurs by formation of smaller ions or smaller radicals or smaller neutral molecules by the breakup off molecular ions.

In the mass spectrometer – the molecular ions, the fragment ions and the fragment radical ions are separated by deflection in a variable magnetic field according to their mass/charge (m/e)ratios.

The mass spectrometer also measures relative abundance of these fragment ions giving rise to the aforementioned 'fingerprint'.

A mass spectrum is printed or drawn by the mass spectrometer. A mass spectrum is a plot of relative abundance against the m/e values of M+· & fragment ions.

For ionization of organic molecules electron beam of 7 to 15 eV is sufficient. However, generally, a high energy electron beam of 70 eV is used in mass spectrometry as standard.

The order of the removal of various electrons

Non bonding electrons > pi- electrons > sigma electrons

8.6.3 Isotopic Peaks

Mass spectra of all organic compounds contain peaks due to the presence of various isotopes,. These peaks appear with mass units dependent on the isotopes present. For example in a molecule composed of carbon and hydrogen a small $(M+1)^+$ ion is observed which is due to the 1% of ^{13}C present in terrestrial carbon. These small peaks appear throughout the spectrum one mass unit greater than the main ions. Since ^2H (deuterium) is less than 0.02% of terrestrial hydrogen this tiny effect can be ignored for all practical purpose in organic chemistry.

Element	Common isptope	% Ambudance	% Abudance of the isotope with 1 mass unit higher		% Abudance of the isotope with 2 mass units higher	
Hydrogen	^1H	100%	^2H	0.016		
Carbon	^{12}C	98.9%	^{13}C	1.1		
Nitrogen	^{14}N	99.6%	^{15}N	0.4		
Oxygen	^{16}O	99.8%	^{17}O	0.04	^{18}O	0.2
Sulpher	^{32}S	95.0%	^{33}S	0.8	^{34}S	4.2
Chlorine	^{35}Cl	75.5%	NO	^{36}Cl	^{37}Cl	24.2
Bromine	^{79}Br	50.5%	NO	^{80}Br	^{81}Br	49.5
Iodine	^{127}I	100%				

^{19}Fluorine, ^{31}Phosphorous, ^{127}Iodine do not have other isotopes. Hence the do not have any contribution to M+1 & M+2 peaks.

Some of the more common fragments which appear as ions and /or lost as radicals or molecules in mass spectra.

m/e	Fragment	m/e	Fragment
1	H	2	H_2
14	CH_2	16	O, NH_2
17	OH	18	H_2O, NH_4
19	F	20	HF
26	C_2H_2, CN	27	C_2H_3, HCN
28	C_2H_4, CO, N_2	29	C_2H_5, CHO

Table *contd....*

m/e	Fragment	m/e	Fragment
30	CH_2NH_2, NO	31	CH_2OH, OCH_3
32	CH_3OH	33	SH
34	H_2S	35	Cl
36	HCL	41	C_3H_5
42	C_3H_6	43	C_3H_7, $COCH_3$
44	CO_2, $C_2H_4NH_2$	45	C_2H_5O, COOH
46	NO_2	51	C_4H_3
55	C_4H_7	57	C_4H_9, C_2H_5CO
59	$COOCH_3$	71	C_5H_{11}
77	C_6H_5	78	C_6H_6
80	HBr	91	C_7H_7
105	COC_6H_5	127	I
128	HI		

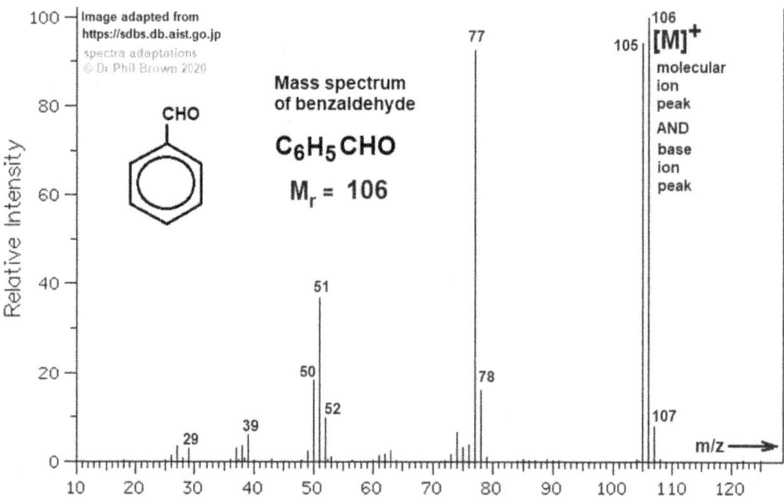

Fig. 8.8 Fragmentation pattern of the mass spectrum of benzaldehyde

8.7 Spectral Analysis

Following the order for interpretation of the spectral data

8.7.1 Molecular Formula Calculations

- The molecular ion M^+ indicates molecular weight.

- If the mass is even there may be zero or even number of Nitrogens and if it is odd there must be odd number Nitrogens in the compounds.

- If mass spectra exhibits M^{+1} & M^{+2} peaks in 3 : 1 ratio (^{35}Cl:^{37}Cl) chlorine may be present and if it is 1:1 (^{79}Br:^{81}Br) bromine may be present.

Obviously isotopic patterns can be very useful as the chlorine and bromine examples illustrate. If two bromines are present in the molecule the molecular ion shows as a triplet. Since the spectrometer measures mass, the two middle combinations show the same mass but twice the abundance giving an ion containing the bromine atoms the pattern 1:2:1 the peaks being 2 mass units apart.

The pattern for bromine is thus given by the coefficients of $(x + 1)^n$ Therefore for 2 bromine atoms in the molecule we have $(x + 1)^2 = x^2 + 2x + 1$ and for 3 bromines $(x + 1)^3 = x^3 + 3x^2 + 3x + 1$ that is (1:3:3:1).

In the case of chlorine the abundances of the isotopes is ^{35}Cl:^{37}Cl = 3:1. The isotopic pattern is therefore given by the coefficients of $(3x + 1)^n$. The calculation of patterns in this way can become difficult but many mass spectrometers now are computer controlled and can produce these calculations.

Questions for Students

1. Calculate the isotopic pattern for a molecule with one chlorine and one bromine. (This is tricky).

2. Calculate the pattern for hexabromobenzene

(**Hint** use the binomial theorem with $(x + 1)^6$ to obtain the seven coefficients).

- Iodine is recognized by the presence of I^+ at m/z 127, combined with a characteristic 127 unit gap in the spectrum corresponding to loss of iodine radical.

- If (M+2) is present with abundance of only 4%, sulphur may be present.

- Number of carbons can be calculated as : $(M + 1)^{+\cdot}$

$$(\text{abundance of } ^{13}C)$$

- Weight of Carbons = No. of Carbons x 12 = X

- If N/Cl/Br/S present, add their mass to mass of carbons.

- Number of Hydrogen's = Molecular weight − (weight of carbons + weight of hetero atoms).

- If necessary add appropriate number of oxygen atoms.

- Therefore the molecular formula may be __________

8.7.2 Index of Hydrogen Deficiency (or) Double Bond Equivalence

Double bond equivalents indicate the total no of rings + total no of double bonds.

If the molecular formula is $C_A H_B N_C O_D$, then the double bond equivalents (DBE) can be calculated from the following equation DBE = A - 1/2 B+1/2 C + 1.

If DBE = 1 it may be one double bond in open chain or monocyclic saturated system.

If DBE = 2 it may be two double bonds in open chain or one double bond in cyclic system or may be a saturated bicyclic system.

If DBE = 3 it may be three double bonds in open chain or two double bonds in cyclic system or tricyclic saturated system.

If DBE = 4 it may be one ring and three double bonds (ie. aromatic ring)

If DBE = 5 it may be one ring and four double bonds (3 in benzene and 1 outside).

Example 1. C_7H_7NO DBE = 5 therefore presence of 1 ring + 4 double bonds

Example II $CH_3 NO_2$ DBE = 1 double bond

Example III. $C_{13} H_9 N_2 O_4 Br S$ DBE = 10 therefore 2 rings & 8 double bonds.

$C_A H_B N_C X_D O_E + 1$

A - 1/2B+ 1/2C + 1/2D + 1

13 - 4.5 +1 + 0.5 + 1 = 10

Example IV. $C_5 H_{10} O_2$ DBE $= 1$ therefore presence of one double bond, an aliphatic compound

$$CH_3 - CH_2 - O - CO - CH_2 - CH_3$$

8.7.3 Interpretation of Spectral Data

(a) **Infrared spectra:** To identify the functional groups using stretching and bending vibrations.

(b) **UV spectral data:** To identify aliphatic or aromatic and characteristic chromophoric groups present.

(c) **Proton and carbon NMR spectral data:** To indicate the number and nature of protons and carbons respectively.

(d) **Mass spectral fragmentation:** One can calculate useful data based on the fragment ions and their relative abundance from the spectrum.

(e) **Proposition of possible structure:** Proposition of the possible structures and confirming the correct structure and elimination of other structures based on the consolidated assignments of IR, UV, NMR and MS.

(f) **Report the structure:** Finally, report the right structure with the help of spectral assignments.

Very high resolution mass spectrometers can measure mass to better than a few parts per million. This allows the elemental composition of the ion concerned to be determined and is an acceptable replacement for elemental analysis especially where very small amounts of sample are available. The reason for this is the mass defects which occur in atoms due to the differing binding energies. For example C_2H_4, CO, N_2 all have an approximate mass of 28. However measurement to high precision easily distinguishes these molecules.

8.8   Sets of Spectra

List: Name and molecular formula & weight of compounds

8.8.1.   Acetophenone, molecular formula C_8H_8O (mol. wt.120)

8.8.2.   3-Phenylbromopropane, mol. formula $C_9H_{11}Br$ (mol.wt. 200)

8.8.3.   Anisole , mol. formula C_7H_8O (mol. wt. 108)

8.8.4.   Methyl 2-furoate, mol. formula $C_6H_6O_3$ (mol. wt. 126)

8.8.5.   *p*-Nitrobenzaldehyde, mol. formula $C_7H_5NO_3$ (mol. wt. 151)

8.8.6.   -Phenylethanol, mol. formula $C_8H_{10}O$ (mol. wt. 122)

8.8.7.   4-Methyl benzaldehyde, mol. formula C_8H_8O (mol. wt. 120)

8.8.8.   Benzyl acetate, mol. formula $C_9H_{10}O_2$ (mol. wt. 150)

8.8.9.   2-Butanone, mol. formula C_4H_8O (mol. wt. 72)

8.8.10.   N-Butylbenzene, mol. formula $C_{10}H_{14}$ (mol. wt. 134)

8.8.11.   *p*-Chloroacetophenone, mol. formula C_8H_7ClO (mol. wt. 154/156)

8.8.12.   Caprolactam, mol. formula $C_6H_{11}NO$ (mol. wt. 113)

8.8.13.   4-Methylpyridine, mol. formula C_6H_7N (mol. wt. 93)

8.8.14.   Diisopropyl ether, mol. formula $C_6H_{14}O$ (mol. wt. 102)

8.8.15.   Diethylphthalate, mol. formula $C_{12}H_{14}O_4$ (mol. wt. 222)

8.8.16.   m-Methoxy benzoic acid, mol formula $C_8H_8O_3$ (mol. wt. 152)

8.8.17.   Aniline mol. formula C_6H_7N (mol. wt. 93)

8.8.18.   Ethylpropionate, mol. formula $C_5H_{10}O_2$, (mol. wt. 102)

8.8.19.   *ortho*-Nitrobenzylalcohol, mol. formula $C_7H_7NO_3$ (mol. wt. 153)

8.8.20.   4-Methyl-2-pentanone, mol. formula $C_6H_{12}O$ (mol. wt. 100

8.8.21.   Cyclohexenone, mol. formula C_6H_8O (mol. wt. 96)

Sets of Spectra - Structure Correlation of Spectra (Spectroscopic Identification of Organic Compounds)

In practice identification of organic compounds begins with analysis, synthesis and spectral data.

Here, we use the obvious features of one spectrum to bring out the more elusive aspects of another spectrum. The power of this approach lies in the complimentary features of the four or five spectra.

For practical reasons, we shall be concerned with the UV absorptions above 200 nm. In fact the application of ultraviolet spectroscopy to structural elucidation appears to be restricted. The strength of the electron spectroscopy depends on the ability to measure the extent of multiple bonds or aromatic conjugation within molecule. In fact the application of ultraviolet spectroscopy to structure investigation appears to be restricted.

In infrared spectroscopy the main preoccupation of an organic chemist is the region 4000-650 cm^{-1}. This is still the most readily examined region which covers the absorptions due to the fundamental vibrations of almost all the common functional groups of organic compounds. The absence of an absorption band from the spectrum may be a very positive indication that a particular functional group is absent from the molecule The clues thus obtained from the infrared spectrum when wisely used can shorten the time required for the elucidation of structure.

In organic chemistry, a large number of nuclei have been studied by proton NMR spectroscopy.As with compounds containing hydrogen (^{1}H NMR), the most valuable parameters are the chemical shifts delta, (measured in ppm) from reference standard and the coupling constant, J(measured in Hz)

By observing the fine structure obtained from such spectra and calculating the chemical shifts and coupling constants, the structure of many samples of organic compounds can be deduced.

One can use mass spectrometry in three important ways. One can measure relative molecular masses (molecular weight) with high accuracy. Also, one can detect within a molecule the places at which it prefers to fragment, from this can be translated the presence of recognizable groupings within the molecule. In many cases this can be done from the fragmentation pattern and comparison with other spectra at hand.

We may not always come up with unequivocal structure but we can at least be able to narrow down the possibilities to several structures and to indicate the steps required to finish the identification. Let us work through the twenty one spectra comprising this chapter.

8.8.1 Problem 1

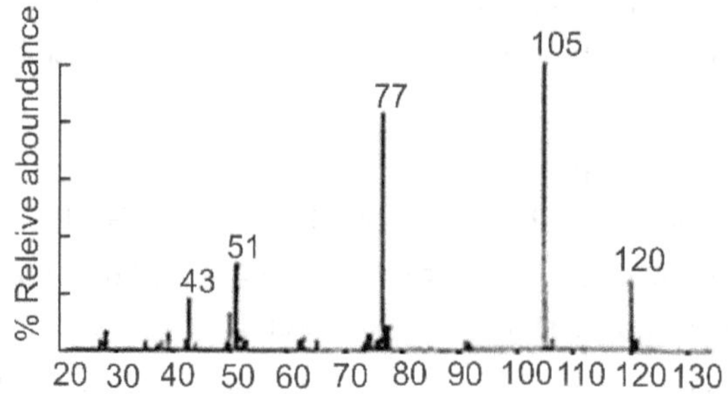

Mass Spectrum

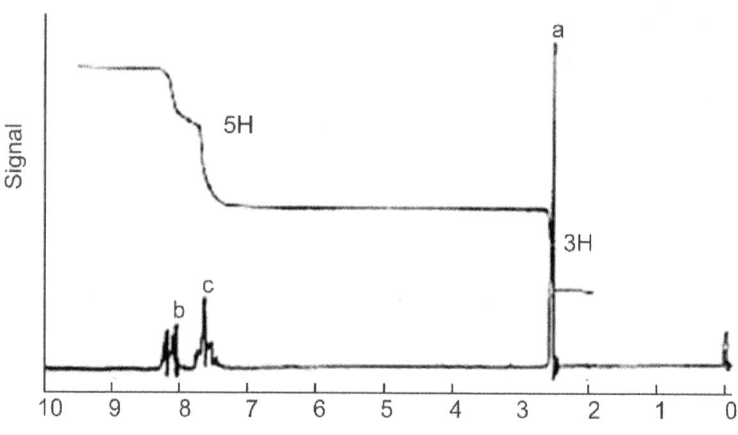

Proton NMR Spectrum

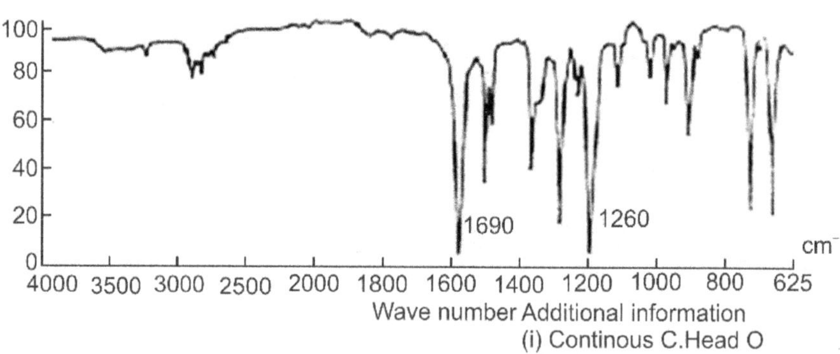

IR Spectrum

25.16 MHZ
0.5 ml : 1.5 ml CDCl$_3$

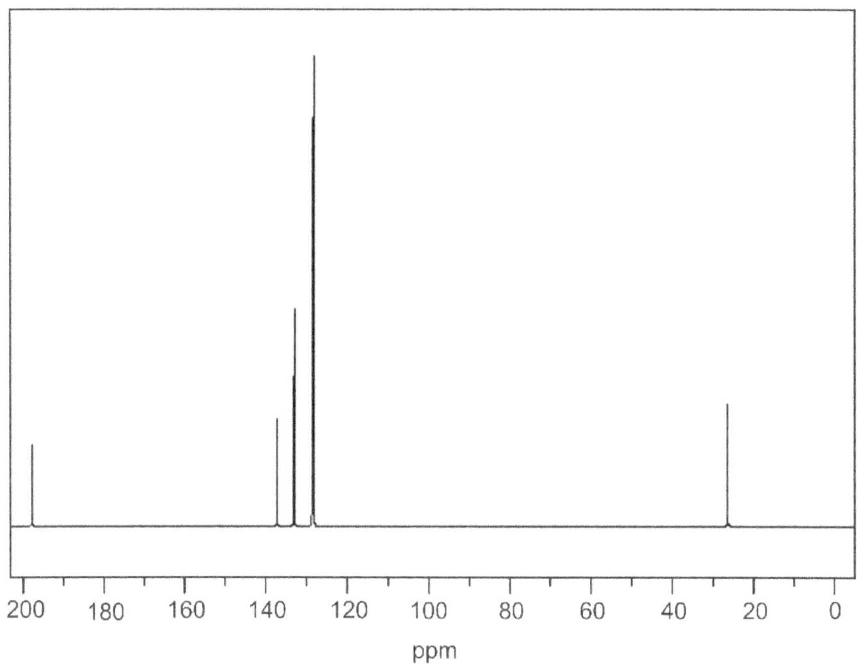

Carbon-13 Spectrum

PPM

8.8.1   Solution

Molecular formula C_8H_8O

We make a note that both the intensity of the parent peak and carbon to hydrogen ratio of the formula indicates aromaticity.

Double bond equivalent (DBE)

Formula DBE =   A - 1/2 B + 1/2 C + 1

Formula C_8H_8O 8 - 4 + 1 = 5

The compound C_8H_8O has 5 elements of unsaturation. Hence one aromatic ring and one double bond may be present.

Possibilities: a) 4 -double bonds OR

b) 1 aromatic ring; 1double bond

Spectral data:

UV λ_{max} (nm)   :   243 & 281

IR v_{max} (cm^{-1})   :   3050 - 1000 (w), 1690 (s),755 (s) & 690 (s)

^{1}H NMR CDCl$_3$ (δ)   :   7.90 - 7.45 & 2.47 (s)

^{13}C NMR (δ)   :   26, 128-137 & 197

MS (m/z)   :   120, 105, 77 & 51

(a) **UV Spectrum:** The presence of conjugated carbonyl group is strongly indicated by weak absorbtion in the longer wave lengh region at λ_{max} 281nm due to n $\rightarrow$ π^* transition.

(b) **Infrared Spectrum:** The presence of conjugated carbonyl is further confirmed by a strong band at 1690 cm^{-1} (non conjugated C=O appears at 1725 cm^{-1}). Here the conjugation is with a phenyl group, hence lower frequency.

Large bands at 755 cm^{-1}(13.5) and 690 cm^{-1} (14.3), suggests a singly substituted benzene ring.

The presence of a benzene ring and a conjugated ketone is established on the basis of lower frequency. Subtractions of singly substituted benzene ring and a carbonyl group from molecular formula gives the following:

$$C_6H_5 \!-\!\!-\!\! \overset{\overset{\textstyle O}{\|}}{C} \!-\!\!-\!\! \quad\quad \!-\!\!-\!\! CH_3$$

(c) **^{1}H NMR Spectrum:** The NMR spectrum is in accordance with the structure. There are five separate ring protons in low field at δ 7.90 – 7.45. The two protons at δ 7.90 *ortho* protons, shifted to down field by the ketone substituent and other three protons at δ 7.45 represent the *meta* and *para* protons of the benzene ring. The three protons of the methyl group as a sharp singlet appeared at δ 2.47.

Now we can write the structural formula. The parent peak is m/z 120. This is the molecular weight.

(d)   **^{13}C NMR Spectrum:** The cluster of four peaks at δ128-137 ppm, suggests the presence of mono-substituted or p-di-substituted benzene. The peaks at δ26 is a typical of CH$_3$group carbon next to carbonyl or benzene and that at signal atδ 197 represents the carbonyl carbon.

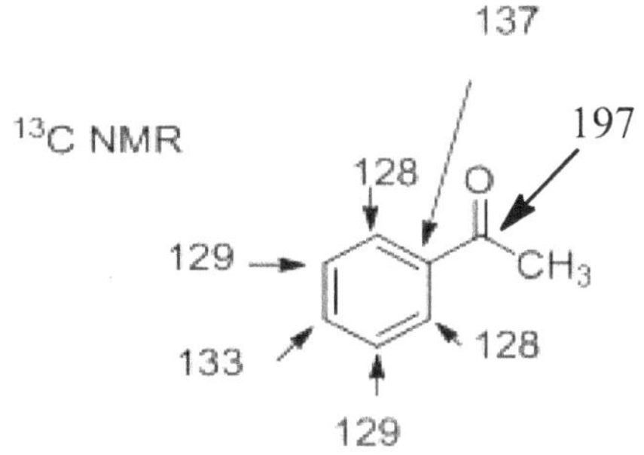

(e) **Mass Spectrum:** We can obtain additional confirmation by examining the mass spectrum and considering the fragmentation pattern. The prominent peak at m/z 105 (M-15) in the mass spectrum is due to the loss of methyl radical leaving a resonance stabilized acylium ion. The second peak, base peak, at m/z 77 is formed by the loss of carbon monoxide. The large peak at m/z 51 represents the loss of neutral molecule acetylene.

These special features and other spectral data discussed above point to the following structure for the compouund.

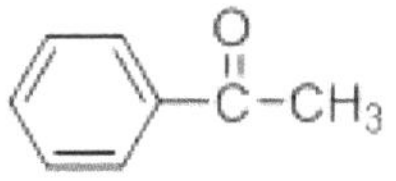

Fragmentation

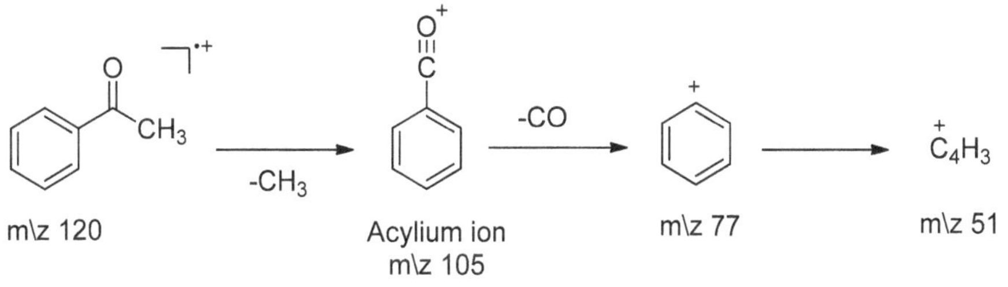

m\z 120

-CH$_3$

Acylium ion
m\z 105

-CO

m\z 77

C$_4$H$_3$

m\z 51

8.8.2 Problem 2

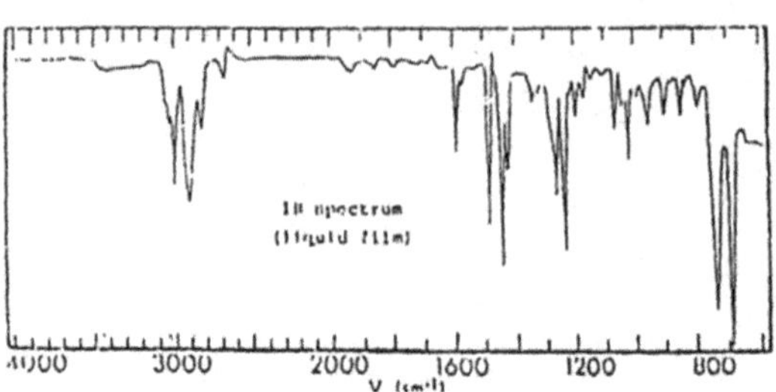

IR Spectrum

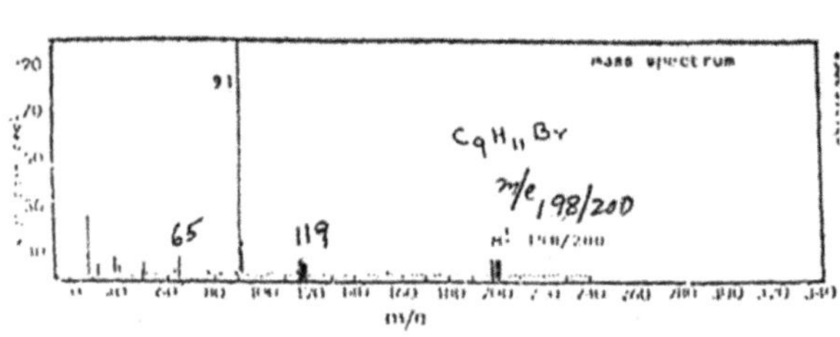

Mass spectrum

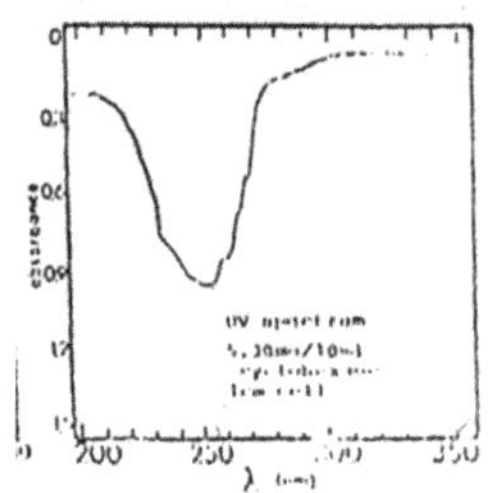

UV Spectrum

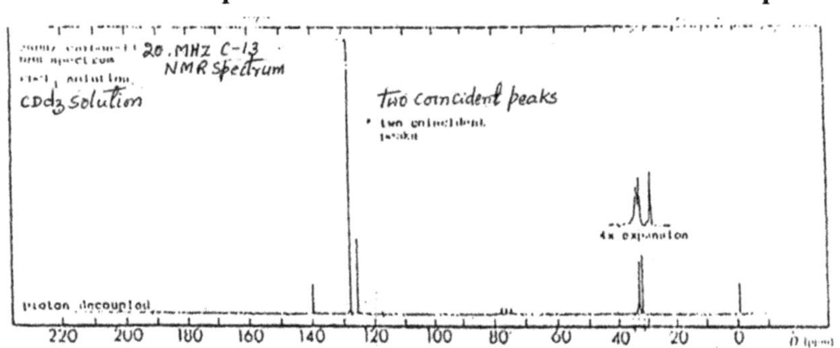

Carbon 13 Spectrum

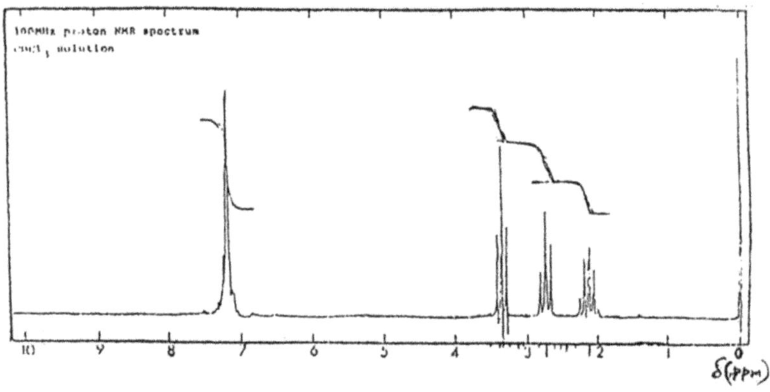

¹H NMR Spectrum

8.8.2 Solution

Molecular formula $C_9H_{11}Br$

The first step is translating the spectral data into a molecular structure. The parent peak is 150; this is the molecular weight. The parent peak is an even number.

Double bond equivalent (DBE)

Formula DBE $=$ A - 1/2 B + 1/2 C + 1

$=$ 9 - 5.5 + 0.5 + 1 = 4

Possibilities are

(a) 4 - double bonds

(b) 1 - aromatic ring, 1- double bond

(c) 1- aromatic ring, 1 – halogen

So possible structure may be

1 – Aromaticring & 1- halogen substitution

Spectral data:

UV λ_{max} (nm)   :   254

IR v_{max} (cm^{-1})   :   2800-3030, 1600, 750 & 697

^{1}H NMR CDCl$_3$ (δ) :   7.30–750 (m), 3.40 (t), 2.0-3.50 (m)

^{13}C NMR (δ)   :   32, 34, 34.3, 126, 128 & 140

MS (m/z)   :   200, (8%) (M+2), 198, 118, 91(base peak), 65, & 39

(a) **UV Spectrum:** Peak at 254 nm – aromatic in nature

(b) **IR Spectrum:** Two large bands at about 750 cm^{-1} (13.35 µ) and 697 cm^{-1}(14.35 µ) suggest singly substituted benzene ring. Peaks at 2800 – 3030 cm^{-1} point to the presence of aromatic C-H stretching and another peak at 1600 cm^{-1} confirms stretching vibration of C = C of the aromatic ring.

(c) ^{1}**HNMR Spectrum:** Aromatic protons appeared at down field δ 7.30 to 7.50 as a multiplet corresponds to benzene protons accounting for 5H on the basis of the integration. The cluster peaks which appeared between δ2.0 to 3.50 integrated for six protons, spin-spin coupling pattern indicated the presence of a –CH$_2$-CH$_2$-CH$_2$-Br chain. The expected triplet of the -CH$_2$Br protons appeared at δ 3.40 ppm.

Another set of peaks in aliphatic region at 2.80 as a triplet for C-3 methylene groups and another methylene group split into a multiplet at 2.20 for C-2 (CH_2-CH_2-CH_2–Br) methylene group protons due to coupling with both sides of the methylene group protons.

(d) ^{13}**CNMR Spectra:** The carbon-13 spectrum shows four peaks in between δ 126 -140ppm which is consistant with phenyl absorption. Note that C-2,6 and C-3,5 are iodentical. There are only four absorption peaks for six carbons of phenyl group because of symmetry. The aliphatic three carbons are at the low field around δ 32, 34 & 34.3 respectively.

(e) Mass Spectrum: The intencity of $M^{+\cdot}$(198) & $(M+2)^{+\cdot}$ (200) peaks are almost equal. This indicates the presence of one bromine atom. The peak at m/z 118 resulting from the loss of 79 mass units from the molecular ion also confirms the presence of one bromine atom in the molecule.

The frequently observed peak at m/z 65 results from elimination of a neutral acetylene molecule from the trophylium ion. The given compound is3-phenylbromopropane

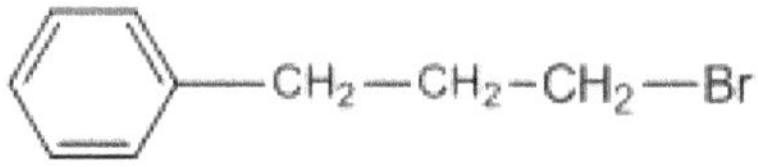

Fragmentation

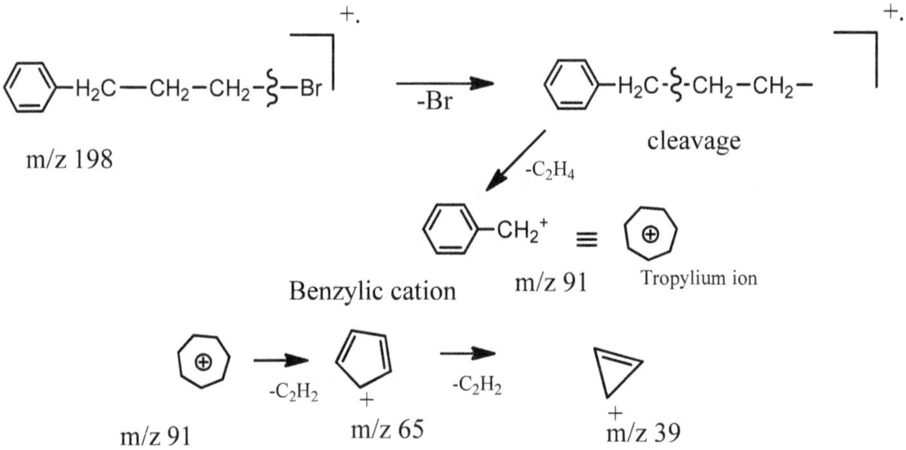

8.8.3 Problem 3

^{13}C NMR Spectra   ppm

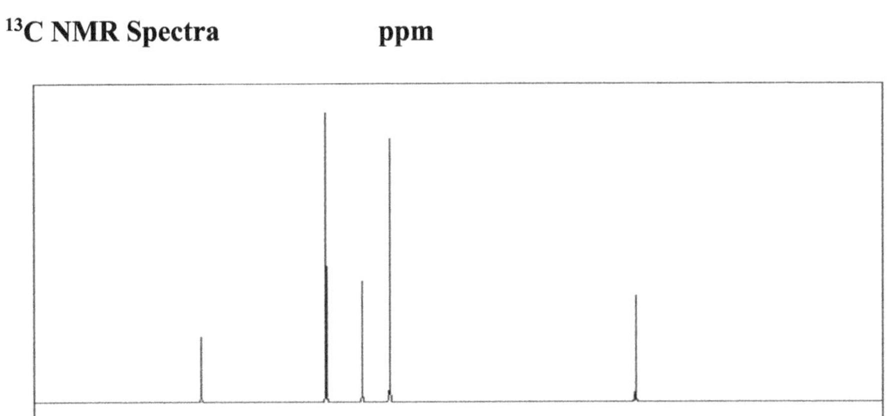

Gas Phase IR:

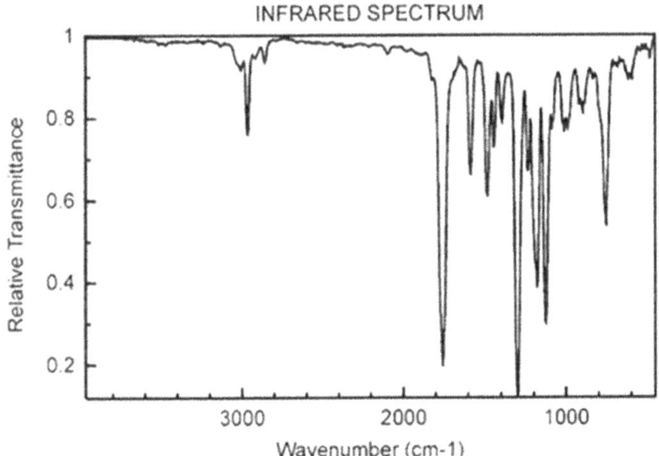

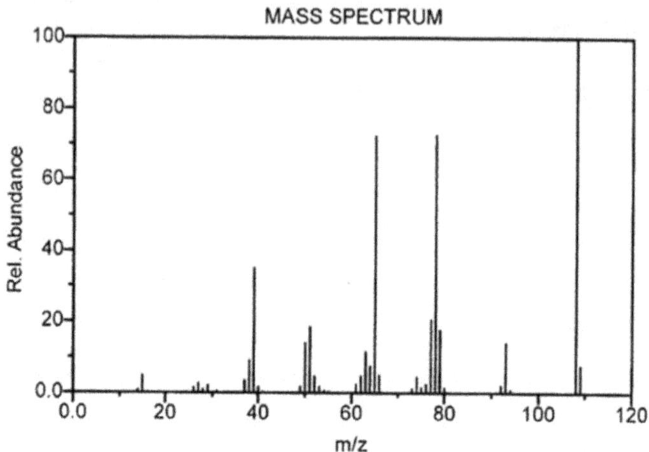

1H NMR Spectra:

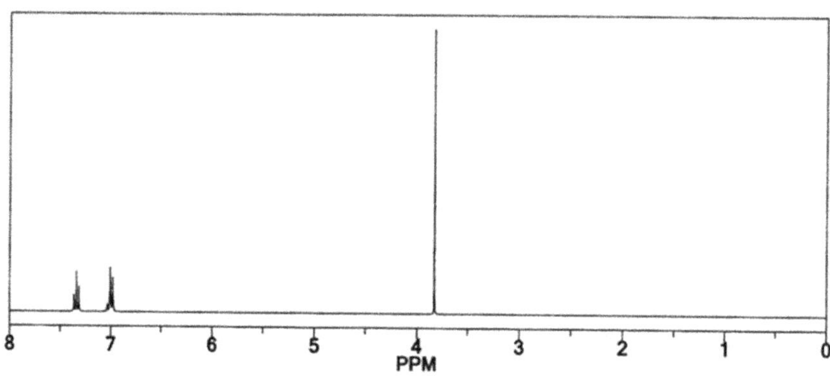

8.8.3 Solution

Molecular Formula C_7H_8O

On the basis of the molecular formula, the carbon to hydrogen ratio indicates that the compound is an aromatic in nature.

If the molcular formula is C_7H_8O, DBE = A - 1/2 B + 1/2 C + 1

$$7 - 4 + 1 = 4$$

The formula shows that four double bonds equivalents are present. It may be a benzene ring.

Spectral Data

UV λ_{max} (nm)   :   222 & 272 nm

IR v_{max} (cm^{-1}) :   3000 – 3050, 1200 – 1275, and 1060 – 1150.

^1H NMR $CDCl_3$ (δ)   :   7.25, 6.8, 6.75 (m), 3.65 (s)

^{13}C NMR (δ)   :   160, 130, 121,114 & 55

MS (m/z)   :   108, 93, 78 & 65 (base peak)

(a) **UV Spectrum:** Ultraviolet spectrum at 272 nm indicates the compound is aromatic.

(b) **IR Spectrum:** When an organic compound contains an oxygen atom then one can expect either O-H or C-O stretching in the region of 3400 - 3550 cm^{-1} (OH) and 1650 -1800 cm^{-1}. But infra-red spectrum showed no peaks in this region. Alternatively, the compound may be ether. Ethers can be recognized by their strong C-O stretching bands at 1200 - 1275 cm^{-1} (aryl ethers) and at 1600 - 1150 cm^{-1} for alkyl ethers. Hence it may be ether.

(c) ^1H **NMR Spectrum:** A multiplet peak appeared in the aromatic region at δ 6.75-7.25 accounts for δ aromatic protons on the basis of integration. Another singlet at δ 3.65 corresponds to $-OCH_3$ (methoxy protons, integrated for 3 protons); *ortho* and *para* protons of anisole are shielded shows upfield signals.

On the basis of the above spectral data, the tentative structure for the compound may be methoxybenzene (Anisole).

(d) ^{13}C **NMR Spectrum:** The carbon-13 spectrum is likely to show the deshielding effect of an electro negative oxgen atom, both on the methyl carbon and the aromatic carbon. The methyl carbon displayes a downfield shift at δ 55 ppm. Among the aromatic carbons, the carbon directly bonded to the oxgen is the most deshielded and appears at δ160 ppm. The *ortho, meta* and *para*-carbon atoms appears at δ 114, 130 and 121ppm respectively.

(e) **Mass spectrum:** The molecular ion peak of anisole is prominent at m/z 108. Primary fission occurs at the bond β to the ring and the loss of CH_3 group gives the ion (M -15), on further loss of C=O to give a base peak at m/z 65. The typical aromatic peaks at m/z 78 & 77 may involve the following fragmentation shown below:

Fragmentation:

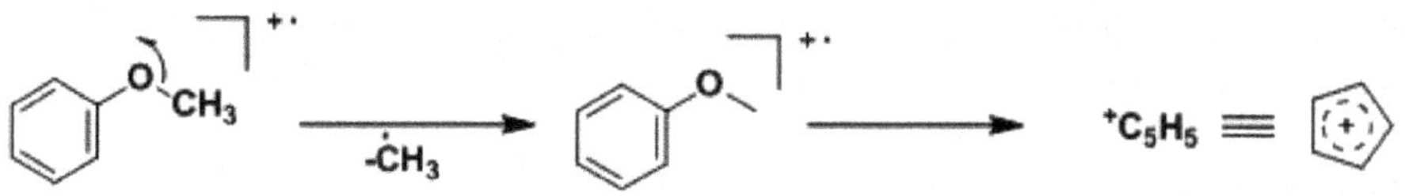

m/z 108 m/z 93 m/z 65

Alternatively

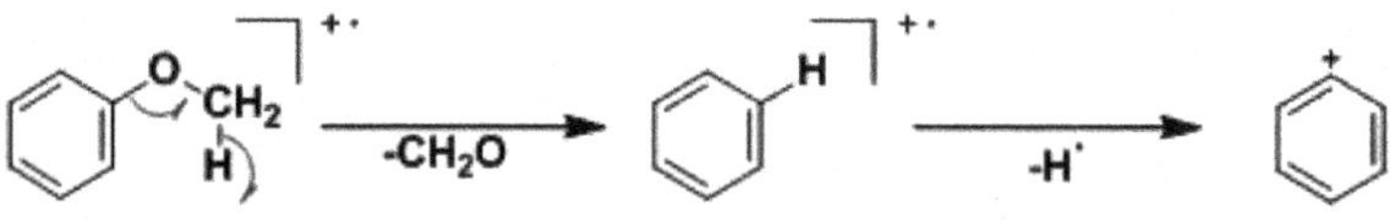

m/z 78 m/z 77

8.8.4 Problem 4

Gas phase IR:

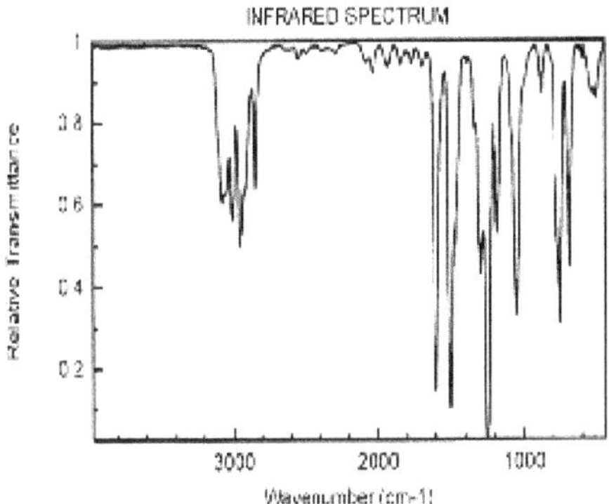

Mass spectra:

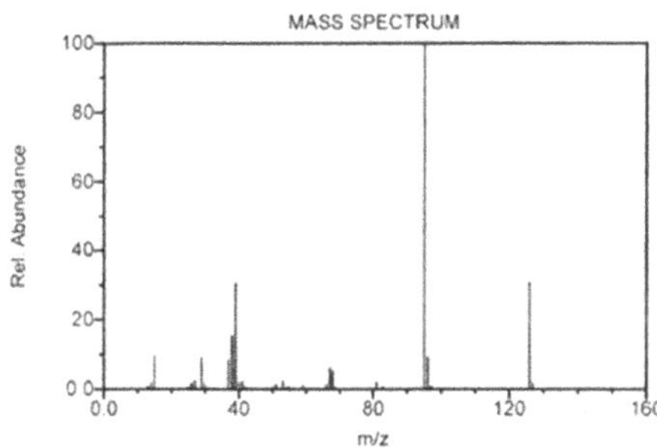

1H NMR Spectra:

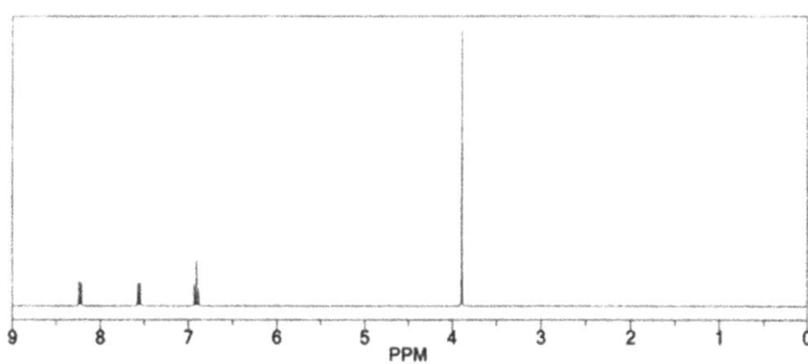

13C NMR Spectra:

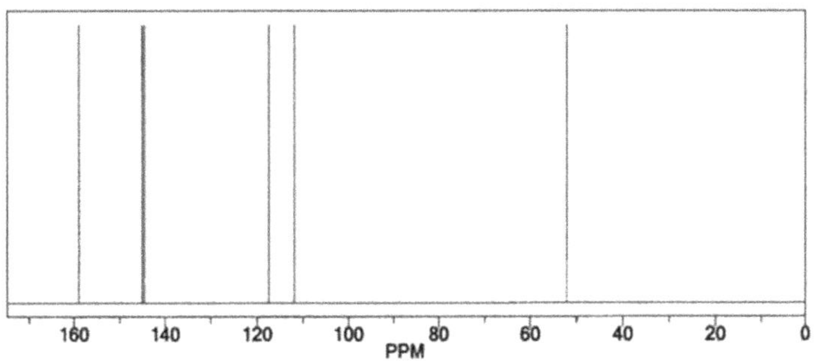

8.8.4 Solution

Molecular Formula $C_6H_6O_3$

The first step is to translate these four spectra into a molecular structure. The formula and general appearance of the infra-red spectrum suggests aromaticity.

Double bond equivalent (DBE) $C_A H_B N_C O_D$ DBE = A - 1/2 B + 1/2 C +1

Formula $C_6H_6O_3$ 6 - 6/2 + 0 + 1 = 4

This means four unsaturated sites.

Spectral data

UV λ_{max} (nm)	:	250
IR ν_{max} (cm^{-1})	:	1730, 1420,&1110
^{1}H NMR CDCl$_3$ (δ)	:	7.55 (1H), 7.10 (1H), 6.85 (1H), & 3.81 (3H).
^{13}C NMR (δ)	:	52, 112, 118, 144, 146 & 159
MS (m/z)	:	126, 95, 68, 53 & 39

(a) **UV Spectrum:** The intense band in the UV spectrum at 250 mu is suggestive of a chromaphore conjugated with an aromatic ring.

(b) **IR Spectrum:** The band at 1730 cm^{-1} in IR spectrum suggests the presence of C=O group. We are dealing with a conjugated chromaphore. Therefore, we can make an option between a ketone or an ester. It may be an ester because the conjugated ketone (Ar-CO-R) bands are usually at lower frequency (conjugated with phenyl or olefinic group and causes absorption in the 1690 – 1660 cm^{-1} region). There are a number of strong bands between 1420 and 1100 cm^{-1} suggesting an ester group, one medium peak at 3100 cm^{-1} and two broad peaks beyond 800 cm^{-1} suggest aromaticity.

(c) ^{1}**H NMR Spectrum:** The NMR spectrum also supports aromatic ring and other aliphatic protons. We see three aromatic protons at low fields and three protons of the methoxy group as a sharp singlet at δ 3.80. The three low field aromatic protons (multiplets) represent at δ 7.55 (1H), 7.10 (1H) and 6.85 (1H) respectively. Each is shifted downfield due to the deshielding effect of the carboxylate substituent. The NMR spectrum is in agreement with the furan and an ester group attached to it. We can write the structure formula:

(d) ^{13}**C NMR** shows six peaks at δ 112, 118,144,146 ppm corresponds to ring carbons, suggests the presence mono substituted aromatic ring. The peak at δ 52 is a typical of mythoxy carbon next to carbonyl and the signal at δ 159 represent the carbonyl carbon.

(e) **Mass Spectrum:** The fragmentation pattern offers loss of 31 amu. The base peak at m/z 95 results from the parent peak m/z 126 by the loss of 31 and this loss practically confirms a methyl ester. The ions at m/z 68, 53 and 39 are additional evidence for the methyl furoate structure.

Fragmentation

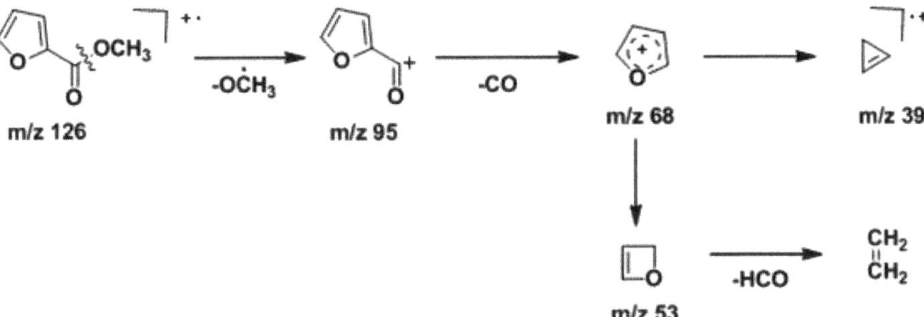

8.8.5 Problem 5

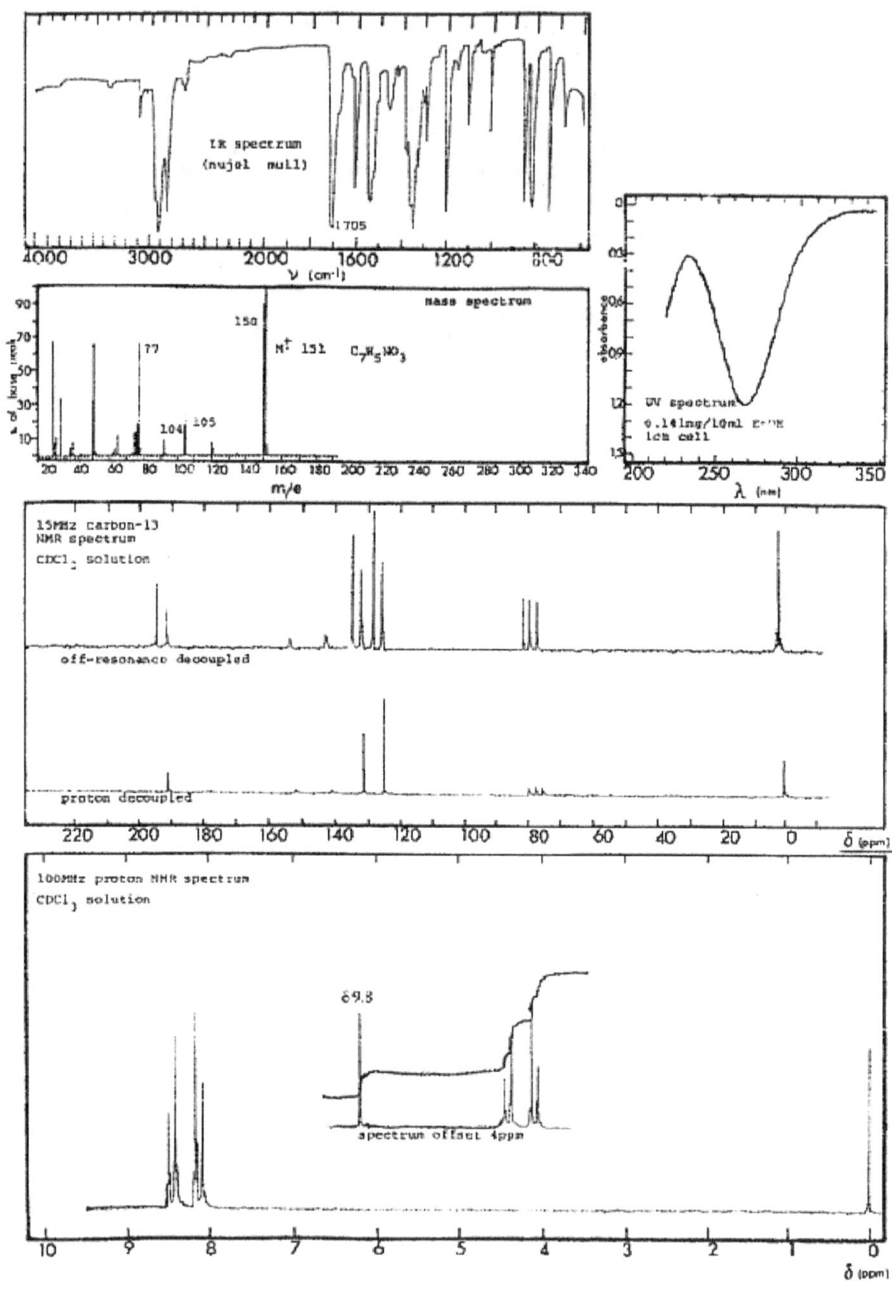

8.8.5 Solution

Molecular Formula $C_7H_5NO_3$

The first step is to establish the structure on the basis of the spectra. The carbon to hydrogen ratio of the molecular formula indicates that the compound may be aromatic.

The parent peak m/z 151 is due to the presence of odd nitrogen present in the compound.

Double bond equivalent (DBE) $C_AH_BN_CO_D$ DBE = A - 1/2 B + 1/2 C +1

Formula $C_7H_5NO_3$ 7 - (5/2 + 0.5) + 1 = 6

The compound has six elements of unsaturation (a benzene ring and two double bonds).

Spectral data

UV λ_{max} (nm)   :   270

IR ν_{max} (cm^{-1})   :   1700, 2830 –2700, 1550 –1500, 1360-1290 & 800

^{1}H NMR CDCl$_3$(δ) :   8.60 – 8.05 & 9.80

^{13}C NMR (δ)   :   124, 130, 140, 150 &190

MS (m/z)   :   151, 150, 122, 105, 77 & 51

  (a) **UV Spectrum:** The absorption at 270 nm is characteristic of a carbonyl group conjugated to the phenyl ring.

  (b) **IR Spectrum:** A strong C=O stretch at 1700 cm^{-1} shows the presence of a conjugated C=O group. This may be a ketone or aldehyde carbonyl. The important difference between an aldehyde and a ketone is that an aldehyde has H-bonded to the carbonyl carbon.This CO-H characteristicstretching absorption bands (doublets) appear between 2830-2700cm^{-1}.Therefore, the carbonyl group (band at 1700 cm^{-1}) is the aldehyde C=O stretch since infra-red spectra also displays aldehydic C-H stretching at 2825 and 2720 cm^{-1}.

     Normal aldehyde C=O stretch is expected to be around 1725 cm^{-1}. Sym and asym stretching of the NO$_2$ absorption appeared near 1550 – 1500cm^{-1} and 1360 – 1290 cm^{-1}. Another absorption band at 800 cm^{-1} indicates it is a disubstituent ring.

  (c) **^{1}H NMR Spectrum:** A down field signal at δ9.80 confirms the presence of an aldehydic proton. The chemical shift value for the 4 protons at 8.05-8.75 is not only in the aromatic region, but their presence as a pair of doublets separated by large *ortho*-coupling constants indicates the presence of two different *para* - substituents.

(d) ^{13}C NMR Spectrum: The ^{13}C NMR spectrum shows five peaks, two peaks at δ 124 &130 ppm suggests the presence of p-disubstited benzene ring and peak at δ 140 & 150 ppm consistent with the phenyl absorbtion. Fifth peak at δ 190 ppm corresponds to aldehyde carbon. The comp[ound can now be formulated as:

$$O_2N-\!\!\!\bigcirc\!\!\!-CHO$$

p-nitrobenzaldehyde

(a) Mass Spectrum: The mass spectrum also confirms the structure and the base peak at m/z 150 is due to the loss of H radical leaving the resonance stabilized acylium ion. Loss of nitro group is typical of nitro aromatics (m/z 105). The phenyl ion peak at m/z 77 (M-28) followed by another larger peak at m/z 51 is the $C_4H_3{}^+$ fragment.

Fragmentation:

8.8.6 Problem 6

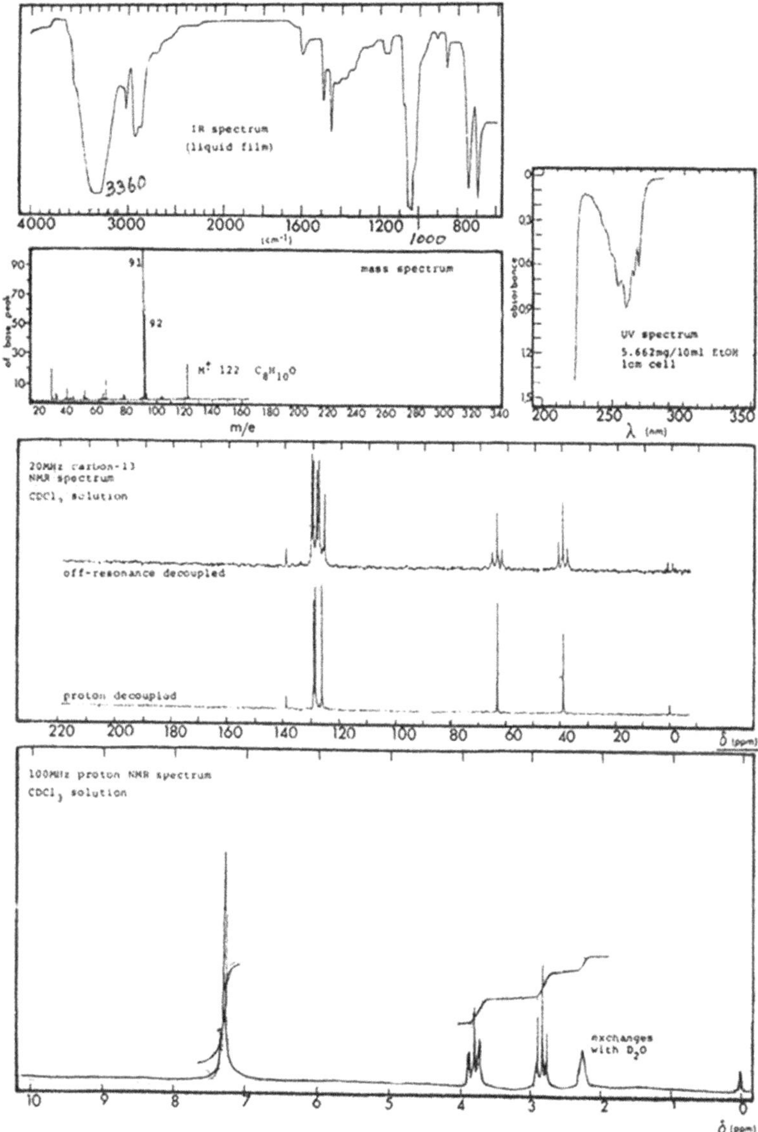

8.8.6 Solution

Molecular Formula: $C_8H_{10}O$

From the molecular formula, the IR and NMR spectra suggest aromaticity substituted with saturated aliphatic protons.

Double bond equivalent (DBE) $C_A H_B N_C O_D$ DBE = A - 1/2 B + 1/2 C +1

Formula $C_8H_{10}O$ DBE = 8 - (10/2 + 0) + 1 = 4

The compound is having four unsaturated sites and it may contain an aromatic ring.

Spectral data

UV λ_{max} (nm)   :   285

IR v_{max} (cm^{-1})   :   3360 (b),1600 (s), 1500 (m), 1050 (s), 750 & 690(s)

^{1}H NMR CDCl$_3$(δ)   :   2.25 (b), 2.85 (t), 3.80 (t) & 7.25 (m).

^{13}C NMR (δ)   :   39, 62, 127 – 139

MS (m/z)   :   122, 92, 91 & 31

(a) **UV Spectrum:**   The band at 285 nm may be due to conjugation or aromatic ring substituted by a group causing bathochromic shift.

(b) **IR Spectrum:** The two bands at 1600 (medium) and 1497 cm^{-1} (medium) are due to stretching vibrations of the C=C bonds of the aromatic ring. Also, mono substituted benzene displays two strong bands at 750 cm^{-1} and 690 cm^{-1}. A strong absorption at 3360 cm^{-1} shows the presence of a hydroxyl group, indicative of intermolecular hydrogen bonding. Alcohols also give strong and broad band due to C-O stretching in between 1000 – 1200 cm^{-1} region. The exact frequency of this band is often used to differentiate between primary, secondary and tertiary alcohols (1^0 alcohol 1050 cm^{-1}, 2^0 alcohols 1100 cm^{-1}, 3^0 alcohols 1040 cm^{-1} and phenols 1230 cm^{-1}), respectively. The band at 1040 cm^{-1} is a characteristic of primary benzyl alcohol.

(c) ^{1}**H NMR Spectrum:** The presence of OH group resonance shifts to high field (i.e., δ 2-4 ppm). The NMR spectra show four sets of signals. A one proton singlet at δ 2.24 (broad) is due to intramolecular hydrogen-bonding, indicates the presence of O-H group (exchange with D$_2$O confirms the O-H proton).

The NMR spectrum shows three signals. The signals are formed in the proton ratio 5:2:2. The spectrum shows two triplets at δ 2.85 and δ 3.80 of equal areas with the same spacing. It is reasonable in writing – CH_2-CH_2-OH and placing the more deshielded methylene group (i.e., -CH_2-CH_2-OH) on the oxygen of the hydroxy group. The benzenoid protons appeared as a multiplet in the region δ 7.10-7.30.

(d) ^{13}C **NMR Spectrum:** The cluster of four peaks at δ127-139 ppm, suggests the presence of monosubstituted benzene ring. The remaining aliphatic carbons, appeared in the upfield in the region of δ 39-62 ppm.

(e) **Mass Spectrum:** Primary alcohols show a peak at m/z 31 due to the loss of CH_2OH. If the aromatic ring is substituted by an alkyl group, characteristic peaks at m/z 91 and m/z 65 are obtained, which are typical of alkyl benzene due to tropylium ion and cyclopentadienyl cation. Also a peak at m/z 92 is due to the rearrangement of benzyl cation. The structure which fits into the data is beta Phenylethanol.

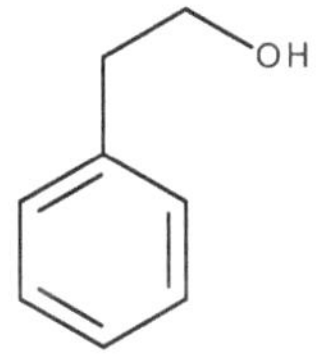

Fragmentation

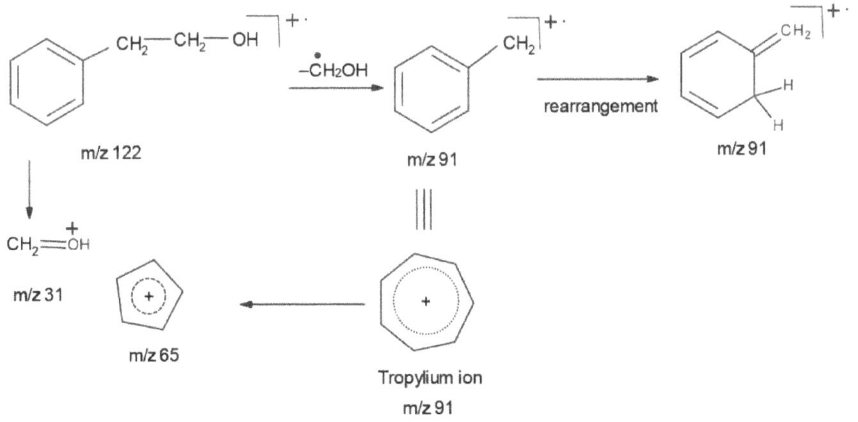

8.8.7 Problem 7

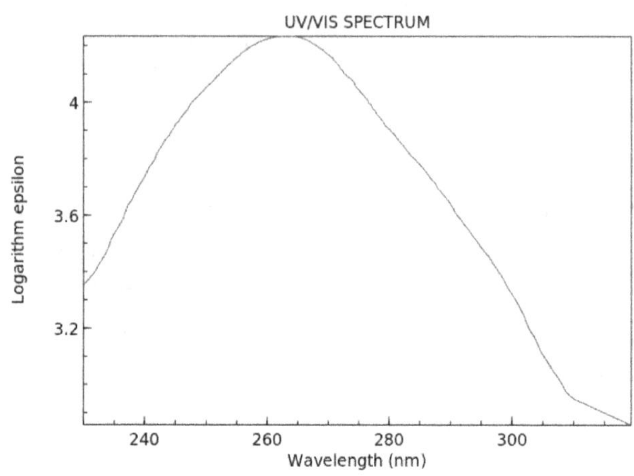

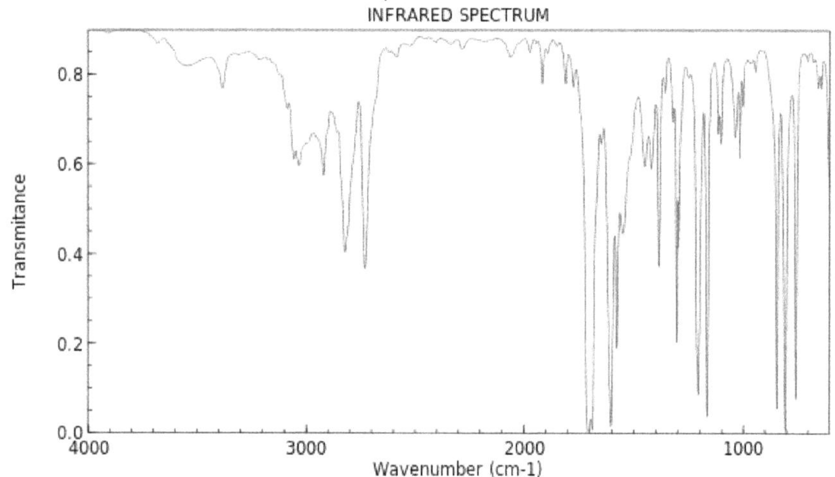

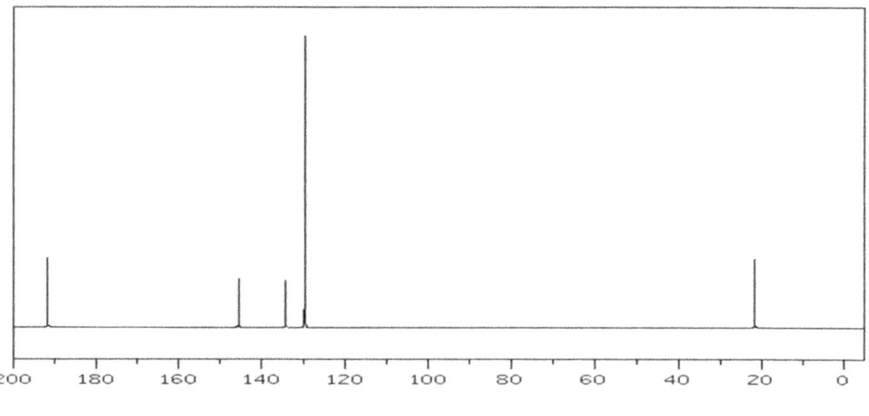

^{13}C NMR (ppm)

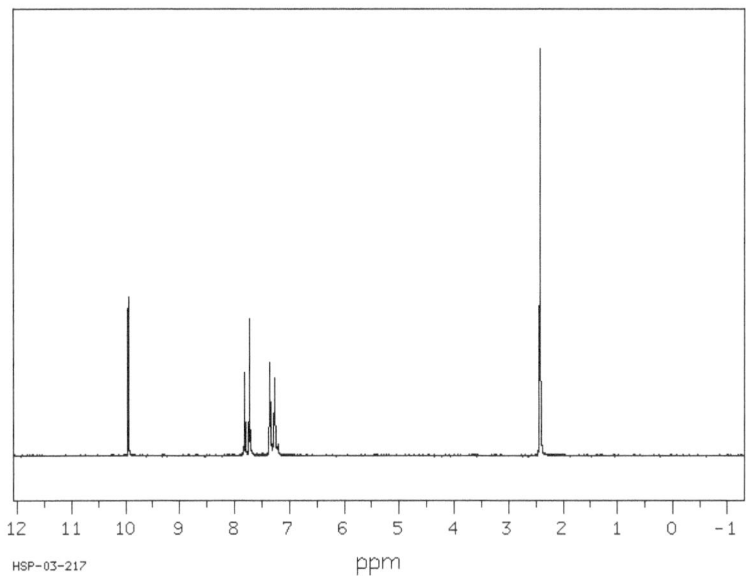

¹H NMR

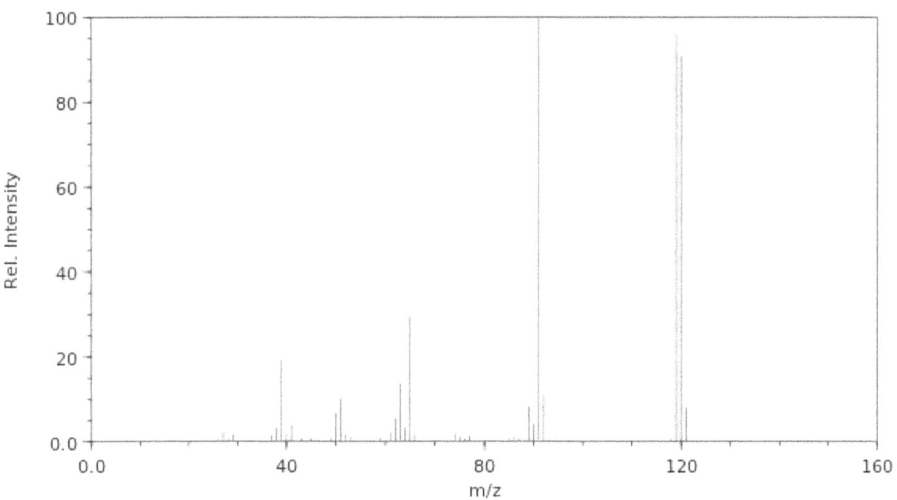

Mass Spectrum

8.8.7 Solution

Molecular Formula C_8H_8O

The first step is to establish the structure on the basis of the spectra. We can make a note that both the intensity and of the parent peak and carbon to hydrogen ratio of the formula indicates aromaticity.

Double bond equivalent (DBE) $C_AH_BN_CO_D$ DBE = A - 1/2 B + 1/2 C +1

Formula C_8H_8O DBE = 8 - 8/2 + 0 + 1 = 5

The compound has five elements of unsaturation. Hence, it may contain one aromatic ring and a double bond.

Spectral data

UV λ_{max} (nm)   :   270

IR v_{max} (cm^{-1}) :   2922, 2733, 1703, & 800 cm^{-1} (one peak)

^{1}H NMR $CDCl_3(\delta)$:   2.4(3H, singlet), 7.3 -7.8 (4H, as a pair of doublets J=8 Hz) & 9.9   (1H, singlet)

^{13}C NMR (δ) :   22, 145, 129 – 145 & 192

MS (m/z)   :   120, 119 & 91

(a) **UV Spectrum:** The absorption at 270 nm is characteristic of a carbonyl group conjugated to the phenyl ring. It refers that there is an aromatic ring with a substitution.

(b) **IR Spectrum:** A strong C=O stretch at 1703 cm^{-1} shows the presence of a conjugated carbonyl group. This –CO-H characteristic stretching absorption bands (doublets) appear between 2922 and 2733 cm^{-1}. In the spectrum, note two bands in the region, suggesting that the compound is indeed an aldehyde.

Normal aldehyde carbonyl stretch is around 1725 cm^{-1} (saturated carbonyl), but the conjugated C=0 stretch is expected to be around 1703 cm^{-1}.

Another absorption band at 810 cm^{-1} indicates it is a di-substituted aromatic ring.

(c) **^{1}H NMR Spectrum**: Down field ^{1}H NMR signal at δ 9.9 (s) confirms the presence of an aldehyde.

The chemical shift value for the four protons at δ 7.3-7.8 is not only in the aromatic region, but also their presence as a pair of doublets separated by a large *ortho*-coupling constant shows the presence of *para*-disubstituted benzene, as the two substituents have different electronegativities.

On the basis of the data available, we can write part of the structure

$$C_6H_5\,CHO + - CH_2 = C_8H_8O$$

The presence of the $-CH_3$ group is evident by the presence of a peak in NMR at δ 2.4 (3H) singlet at *para* to the aldehyde group on the basis of NMR spectra.

(d) **^{13}CNMR:** The carbon 13 spectrum shows 4 peaks two peaks around δ 129 – 145 ppm, which is consistent with phenyl absorption. The peak at δ 22 ppm, is a typical of CH_3 group carbon and the signal δ 192 ppm, represents the alddehyde carbonyl carbon.

(e) **Mass Spectrum:** The mass spectrum also supports the structure. Aromatic aldehydes are characterized by an $(M^{+\cdot} -1)$ peak at m/z 119 is due to the loss of proton radical leaving the resonance stabilized acylium ion.The second major peak at m/z 91 is tropylium cation formed by the loss of carbon monoxide(m-28).

The compound is thus *p*-tolualdehyde (4-methyl benzaldehyde).

$$H_3C-\langle\!\!\bigcirc\!\!\rangle-CHO$$

Fragmentaion:

8.8.8 Problem 8

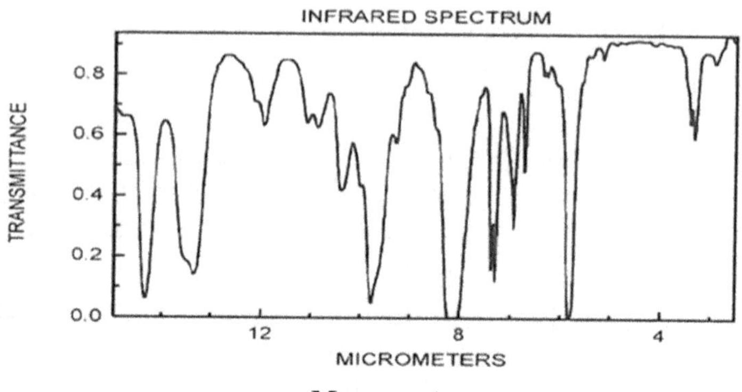

Mass spectrum

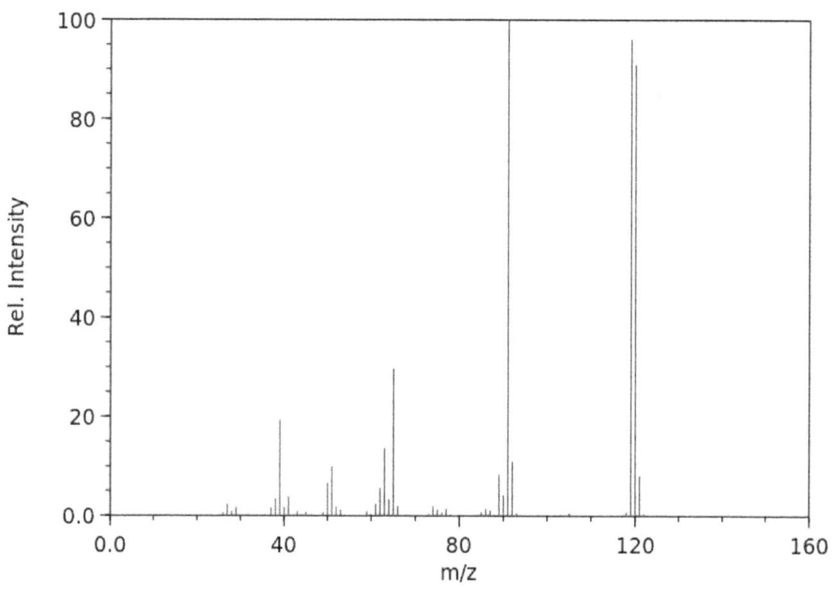

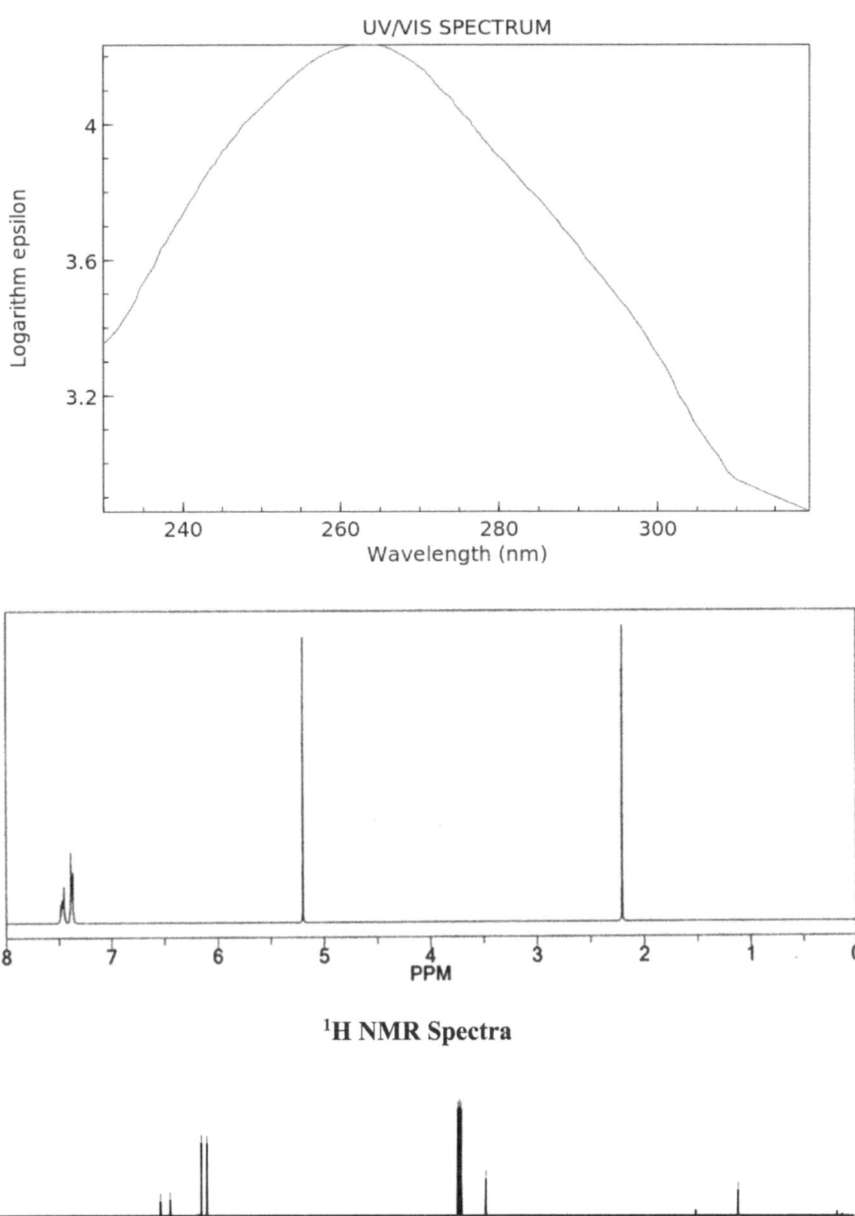

^{1}H NMR Spectra

^{13}C NMR Spectra

8.8.8 Solution

Molecular Formula $C_9H_{10}O_2$

The first step is to translate these spectra into a molecular structure.

It is observed that both the intensity of the parent peak and the C to H ratio of the molecular formula indicates aromaticity.

Double bond equivalent (DBE) $C_AH_BN_CO_D$ DBE $= A - 1/2\,B + 1/2\,C + 1$

Formula $C_9H_{10}O_2$ $9 - (10/2 + 0) + 1 = 5$

The formula shows that five double bond equivalents are present.

Spectral data

UV λ_{max} (nm)	:	268, 264
IR v_{max} (cm^{-1})	:	3050 – 2975, 1745, 1225, 749 & 697
^{1}H NMRCDCl$_3$(δ)	:	7.22 (m), 5.00 (s), & 1.96 (s)
^{13}C NMR (δ)	:	21, 66, 128, 128.6, 136 & 170
MS (m/z)	:	91 (base peak), 77 & 43.

(a) **UV Spectrum:** An ultraviolet band at 268 & 264 nm is due to $\pi - \pi^*$ transition due to aromatic ring.

(b) **IR Spectrum:** In the infra-red spectrum, the strong band at 1745 cm^{-1} is due to C=O stretching, suggestive of an acetate group. One has to look for confirmation in the C–O–C stretching region (1300–1000 cm^{-1}), a strong band at about 1225 cm^{-1} characteristic of acetate (CH$_3$-COO-).

Strong bands at 700 - 750cm^{-1} are indicting mono substituded benzene.

Furthermore, we note from the position of the carbonyl band that the C=O moiety is not conjugated with the ring (conjugation lower the frequency). This is confirmed by the wavelength and intensity of the UV absorption peaks.

The presence of the benzene ring and an acetate group is established.

Subtraction of singly substituted benzene ring and an acetate group from the molecular formula gives the following:

$$C_6H_5 + CH_3COO- = C_8H_8O_2$$

The missing link is the -CH$_2$ group

(c) **^{1}H NMR Spectrum:** NMR spectrum produces almost conclusive evidence for the final structure.

From the NMR data, we can see three types of protons. The spectrum shows a five proton singlet at δ 7.22 which must be due to aromatic protons. The singlet nature of this peak further shows that the compound is a monosubstituted benzene and the substituent has about the same electronegativity as carbon. A singlet in the PMR spectrum at δ 1.96 shows the presence of a methyl group on carbonyl carbon (CH_3-C=O) of two protons at δ 5.00 represents the methylene group between a phenyl and an ester group. The singlet of three protons at δ 1.96 represents the methyl group. The structure represents by these spectra is benzyl acetate.

(d) **^{13}C NMR Spectrum:** In the carbon 13 spectrum one sees only seven peaks, although the compound has nine carbon atoms. A peak at δ21 (CH_3 C0) followed by a peak at δ 66 ppm clearly reveals the resonance

for the two carbon atoms of the substituent of the phenyl ring. Phenyl C-2,6 and C-3,5 are identical and peaks appeared at δ 128, 128.6 & 136. The carbonyl carbon being the most deshielded comes atδ 170 ppm. The CMR data further supports the given compound may be benzyl acetate.

(e) **Mass Spectrum:** The mass spectrum of benzyl acetate is fairly typical of that of mono-substituted benzene, in that the molecular ion is fairly abundant and relatively few fragment ions are observed.

Benzyl acetate eliminates the neutral molecule ketene to form the base peak m/z 108. The m/z 43 peak ($CH_3 - C = O^+$) is a prominent one. It has been shown that most cases the ion of mass 91 is a tropylium rather than a benzylic cation. The frequently observed peak at m/z 65 results from the elimination of neutral acetylene molecule from the tropylium ion.

The fragment pattern may be rationalized as indicated below:

Fragmentation:

m/z 108
base peak

-OH

m/z 65   m/z 91

8.8.9 Problem 9

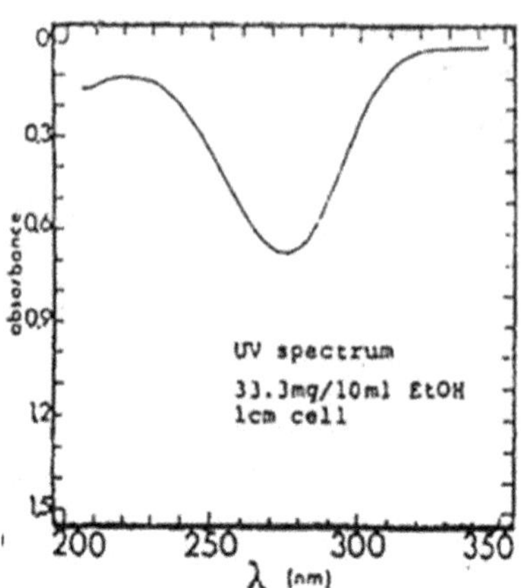

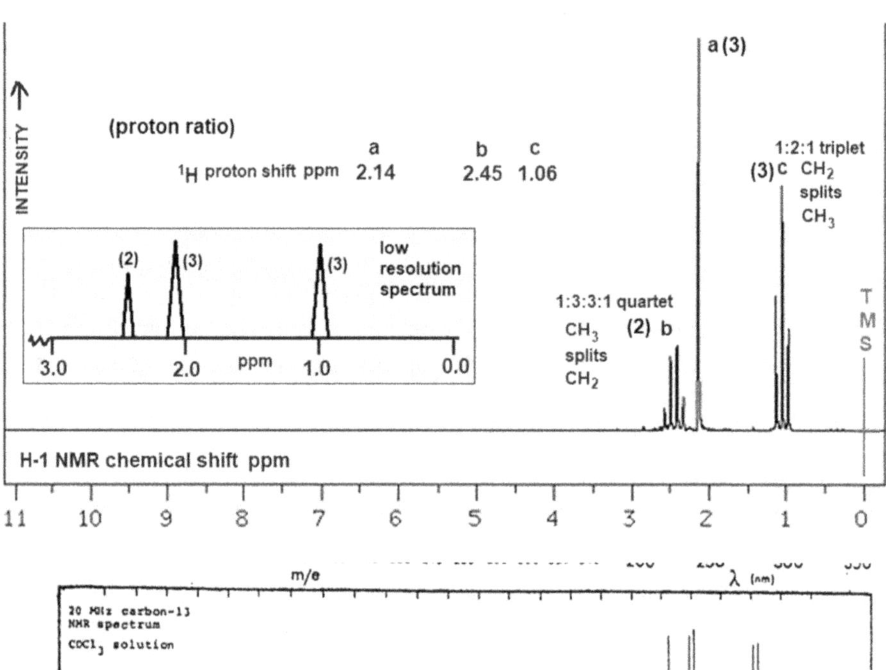

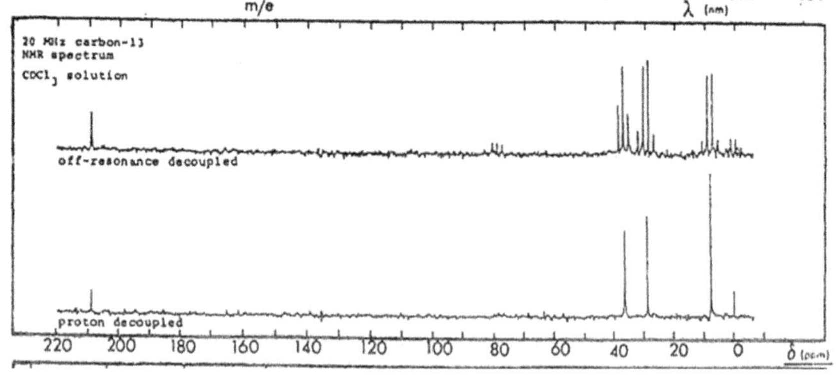

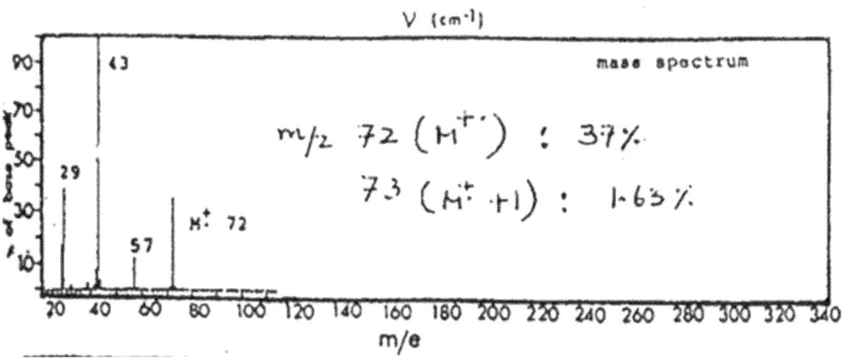

m/z 72 (M⁺·) : 37%

73 (M⁺ +1) : 1·63%

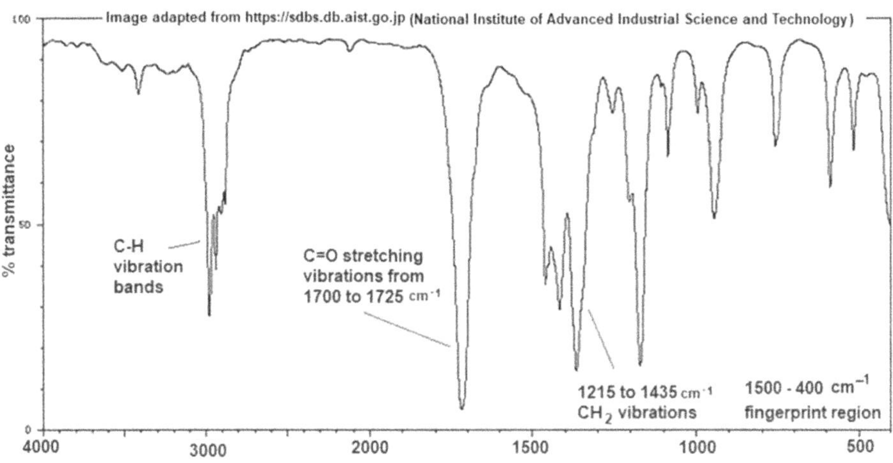

8.8.9 Solution

Molecular Formula C_4H_8O

On the basis of the molecular formula the carbon to hydrogen ratio indicates the compound is aliphatic in nature.

Double bond equivalent (DBE) Formula $C_AH_BN_CO_D$ DBE $= A - 1/2\ B + 1/2\ C + 1$

Formula $C_4\ H_8\ O$ 4 - 8/2 + 0 + 1 = 1

The molecular formula has index of hydrogen deficiency of one and therefore one double bond equivalent.

Spectral data

UV λ_{max} (nm)	:	295
IRν_{max} (cm^{-1})	:	1715 (strong)
^{1}H NMR CDCl$_3$ (δ)	:	2.4 (q), 2.09 (s), 1.04 (t)
^{13}C NMR (δ)	:	209, 37, 29, 8.0
MS (m/z)	:	72, 57 (weak), 43 (base peak), 29

Interpretation:

(a) **UV & IR Spectrum**: The UV spectrum has λ_{max} 295 nm, the absorption is therefore a typical of the n – absorption of a saturated ketone or aldehyde. One should not place too much faith in this information alone, however, in this case, the presence of a carbonyl group in the IR spectrum with its very strong band at 1715 cm^{-1}. Furthermore, it is clearly a ketone and not an aldehyde, since the latter (aldehyde) would have (a) a carbonyl band at slightly higher frequency, for example:-

O
R — C — R
(1715 cm^l)
(1700 cm^l)

Further absorption between 2900 and 2700 cm^{-1} (c) absorption in the δ 9-10 region of the 1H NMR spectrum. With only one double bond equivalent one heteroatom to account for, we have now found them both in the ketone carbonyl group.

There can be no further functionality, so all we have to do is arrange the carbon skeleton.

(b) **^{1}H NMR Spectrum:** We now look at the NMR spectrum, where we can see a signal at δ 2.4 (quartet), 2.09 (singlet), and 1.04 (triplet). The integration accounts for the presence total of eight hydrogen atoms in the molecule. The three-proton signal at δ 2.09 will be a methyl group; since it is a singlet, it must be attached to an atom having no protons on it. Its position of resonance is compatible with its being a CH_3-CO– group. The 1:3:3:1 quartet and the 1:2:1 triplet integrating for two and three protons respectively is typical of an ethyl group. Since the former is only a quartet, the CH_2 group must be attached to an atom having no hydrogen atoms on it. Clearly, it is joined directly to the carbonyl carbon, and the whole structure is, of course, methyl ethyl ketone -CH_3–CO–CH_2–CH_3.

(c) **^{13}C NMR Spectrum:** The ^{13}C NMR shows that all four carbon atoms are different; one at δ 209 is clearly the carbonyl carbon, being at a very low field; the peak at δ 8.0 is the isolated methyl group next to the ketone, the peak at δ 29 is the methylene group and the other peak at δ 37 is the methyl group carbon.

(d) **Mass spectrum:** The mass spectrum too, is definitive: The base peak at m/z 43 by the loss of ethyl radical and m/z 57 by the loss of methyl group from the molecular ion m/z 72) and hence detects the methyl and ethyl groups.

On the basis of spectral data the unknown compound is 2- butanone.

Fragmentation

8.8.10   Problem 10

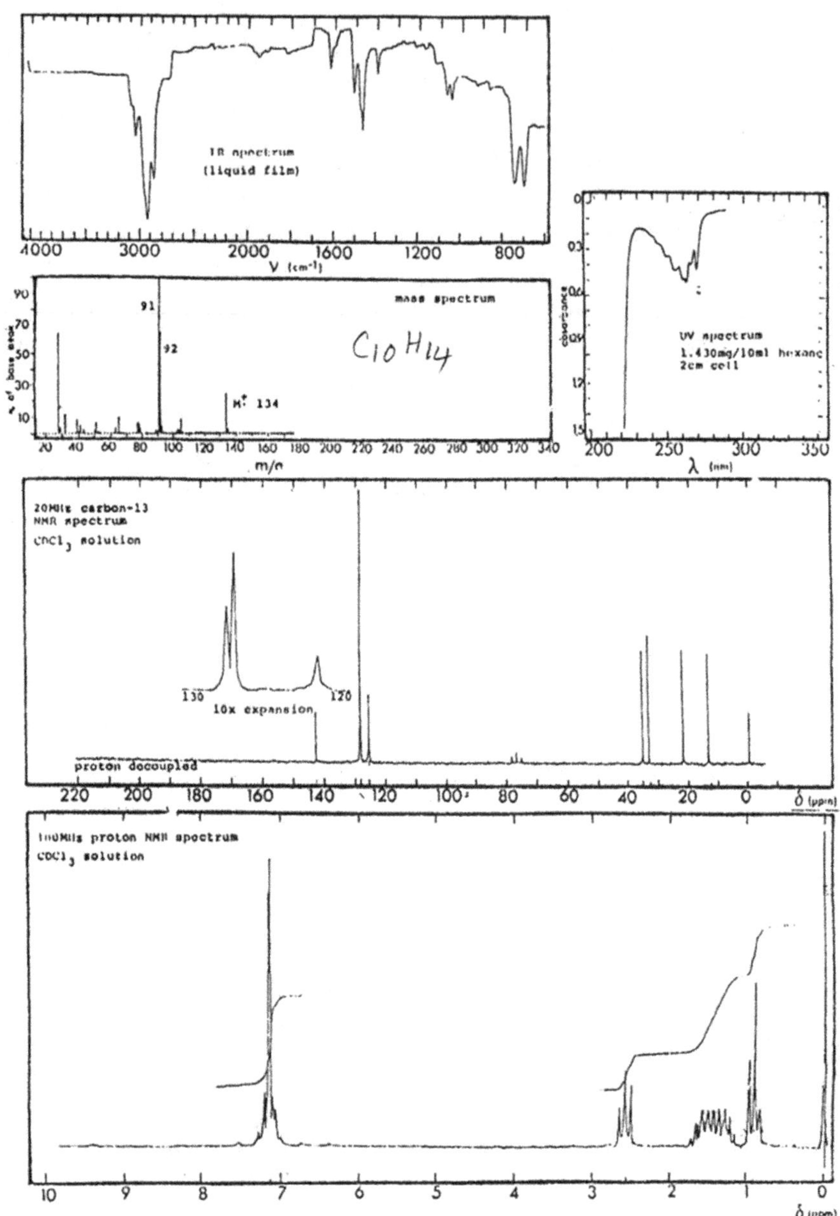

8.8.10 Solution

Molecular Formula $C_{10}H_{14}$

On the basis of molecular formula, the C to H ratio of the formula indicates that the given compound may be an aromatic. The first step is to establish the structure on the basis of spectral data.

Double bond equivalent (DBE) $C_A H_B N_C O_D$ DBE = A - 1/2 B + 1/2 C +1

Formula $C_{10}H_{14}$ $= 10 - 14/2 + 0 + 1 = 4$

Therefore, three double bonds and one ring present in the molecule.

Spectral data

UV λ_{max} (nm)	:	256, 275
IR v_{max} (cm^{-1})	:	1600, 1480 & 800
^{1}H NMRCDCl$_3$(δ)	:	7.1, 2.5, 1.8, 1.2 & 0.9
^{13}C NMR CDCl$_3$(δ)	:	13, 22, 33, 35, 125, 128 & 142 ppm
MS (m/z)	:	134, 91 & 65

(a) **UV Spectrum:** Benzene displays one absorption band at λ_{max} 256 but substitution of alkyl group on the benzene ring produces a bathochromic shift around λ_{max} 275, hence there may be a substitution of alkyl group.

(b) **Infrared Spectrum:** The formula and general appearance of the IR spectrum suggest aromaticity. We note that two strong peaks between 700 to 750 cm^{-1} suggest a mono-substituted benzene ring and two peaks at 1600 and 1480 cm^{-1} suggest that C = C stretching frequency in the benzene ring.

(c) **^{1}H NMR Spectrum:** On the basis of ^{1}H spectra, it is a mono-substituted benzene. As the substituents are neither shielding nor deshielding (eg., alkyl groups). Therefore, all the benzenoid hydrogens are expected to have the same chemical shift and will appear as a multiplet in the aromatic region around δ 7.1ppm.

The only well defined signal is the expected triplet of the $C_6 H_5 - CH_2$- group at δ 2.5 ppm. The terminal methyl group gives a highly distorted triplet at δ 0.9 ppm.

Finally, the protons of the aliphatic $-CH_2$- groups at C-2 & C-3 appear as a broad multiplet between δ 1.2 & 1.8 ppm.

Since the difference in chemical shifts is so small that the splitting patterns overlap.

(d) **^{13}C NMR Spectrum:** The spectrum shows peaks appeared in between δ 127 -141ppmfor ring carbons, as shown by a group of three absorptions in the region. The remaining carbons of aliphatic four appeared in the upfield in the region δ 16–34ppm.

(e) **Mass Spectrum:** The mass spectrum shows characteristic peak at m/z 91 and 65 typical of alkyl benzenes (i.e., the tropylium ion and cyclopentadienyl cation).

In alkyl benzene the dominant fragmentation is at the benzylic bond, since the benzylic cation that is formed is resonance stabilized. The base peak is at m/z 91.

β – Cleavage in the side chain to give benzylic cation which rearranges to more stable tropylium ion m/z 91 by the loss of hydrogen atom.

Finally, the given unknown on the basis of combined spectral data is n-butyl benzene.

Fragmentation

8.8.11   Problem 11

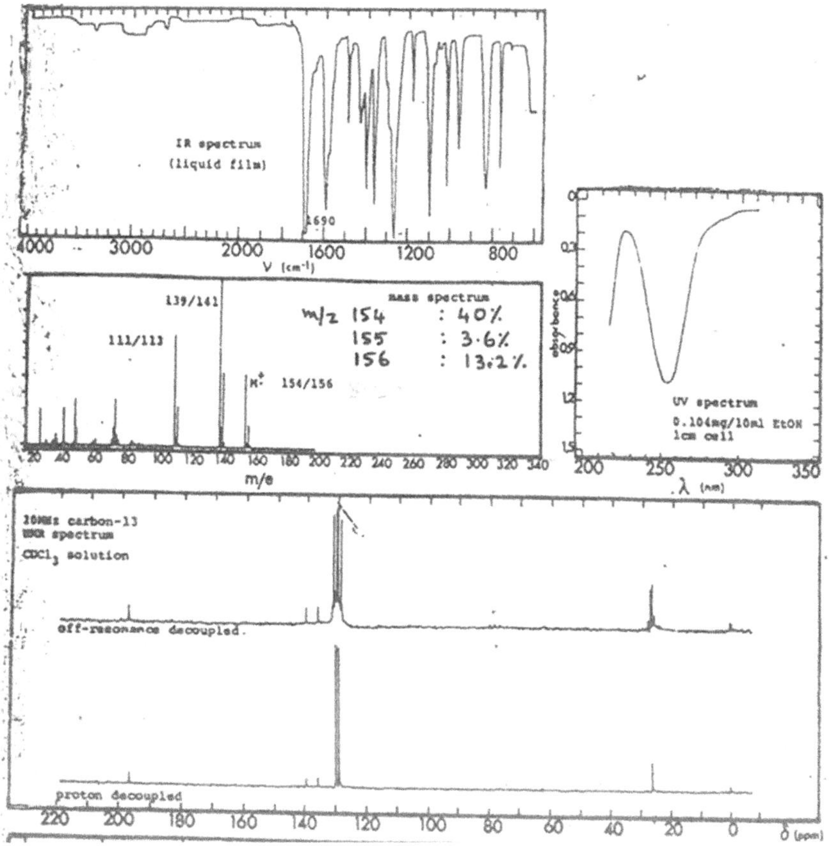

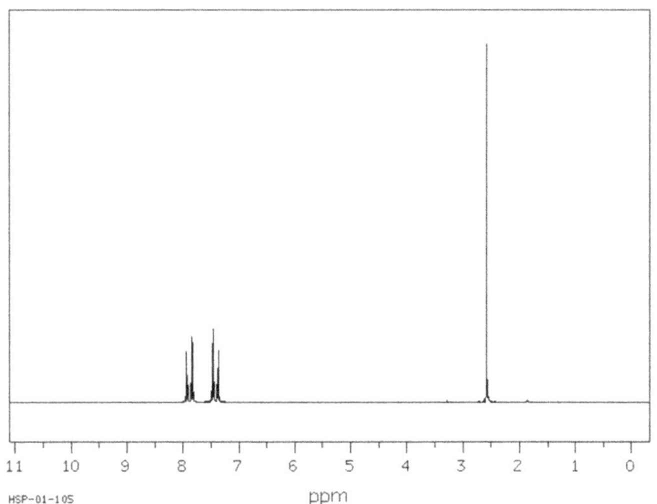

¹H NMR Spectrum

8.8.11 Solution

Molecular Formula C_8H_7ClO

Molecular weight $M^{+\cdot} = 154$

On the basis of the molecular formula, the carbon to hydrogen ratio of the formula indicates that the compound may be aromatic in nature.

The molecular ion at m/z 154 & 156 due to $P + 2$ specify the presence of one halogen atom.

Double bond equivalent (DBE): $C_A H_B X_C O_D$ DBE = A - 1/2 B + 1/2 X + 1

Molecular formula $C_8H_7ClO = 8 - 7/2 + 1/2 + 1 = 5$

Possibilities are

  (a) one ring

  (b) 4-double bonds in the system

Spectral data:

UV λ_{max} (nm)	:	250
IR v_{max} (cm^{-1})	:	2820, 2720, 1690, 1600,& 825
^1H NMRCDCl$_3$(δ)	:	7.9, 7.5, & 2.65
^{13}C NMR (δ)	:	27, 128, 130, 132, 136 & 197
MS (m/z)	:	156, 154,141,139,113, 111& 74

  (a) **UV Spectrum:** The intense band in the ultraviolet spectrum at 250 mμ is suggestive of a chromophore conjugated with an aromatic ring.

  (b) **Infrared Spectrum:** Carbonyl peak at 1690 cm^{-1} indicates the presence of a carbonyl group. We are probably dealing with conjugated ketone or an aldehyde. Absence of peaks near 2820 and 2720 cm^{-1} indicates the compound is not an aldehyde.

  We note strong peaks at 1600 cm^{-1} (conjugated C = C stretching) and a single peak at 825 cm^{-1} suggested 1,4-disubstituted aromatic ring.

  (c) ^1H **NMR Spectrum**: We consider the singlet in NMR spectra in the aliphatic region at δ2.65 indicates the presence of a – CH$_3$ group on the basis of the integration. We see two separate ring protons at low field double doublets (symmetrical pattern) for four aromatic protons at δ 7.5 (2H) and δ 7.9 (2H). On the basis of the spectral data discussed above, we can write a partial structure formula.

Now, we have to fix the position of the chlorine atom. The fact that the four protons appeared as doublet and the peak in IR spectrum at around 825 cm^{-1} indicates the chlorine atom is in the *para* position to the ketone group.

(d) ^{13}C NMR Spectrum: The spectrum shows the ring carbons are considered with respect to benzene (in which all six carbons) results in four absorptions between δ136 (C-2), 132 (C-3), 130 (C-4), 128 ppm (C-5). Methyl carbon next to carbonyl displays itself at δ 27 ppm (down field shift) and the carbonyl carbon being the most deshielded appears at δ 197 ppm. This data further supports the proposed structure.

(e) Mass Spectrum: The fragmentation pattern affords ample confirmation, characteristic chlorine compound.

Molecular ion at m/z 154 (relative abundance of normal isotope ^{35}Cl and m/z156 (relative abundance 1/3 of heavier isotope ^{37}Cl).

The base peak appeared at m/z 139/141 by the elimination of methyl group from the parent ion. Loss of CO from this fragment gives the phenyl ion m/z 111/113, followed by loss of chlorine atom to give another peak at m/z 77.

Hence, the given molecule is *para*-chloroacetophenone.

MS

8.8.12 Problem 12

Condensed phase IR:

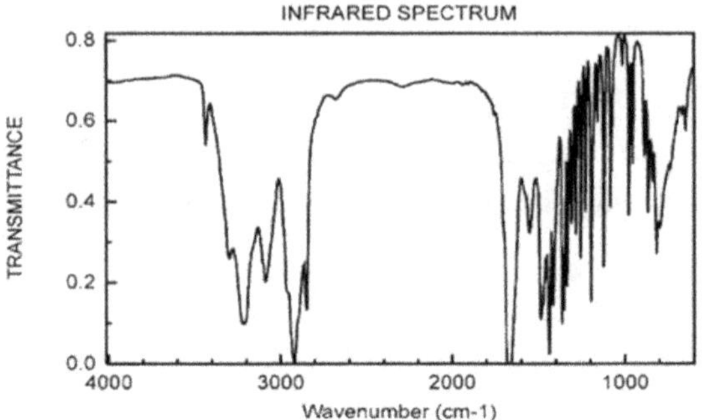

Mass spectra:

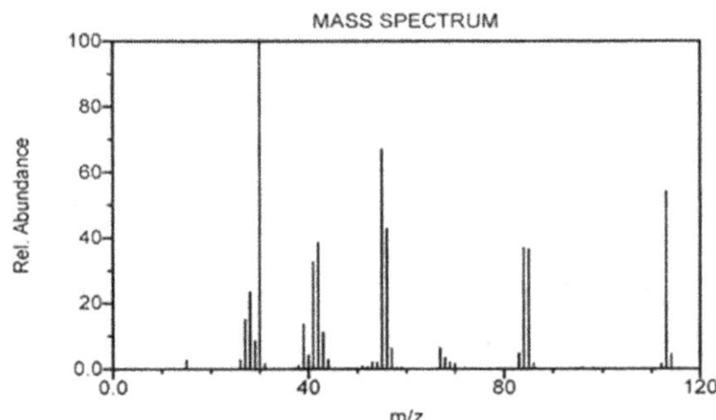

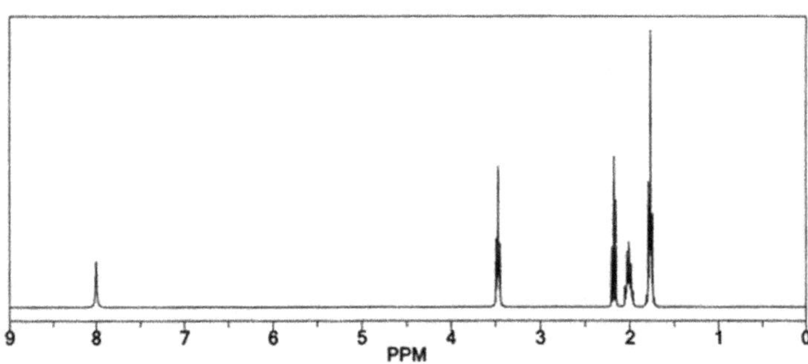

13C NMR Spectra:

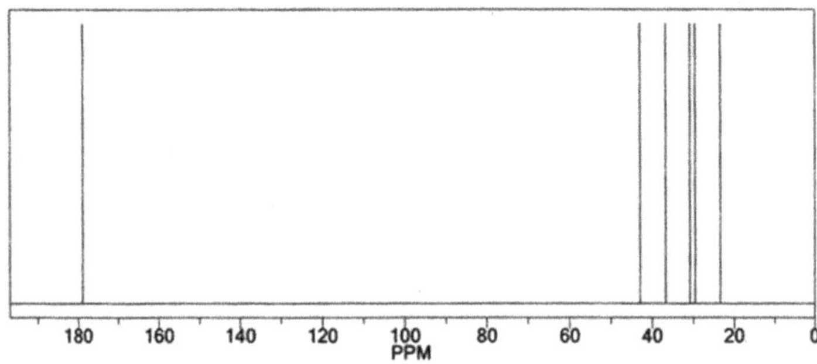

8.8.12 Solution

Molecular Formula $C_6H_{11}NO$

We can make a note on the basis of the molecular formula, the carbon to hydrogen ratio of the formula indicates that the compound is an aliphatic in nature.

Double bond equivalent (DBE) $C_A H_B N_C O_D$ DBE = A - 1/2 B + 1/2 C +1

Formula DBE   = 6 - 11/2 + 1/2 C +1 ; 6 - 5.5 + 0.5 + 1 = 2

Hence, there are two double bonds or one double bond and a ring. That is there are two unsaturated sites.

Spectral data

UV λ_{max} (nm)	:	210
IR v_{max} (cm^{-1})	:	3450 – 3077 (2.90 & 3.25 µ), 3200 (3.12 µ), 1670 (5.99 µ), 1620 –1590 (6.17 & 6.29 µ)
^{1}H NMR CDCl$_3$(δ)	:	8.2, 3.2, 2.3 & 1.7
^{13}C NMR (δ)	:	23, 29, 31, 37, 43 & 179
MS m/z	:	113, 55 & 30

(a) **UV spectrum:** It did not exhibit any absorption in the UV-Visible region. Therefore the compound is likely to be aliphatic in nature.

(b) **IR Spectrum:** The strong carbonyl band at 1670 cm^{-1} in the IR spectrum together with the series of bands in the region between 3450 - 3077 cm^{-1} suggest an amide group but the absence of an amide II band at 1620-1590 cm^{-1} indicates lactam seems more likely. Strong peak near 3200 cm^{-1} due to N–H stretching also suggests the presence of a lactam.

(c) **^{1}H NMR Spectrum:** The molecular weight permits us to write a six carbon lactam structure. We can allot the broad peak at δ 8.2 in the NMR spectrum due to hydrogen on a lactam nitrogen.

i.e.,   - CO – NH–

The broad triplet at δ 3.2 in the NMR spectrum must represent a –CH$_2$ group next to the nitrogen atom and somewhat less deshielded CH$_2$ group on the C=O group at δ 2.3. Part of the structure can be written as

$$\overset{\displaystyle O}{\overset{\displaystyle \|}{}}$$

$$-\!\!-CH_2-\!\!-NH-\!\!-\overset{O}{\overset{\|}{C}}-\!\!-CH_2-\!\!-$$

We have accounted for 3 carbons and the remaining three– $(CH_2)_3$– protons appeared as multiplet centered a δ 1.70 for the six protons on the basis of integration. The compound can now be formulated as caprolactam.

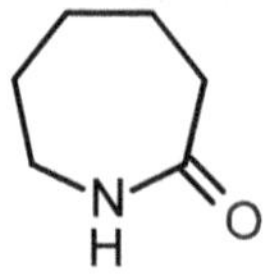

(d) **^{13}C NMR Spectrum:** The ^{13}C NMR spectrum shows six signals at δ 23, 29, 31, 37, 43 and the carbonyl carbon being the most deshielded comes at δ 179 ppm. This assignment further confirms the structure.

(e) **Mass Spectrum**: The base peak at m/z 30 may result as follows:

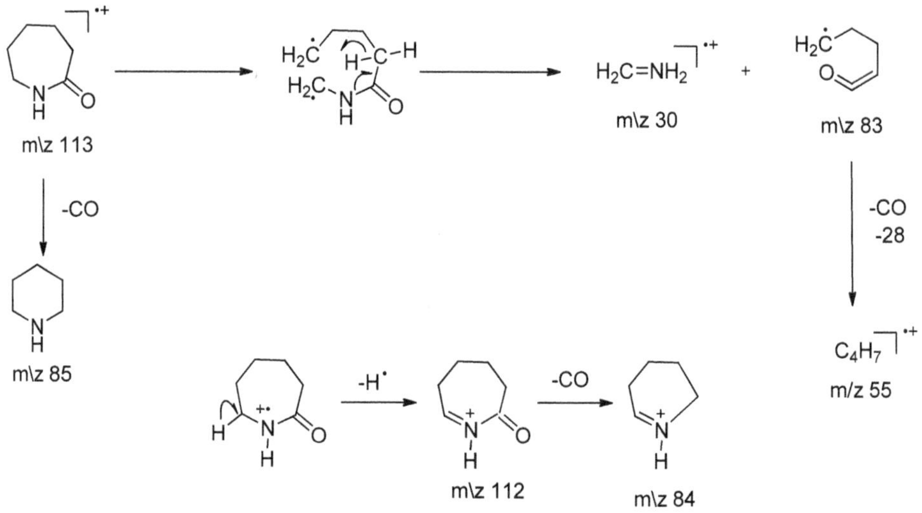
m\z 113
H₂C
H–C–H
H₂C
H₂C=NH₂
+
m\z 30
m\z 83
-CO
-28
C₄H₇
m/z 55
-CO
m\z 85
-H•
-CO
m\z 112
m\z 84

8.8.13 Problem 13

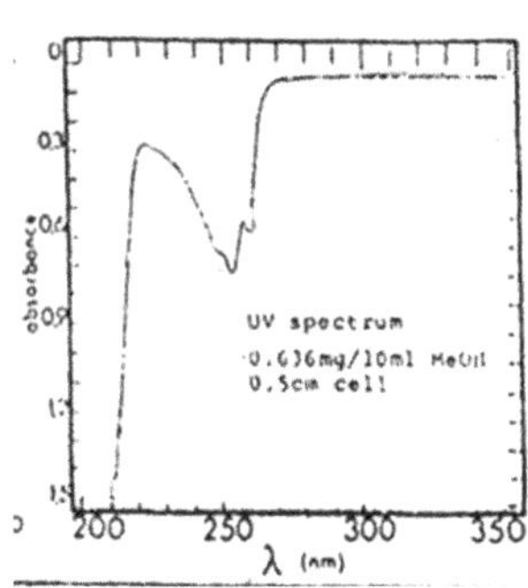

UV Spectrum

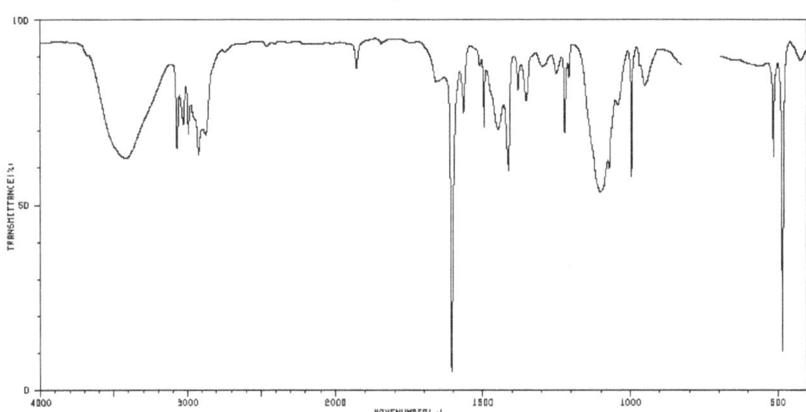

IR Spectrum

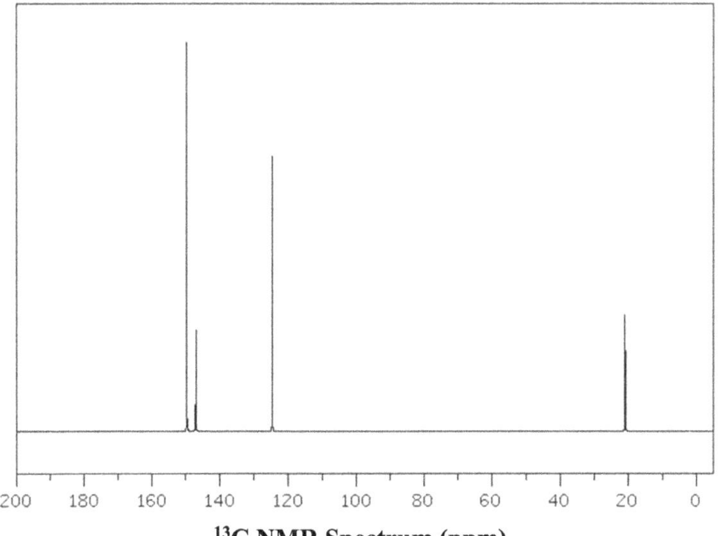

^{13}C NMR Spectrum (ppm)

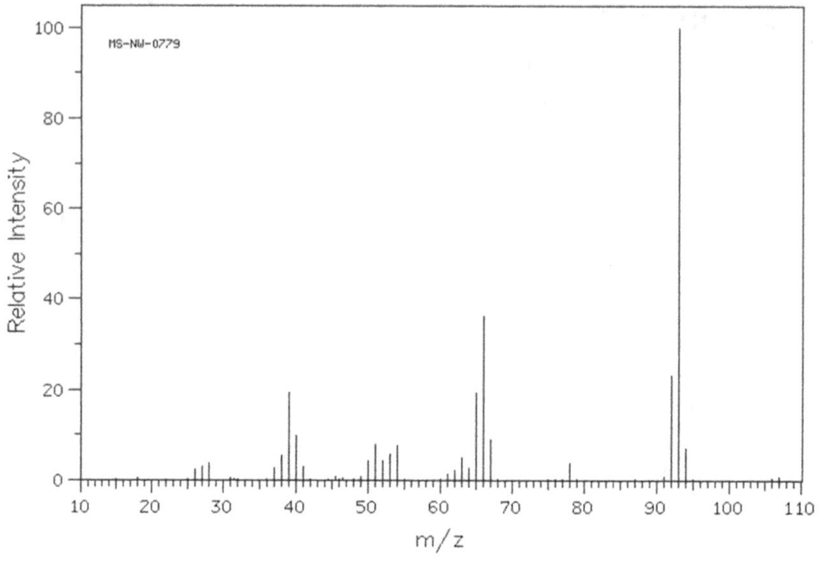

Mass spectra

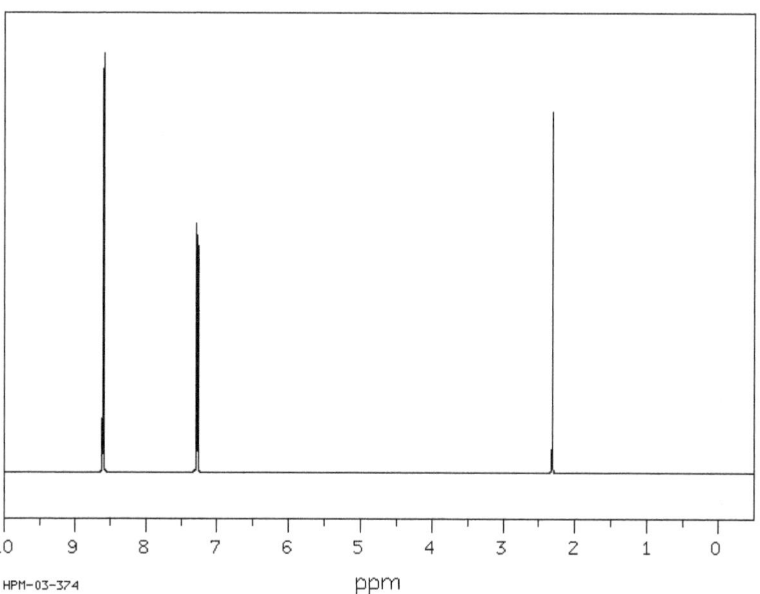

¹H NMR Sprectrum

8.8.13 Solution

Molecular formula C_6H_7N

It is noted both the intensity of the parent peak and the carbon to hydrogen ratio of the molecular formula and the NMR data indicate aromatic nature of the molecule.

For the compound with molecular formula $C_AH_BN_CO_D$

DBE $= A - 1/2B + 1/2C + 1$ $6 - 7/2 + 1/2 + 1 = 4$

The formula shows that for double bond equivalents are present.

Spectral data:

UV λ_{max} (nm)   : 262

IR v_{max} (cm⁻)   : 3077–3000 (m), 1600 (s), 1400 - 1450 (m) & 795

^1H NMRCDCl$_3(\delta)$: 7.18 (dd), 8.50 (dd) & 2.3

^{13}C NMR Spectra (δ): 20, 125 – 150

MS (m/z)   : 93, 68 & 30

(a) **UV Spectrum:** The absorption maximum at λ_{max} 262 is due to π- π^* transition reflects the presence of aromatic system.

(b) **Infrared Spectrum:** A characteristic feature of the IR spectra of aromatic compounds is the presence of a large number of sharp bands and particularly diagnostic are those near in the region $3077 - 3000$ cm^{-1} due to $= C - H$ stretching vibrations.

In particular, those in the $1600–1450$ cm^{-1} region which result from the in-plane skeletal vibrations of the aromatic and $–C=N$ stretching.

A strong band appeared at 795 cm^{-1} which is entirely consistent with the presence of four adjacent hydrogens leading to a single absorption. It may be nitrogen containing heterocyclic molecule, which accounts for the presence of nitrogen atom in the molecular formula.

In the case of mono-substituted benzene, the five adjacent hydrogen atoms give raise to two absorption bands in the region of $770–690$ cm^{-1}. In this respect, it closely resembles amino-substituted benzene.

(c) **^1H NMR Spectrum**: The ^1H NMR spectrum reveals two kinds of protons. The singlet at δ 2.3 is due to methyl group attached to aromatic ring. The four protons symmetrical pattern at δ 7.18 and δ 8.50 as double of doublet reveals that it may be a pyridine ring having substitution at C-4 position. The peak at δ 8.50 down field shift refers to protons alfa to nitrogen atom.

(d) ^{13}C NMR Spectrum: The ^{13}C NMR spectrum of compound shows four absorptions resulting from the four distinct locations of carbon in the molecule. The methyl carbon appears at δ 20, shifted down field slightly compared to a methyl carbon bonded to a saturated carbon. The five carbons of the aromatic ring results in three absorptions in between δ 125-150 arising from the carbon directly bonded to the methyl group and those in the *ortho* and *meta* positions (i.e., two equivalents) with respect to nitrogen atom.

Based on the spectral data, the given unknown compound can now be formulated as 4-methyl pyridine.

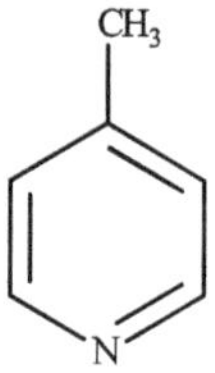

(e) Mass Spectrum: Typical fragmentation pattern of 4-methyl pyridine

Fragmentation pattern

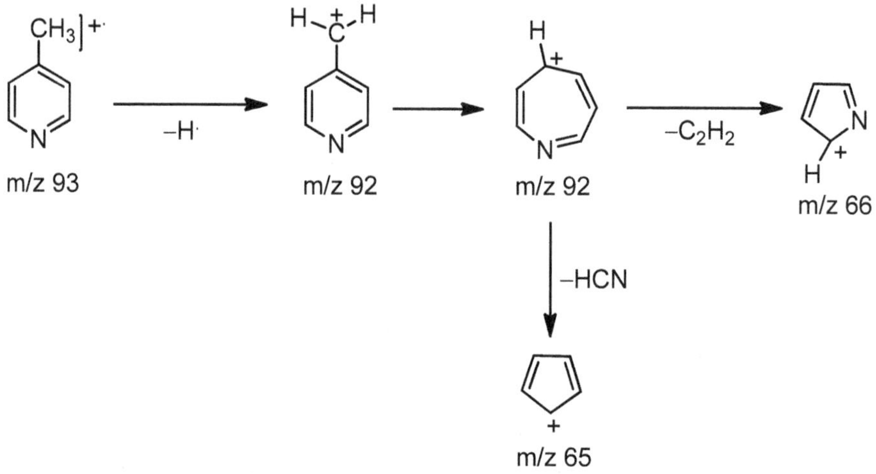

8.8.14   Problem 14

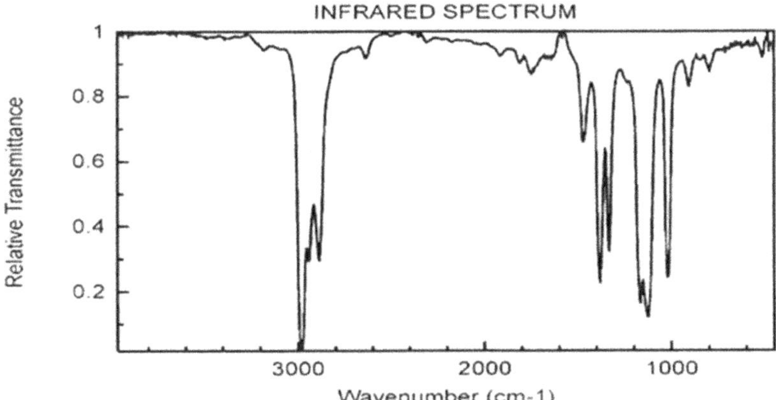

Gas Phase IR Spectrum

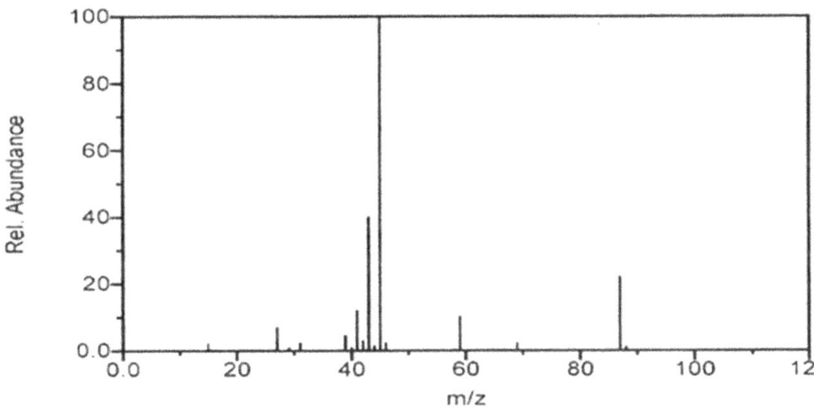

Mass Spectra

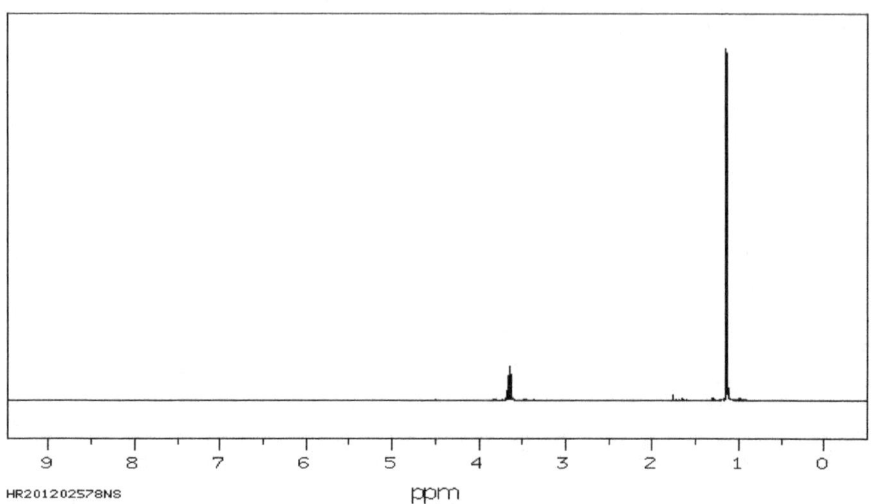

^{1}H NMR Spectrum

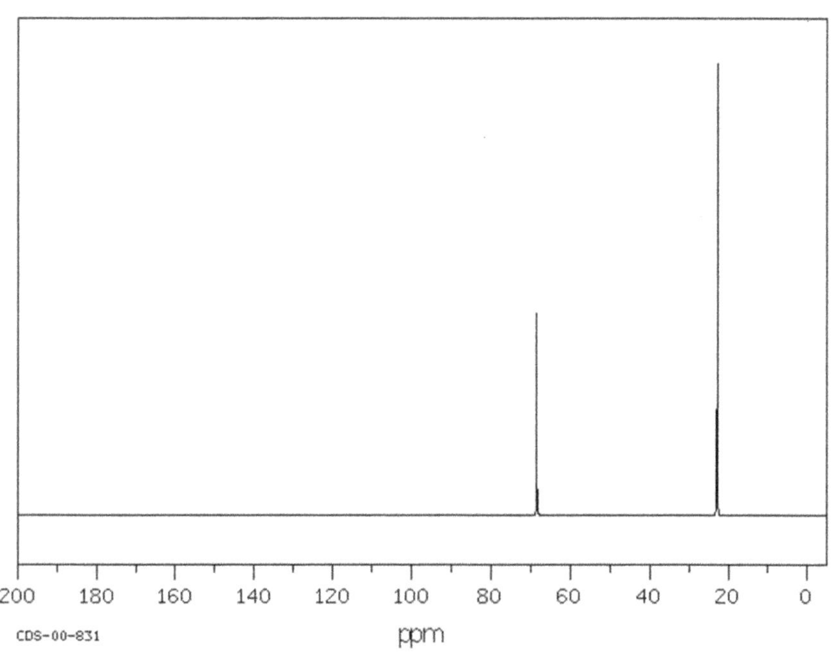

^{13}C NMR Spectrum

8.8.14 Solution

Molecular Formula $C_6H_{14}O$

The compound is aliphatic on the basis of C–to-H ratio, transparent above 210 $m\mu$ and the absence of proton signals at low field in the proton NMR spectra.

Double bond equivalent (DBE): $C_AH_BN_CO_D$ $DBE = A - 1/2\,B + 1/2\,C + 1$

Here $C_6H_{14}O$ $= 6 - 14/2 + 0 + 1 = Zero$

Hence, it's a saturated aliphatic compound.

Spectral data:

UV λ_{max} (nm)   :   210

IR v_{max} (cm⁻)   : $1130 - 1110$ (8.7 to 9.0 μ), 1380, 1370

(7.23 & 7.28 μ), and 1170 (8.55 μ)

^{1}H NMRCDCl$_3$(δ)   : $3.4 - 3.8$ (septet) & 1.1 (d)

^{13}C NMR (δ)   :   23 & 73

MS (m/z)   :   87, 45 (base peak) & 43

(a) **UV Spectrum:** Transparent above 210 μ suggests an aliphatic compound.

(b) **IR Spectrum:** There is no C=O or –OH absorption in the IR spectra.

Since the formula contains an oxygen atom, we can consider that it may be ether, and we observe a peak (split) at about $1130 - 1110$ cm^{-1} which corresponds to C–O absorption that confirms it is ether. The IR shows doublets at 1380 & 1370 cm^{-1} suggests the presence of isopropyl group.

The rather high frequency C–H (str) peaks at $2975 - 2850$ cm^{-1} further confirms the presence of the oxygen atom.

(c) ^{1}**H NMR Spectrum:** The proton spectrum is very clear to support the structure. The doublet at δ 1.1 (methyl group) and symmetrical septet aroundδ 3.65 (CH – groups) in the integrated ratio of 6:1 confirms the presence of isopropyl group. Evidently the molecule is symmetrical about the oxygen atom.

The compound can now be formulated as diisopropyl ether.

(d) **^{13}C NMR Spectrum:** The Carbon 13 NMR shows only two signals around δ 23 and 73 ppm although the compound has six carbon atoms. One can assume four methyl carbons are identical. The remaining two isopropyl carbon atoms directly bonded to the oxygen atom shifted to downfield appearing at δ73. This carbon 13 data also supports the diisopropyl ether structure.

(e) **Mass Spectrum:** Ethers show a weak or negligible molecular ion. In this case the M$^{+\cdot}$ was not observed. Fragment at the bond β to the oxygen (i.e. α,-cleavage) may occur. The base peak at m/z 45 in the MS results from double cleavage with rearrangement of a hydrogen atom. Removal of $-$ CH_3 group accounts for the mass of 87 peak and C $-$ O cleavage with retention of the charge on the alkyl part results in the large mass at m/z 43 peak.

The base peak at m/z 45 can be explained by ? cleavage in m/z 87 ion

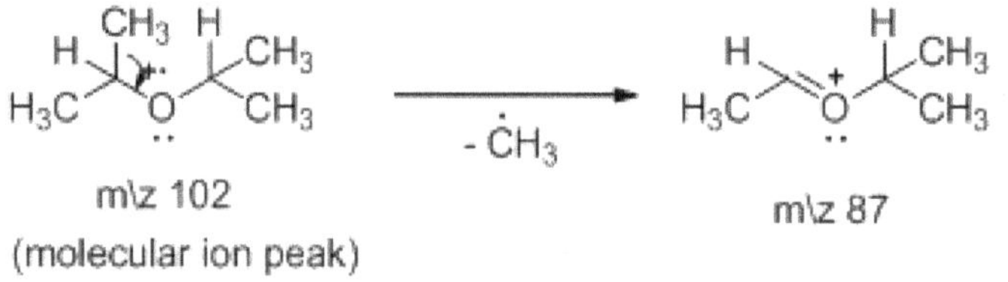

m\z 102
(molecular ion peak)

$-\dot{C}H_3$

m\z 87

Alternatively

Heterolytic Cleavage

m\z 43

Homolytic Cleavage

m\z 59

m\z 43

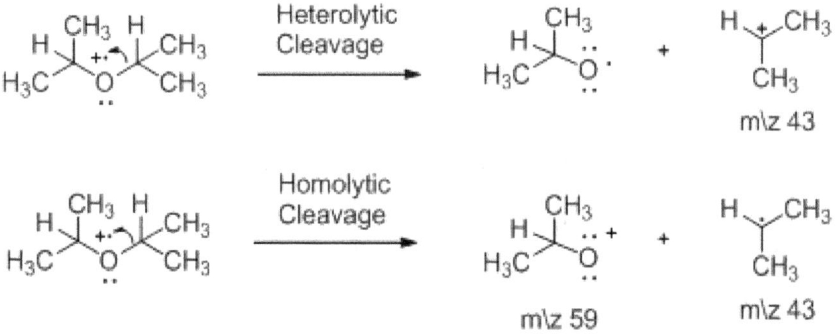

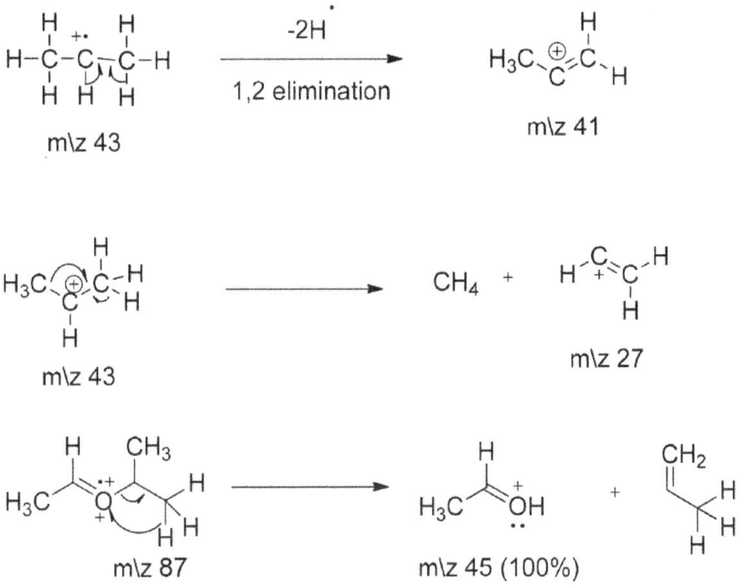

m\z 43

-2H

1,2 elimination

m\z 41

m\z 43

CH_4 +

m\z 27

m\z 87

m\z 45 (100%)

+

8.8.15 Problem 15

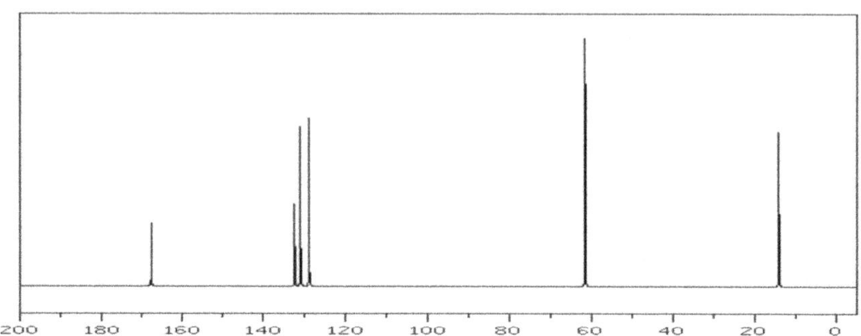

Carbon 13 NMR (ppm)

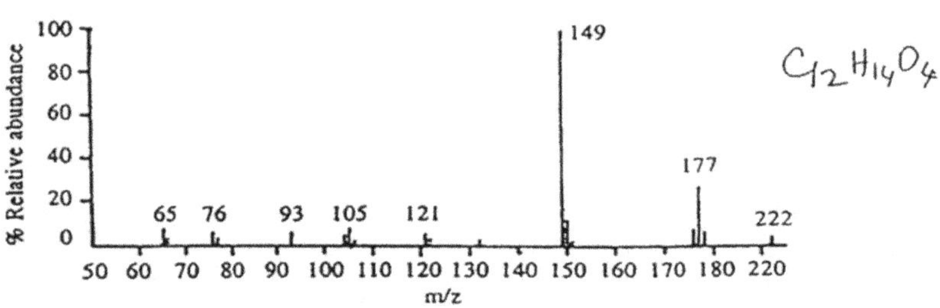

Mass Spectrum

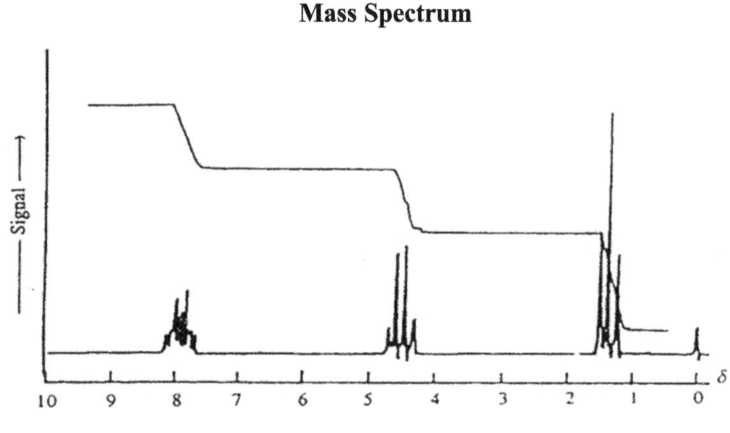

¹H-NMR Spectrum

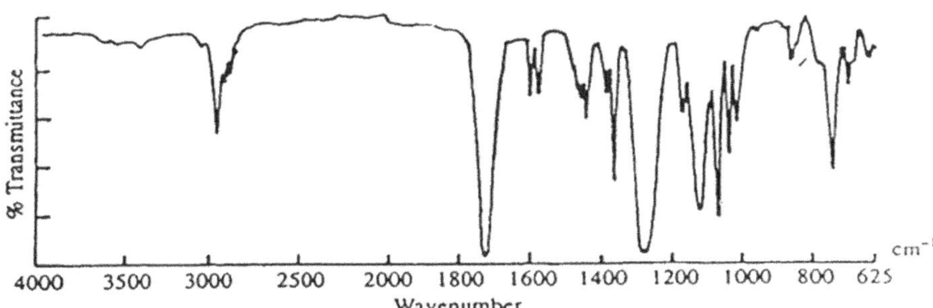

IR Spectrum

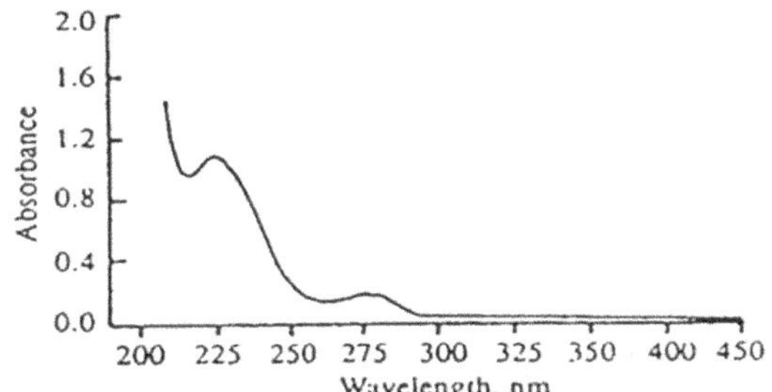

UV Spectrum

Additional information
(i) Contains C, H and O
(ii) MS : M$^+$ 222

8.8.15 Solution

Molecular Formula $C_{12}H_{14}O_4$

The carbon to hydrogen ratio of the molecular formula indicates aromaticity.

Double bond equivalent (DBE): $C_A H_B N_C O_D$ DBE = A - 1/2B + 1/2C + 1

Formula $C_{12}H_{14}O_4$ = 12 - 14/2 + 0 + 1 = 6

    The compound contains six elements of unsaturation. A benzene ring and two double bonds may be present.

Spectral data:

UV λ_{max} (nm)   :   235 & 280

IR v_{max} (cm⁻)   :   3000, 1770, 1250 (s) & 760

^{1}H NMR CDCl$_3$(δ)   :   7.90 (m), 4.50 (d), 1.3 (t)

^{13}C NMR (δ)   :   14, 61, 129-133 & 169

MS m/z   :   222, 177, 149, 121, 105 & 77

(a) **UV Spectrum: The low intensity absorption at** at 280 nm is a characteristic of *n* -π* transition. Therefore the compound is inferred to contain a carbonyl group.

(b) **Infrared Spectrum:** The IR spectrum shows a carbonyl band at 1770 cm^{-1} and a large band appeared at 1250 cm^{-1} characteristic of C–O–C stretching. This coupled with the presence of two oxygen atoms in the formula, suggests an ester. A simple peak at about 760 cm^{-1} indicates a 1, 2 - disubstituted benzene ring.

(c) **^{1}H NMR Spectrum:** It reveals the environment of hydrogen atoms in the molecule. The down field shift of protons appeared at δ 7.90 as a multiplet corresponds to benzene ring accounts for 4 hydrogen atoms on the basis of integration. Deshielded –CH$_2$- protons appeared as a quartet at δ 4.5 and at the same time shielded –CH$_3$- protons appeared as a triplet at δ 1.3 indicates an ethyl ester.

These spectral features point to the following structural units –

We can propose the structure as diethylphthalate.

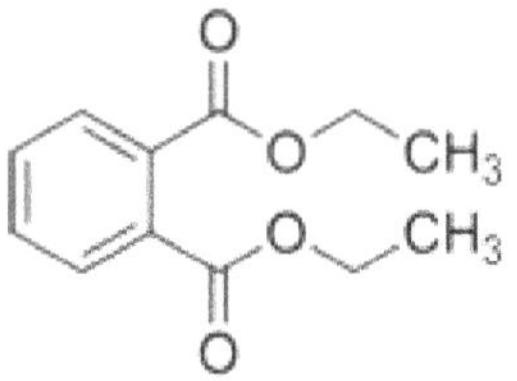

(d) ^{13}**C NMR Spectrum:** Carbon-13nmr spectrum shows a cluster of peaks appeared in between

δ 129 –133 ppm, suggests a disubtututed benzene ring. The remaining two ethoxy carbon atoms, being identical, subtutuent on the benzene ring directly bonded to the oxygen atom shifted to downfield appearing at δ 14 & 61 ppm. The two carbonyl carbons, being identical, and the most deshielded comes at δ 169 ppm.

(e) Mass Spectrum: We can draw additional confirmation by analyzing the mass spectrum and considering the fragmentation pattern. The base peak at m/z 149 (100%) fragment is formed by the loss of ethoxy radical followed by carbon monoxide. The third peak at m/z 121 arises from McLafferty rearrangement to give benzoic acid radical cation. The peaks at m/z 105, 95, 77 and 65 are additional evidence for the benzene ring.

Fragmentation pattern:

m/z 222 $\xrightarrow[-45]{-CH_3CH_2O}$ m/z 177 $\xrightarrow{-CO}$ m/z 149   base peak $\xrightarrow{H_2C=CH_2}$ m/z 121 $\longleftarrow$ m/z 105 $\xleftarrow{-CO}$ m/z 77

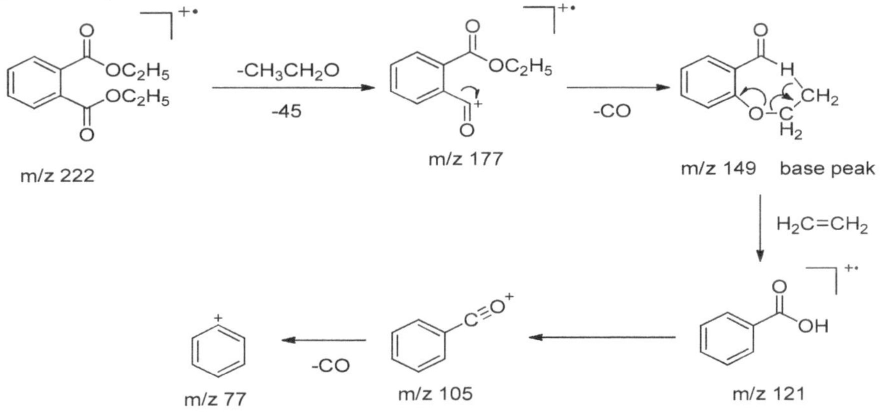

8.8.16   Problem 16

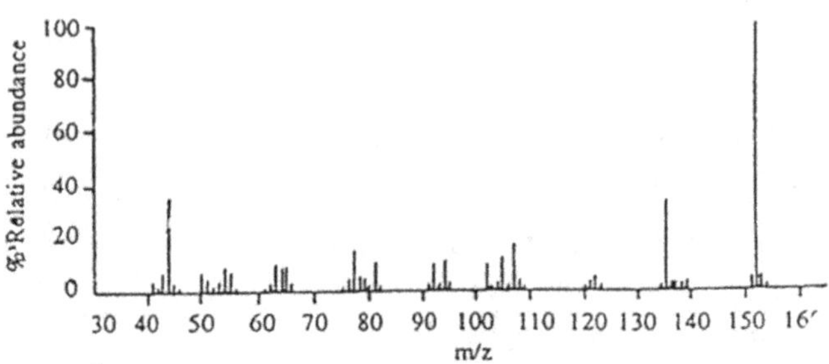

Mass Spectrum

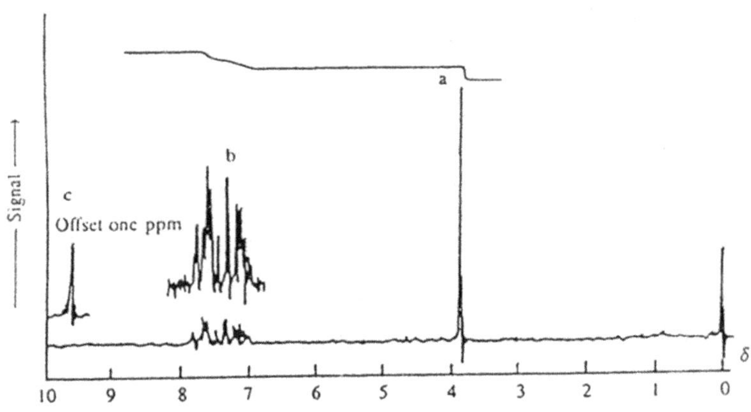

¹H NMR Spectrum

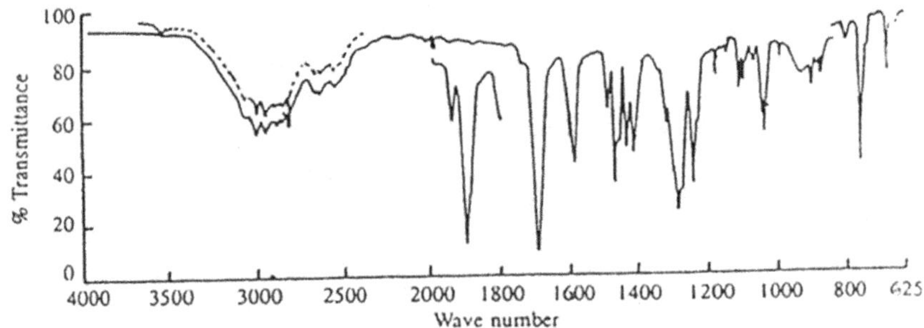

IR Spectrum

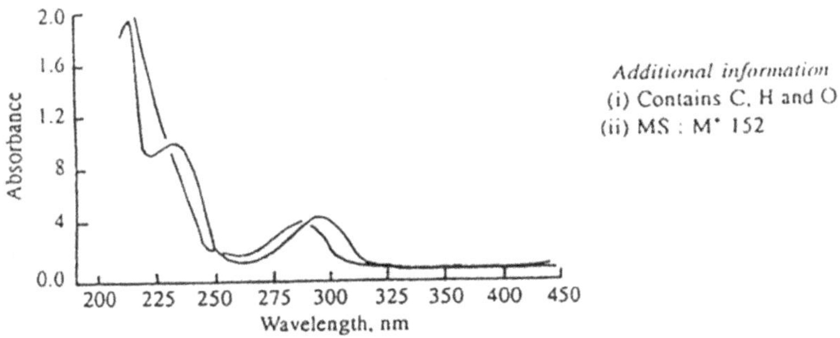

UV Spectrum

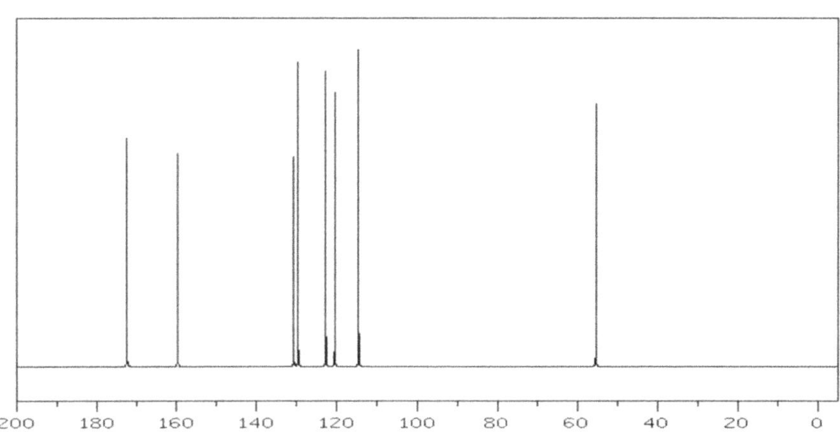

Carbon 13 NMR Spectrum (ppm)

8.8.16 Solution

Molecular Formula $C_8H_8O_3$

The first step is to translate these given spectra into a molecular structure. On the basis of intensity of the parent peak and the C- to - H ratio of the formula indicates aromaticity.

From the molecular formula, it is always useful to calculate number of double bonds equivalents.

Double bond equivalent (DBE) $C_AH_BN_CO_D$ DBE = A - 1/2B + 1/2C + 1

DBE formula $C_8H_8O_3$ = 8 - 8/2 + 0 + 1 = 5

Thus, $C_8H_8O_3$ contains 5 double bond equivalents, probably it may contain an aromatic ring and one double bond.

Spectral data

UV λ_{max} (nm)	:	290
IR v_{max} (cm^{-1})	:	3000, 1700, 1250, 1020 & 750
^{1}H NMR CDCl$_3$(δ)	:	11.2 (s),7.80 (m),3.80 (s)
^{13}C NMR (δ)	:	55, 114-130, 160 & 172
Mass m/z	:	152, 135,107, 77 & 45

Detail analysis :

(a) **UV Spectrum:** The absorption at 290 nm is the characteristic of a conjugated carbonyl group (i.e.,**n – π*** transition).

(b) **IR Spectrum:** A strong carbonyl stretch at 1700 cm^{-1} shows the presence of conjugation with the aromatic ring; non-conjugated C=O stretch is around 1720 cm^{-1}. This may be a carboxylic acid because a broad stretching absorption centered around 3000 cm^{-1}is the result of the strongly hydrogen bonded O-H stretching vibrations. Other bonds, which assist in the identification of carboxyl groups are the coupled vibrations involving the C-O stretching and O–H in plane deformation vibrations appeared around 1440–1390 and 1250 cm^{-1}.

(c) **^{1}H NMR Spectrum:** A peak at δ 11.20 (s) is due to the presence of COOH proton and a singlet at δ 3.80 (3H, deshielded) suggests the presence of a methoxy group. The remaining signals around δ 7.10 – 7.80 (m) can be assigned to the aromatic protons.Thus, the major structural features which stand clear are the presence of a carboxyl group, a phenyl ring and a methoxyl group.

On the basis of UV, IR data and considering the splitting pattern in NMR spectra, it is not 1,2 or 1,4 – disubstituted but 1,3-disubstituted; so we can write the tentative structure as follows:

(d) **^{13}C NMR Spectrum:** Carbon 13 NMR spectrum shows eight signals although the compound has ten carbon atoms. Phenyl C-2, C-6, are equalant as well as C-3 and C-5. One can therefore differentiate among ortho, meta and para-methoxy benzoic acid mearly the number of carbon-13 sinals. The cluster of peaks appeared in between δ 114-130 ppm, and another peak at δ 160 ppm represents phenyl ring carbons. The peak at δ 55 represents the methoxy carbon and the acid cabonyl being the most dshielded comes at δ 172.

(e) **Mass Spectrum:** The molecular ion for this compound ($C_8H_8O_3$) has m/z 152. The peak at m/z 135 is due to the loss of hydroxyl radical leaving a resonance stabilized acylinium ion. The third peak at m/z 107, methoxy phenyl cation, is formed by the loss of carbon monoxide. Further loss of methoxy group ($-OCH_3$) to give the phenyl ion at m/z 77.

Finally, the given unknown compound, based on the combined spectral data, is m-methoxybenzoic acid.

COOH
OCH3

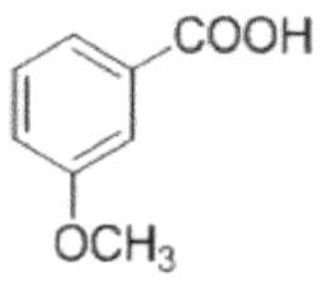

3-methoxy benzoic acid

Fragmentaion:

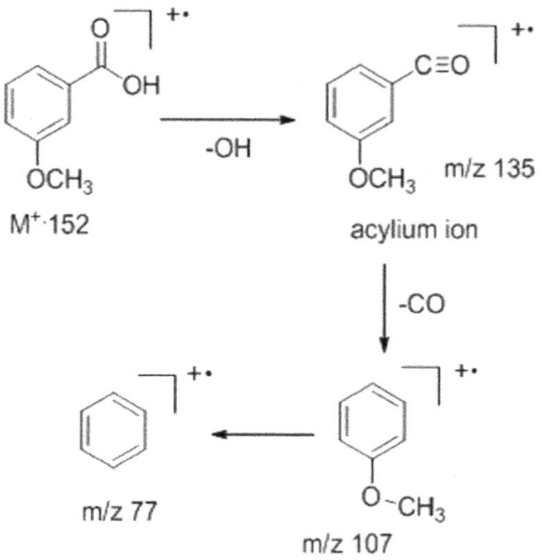

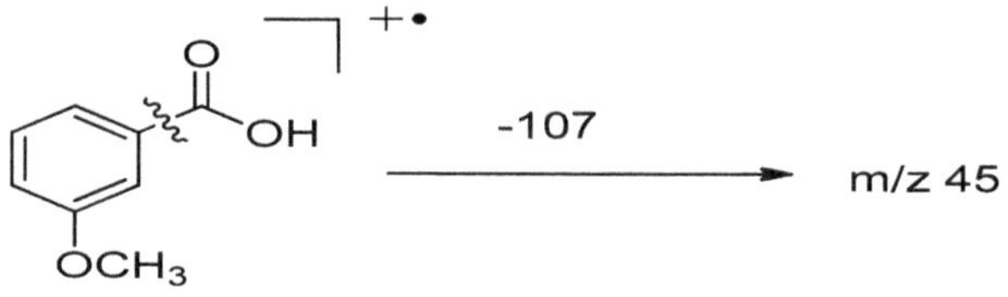

OH
-OH
C≡O
OCH3
M+· 152
m/z 135
acylium ion
-CO
m/z 77
O-CH3
m/z 107

OH
OCH3
-107
m/z 45

8.8.17 Problem 17

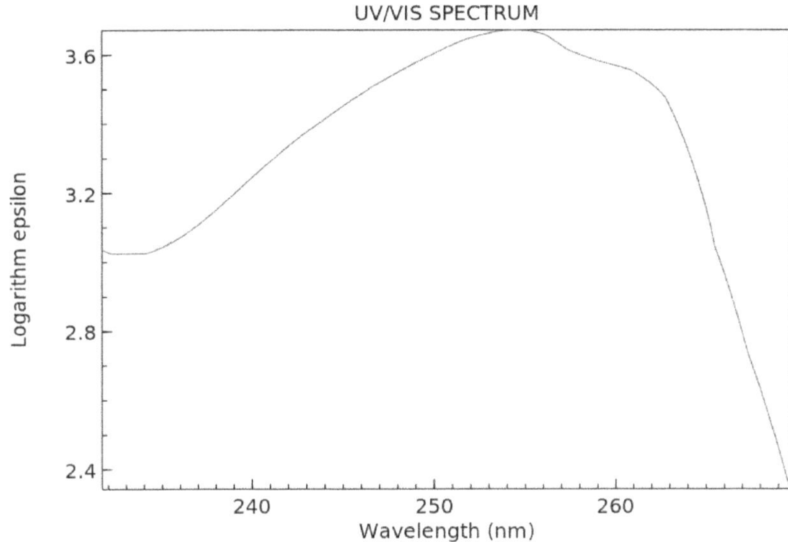

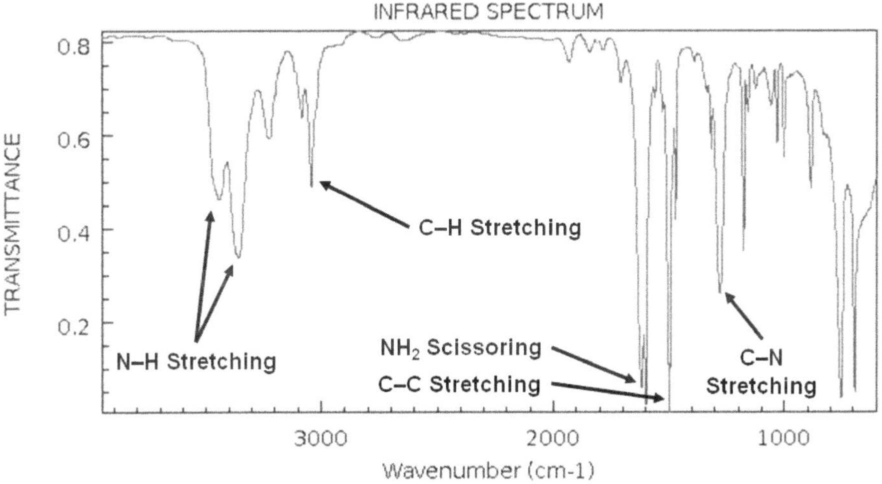

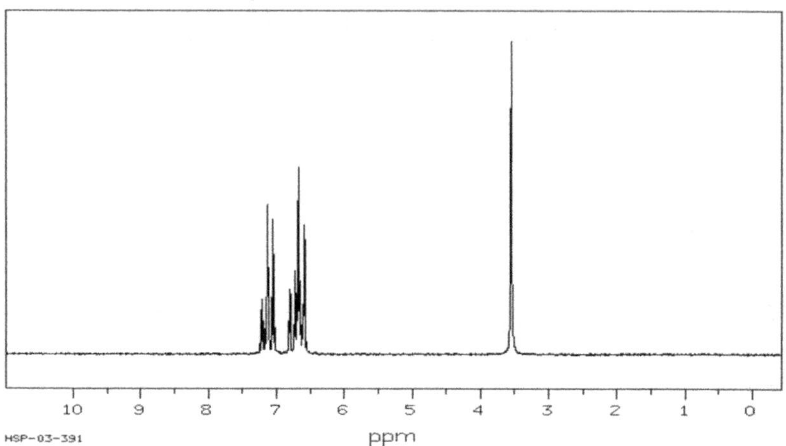

^{1}H NMR Spectrum

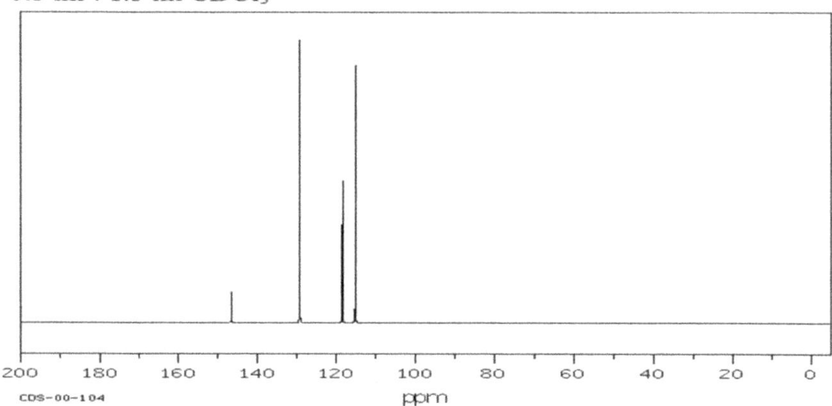

Carbon-13 Spectrum

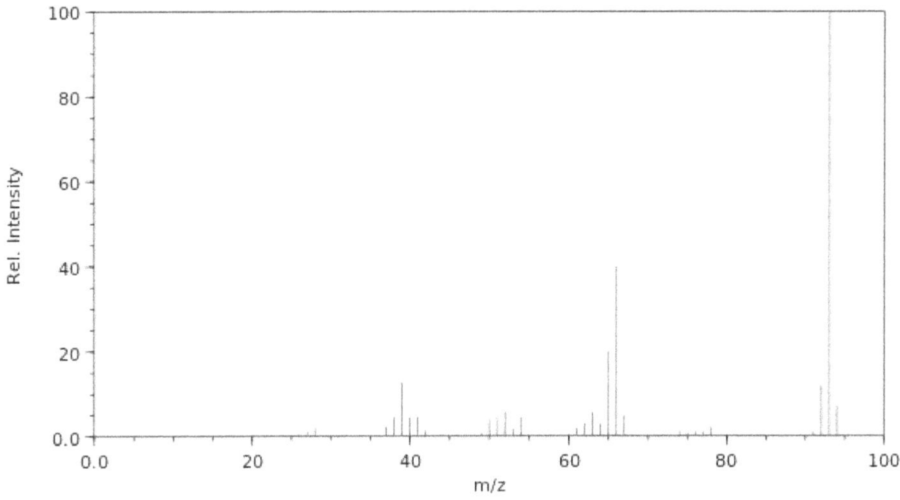

Mass Spectrum

8.8.17 Solution

Molecular Formula C_6H_7N

On the basis of the molecular formula, the C- to - H ratio of the formula indicates that that the compound is aromatic in nature.

Double bond equivalent (DBE) $C_AH_BN_CO_D$ DBE $= A - 1/2B + 1/2C + 1$

DBE $C_6 H_7 N$ $= 6 - 7/2 + 1/2 + 1 = 4$

Three double bonds and one ring may be present.

Spectral data:

UV λ_{max} (nm)	:	230, 280
IR v_{max} (cm⁻)	:	3440-3350, 1620, 1500, 1468 & 650 -900
^{1}H NMR CDCl$_3$(δ)	:	6.5-7.5 (m),& 3.75
^{13}C NMR CDCl$_3$(δ)	:	115, 118, 129 &146
Mass m/z	:	93, 66, 65 & 27

Detail analysis:

(a) **UV Spectrum:** Indicates the presence of aromatic ring because there is an extension of auxochrome, hence longer wave length around 280 nm in the UV spectra.

(b) **IR Spectrum:** When the formula contains a nitrogen atom then one can expect N-H stretching; a spectrum shows the presence of doublets around $3350 - 3440$ cm^{-1} indicates presence of a primary amine. Strong bands at 650-900 cm^{-1} and another peak at 1620 cm^{-1} for N-H bending further confirm presence of a primary amine group.

(c) **^{1}H NMR Spectrum:** The down field shift aromatic protons appeared at δ 6.5- 7.5 as a multiplet corresponds to benzene ring protons which accounts for 5H on the basis of integration. Singlet at δ 3.75 accounts 2H for amino group protons.

(d) **^{13}C NMR Spectrum:** The carbon-13 spectrum one sees only four signal since only four non equivalent carbons present. Phenyl C-2 & C-6 carbons appeared at δ 115 are equivalent as well as C-3 and C-5. carbons absorbs at δ 129. The remaining peaks at δ147 and δ 118 represents C-1 & C-4 carbons respectively.

(e) **Mass Spectrum m/z:** The peak for the molecular ion (odd number) is strong in aromatic monoamines. The loss of one of the amino H atoms gives an intense M - 1 peak. Loss of molecule of HCN, and a subsequent loss of a hydrogen atom gives strong peaks at m/z 66 and 65.

We can now write the complete structure aniline on the basis of all spectral data.

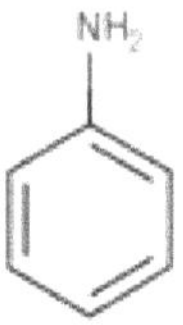
NH2

Fragmentation

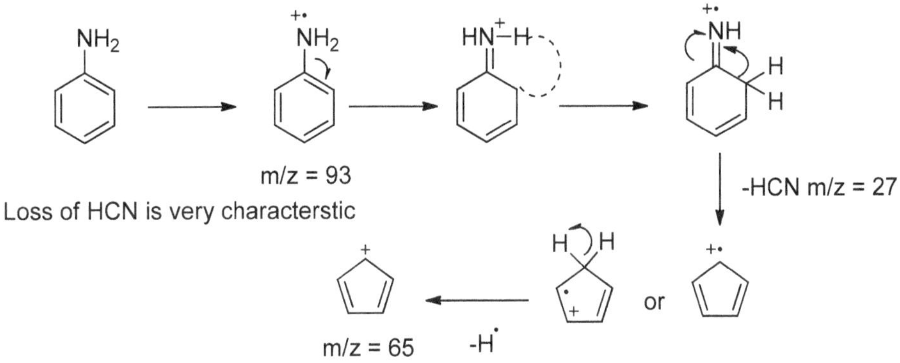
NH2
+•
NH2
HN+–H
+•
NH
H
H
m/z = 93
-HCN m/z = 27
Loss of HCN is very characterstic
H
H
+
+•
or
m/z = 65
-H•

8.8.18 Problem 18

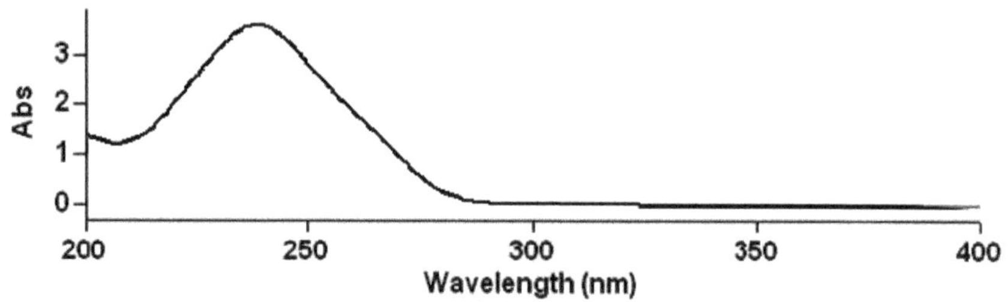

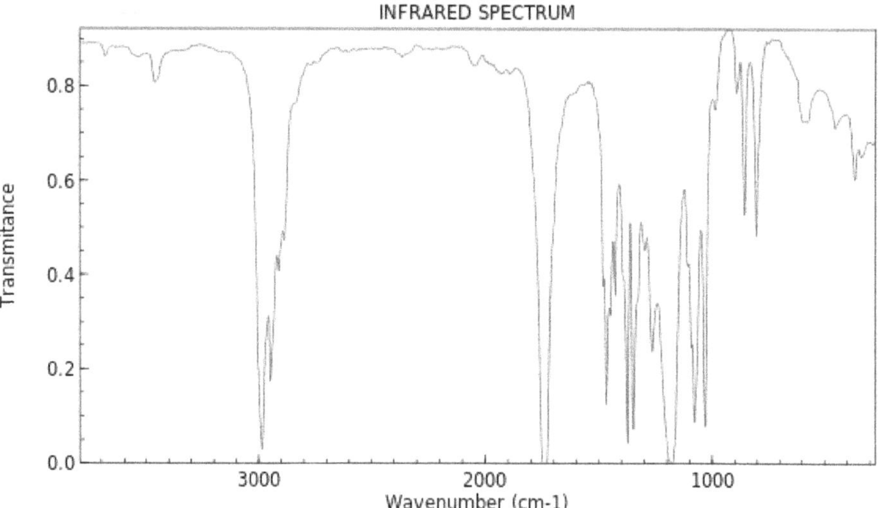

IR Spectrum

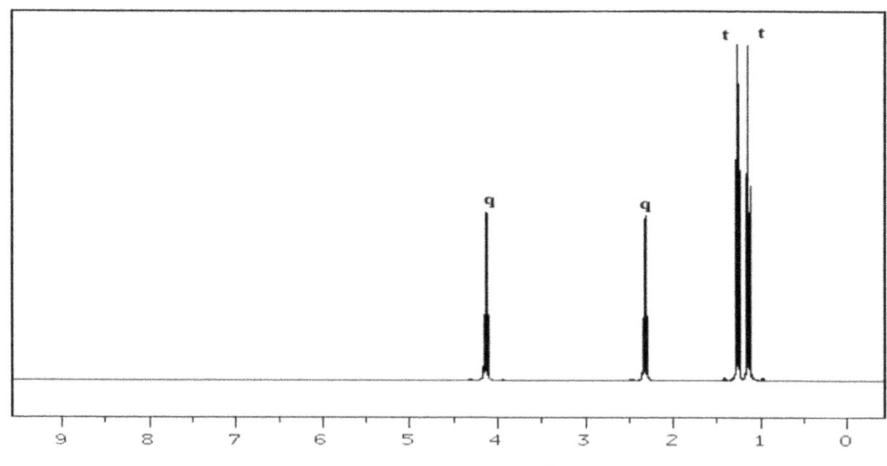

400 MHz NMR Spectrum (δ values)

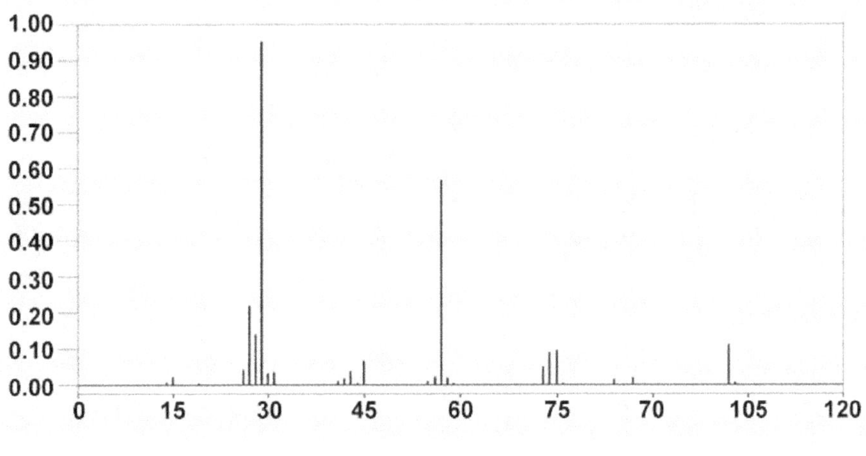

Mass Spectrum

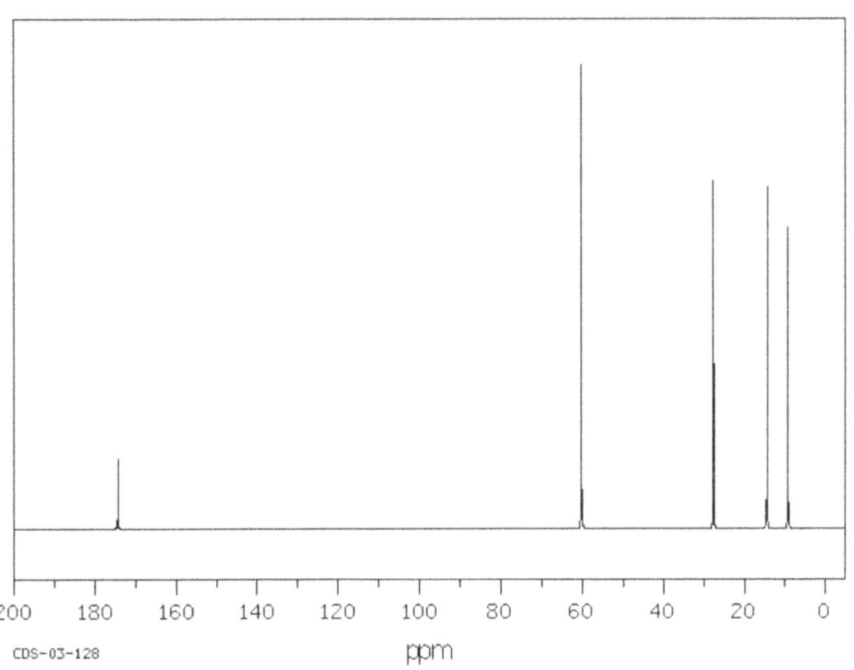

^{13}Carbon Spectrum

8.8.18 Solution

Molecular formula $C_5H_{10}O_2$

DBE: $C_AH_BN_COD$ DBE = A - 1/2B + 1/2C + 1

$$= 5 - 10/2 + 0 + 1 = 1$$

The compound is aliphatic on the basis of the C-to-H ratio, the lack of absorption in the ultraviolet spectrum and the absence of peaks at low field in the NMR spectrum.

Spectral data:

IR vmax(cm^{-1})	:	3000, 1740 (s), 1200
^{1}H NMR CDCl$_3$ (δ)	:	1.2 (t), 1.4 (t),2.4 (q) & 4.2 (q)
^{13}C NMR (δ)	:	09, 14, 21, 60 & 174
MS (m/z)	:	102, 57 & 29

(a) **IR Spectrum:** The infrared spectrum shows a C = O band at about 1740 cm^{-1}, this together with the presence of two oxygen atoms in the formula, suggests an ester. We look for confirmation in the C–O–C stretching region and note the large broad band at about 1200 cm^{-1} characteristic of acetate.

(b) **^{1}H NMR Spectra:** The NMR spectrum suggests no characteristic peaks in the aromatic region. The triplet of three protons (shielded) at δ 1.2 represent the methyl group adjacent to a (-OC-CH$_2$-CH$_3$). Another triplet of three protons (shielded) at δ 1.4 represents another methyl group H$_3$C – CH$_2$-O-) with a -CH$_2$- as an adjacent group. The quartet of two protons at δ 2.4 represents the deshielded methylene group (between OC-CH$_2$-CH$_3$). The remaining the strongly deshielded position of the -CH$_2$- absorption a quartet at δ 4.2 (CH$_3$-CH$_2$-O), places the CH$_2$ on the oxygen atom.

(c) **^{13}C NMR Spectrum:** The peak at δ 174 is due to the carbonyl carbon which accounts for index of hydrogen deficiency. The peak at δ 60 represents CH$_2$ group carbon bonded to the electro negative carbonyl group. The remaining three peaks, aliphatic carbons, appear at δ 21, 14 &09 ppm.

(d) **Mass spectrum:** We can obtain additional confirmation on the basis of mass spectral fragmentation pattern. The molecular ion is prominent at m/z 102. The large peak at m/z 57 strongly suggests the CH$_3$-CH$_2$-CO group in view of the C=O peak in the IR. Another peak at m/z 29 represents ethyl radical. Subtraction of both the above groups from the formula leaves a mass of 16. Consideration of the infrared spectrum leaves vary little question as to how to handle this oxygen atom.

Result: On the basis of the spectral analysis the compound is ethyl propionate.

$$H_3C–CH_2–CO–O–CH_2–CH_3$$

Fragmentation

$$CH_3–CH_2–O–\overset{\overset{O}{\|}}{C}–CH_2–CH_3 \quad \longrightarrow \quad H_3C–\overset{+}{C}H_2 \quad m/z\ 29$$

m/z 102

$$CH_3–CH_2\overset{\overset{O}{\|}}{C}{}^{+} \quad m/z\ 57$$

8.8.19 Problem 19

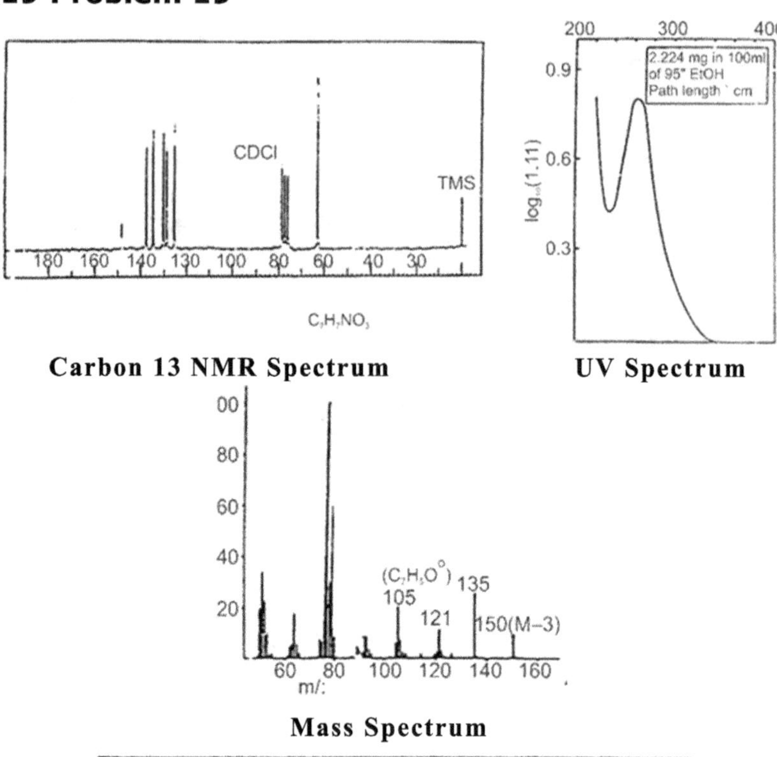

Carbon 13 NMR Spectrum   **UV Spectrum**

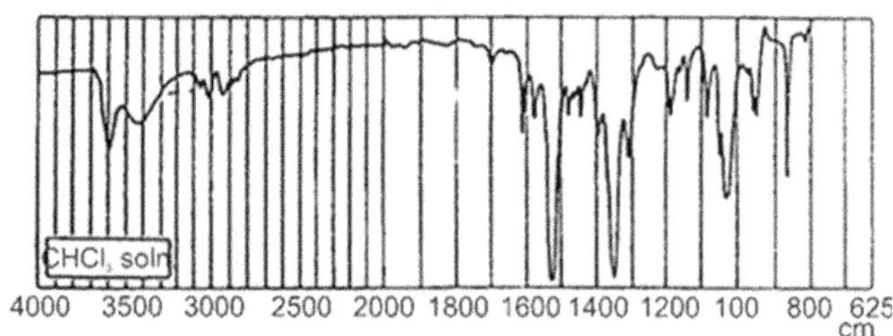

Mass Spectrum

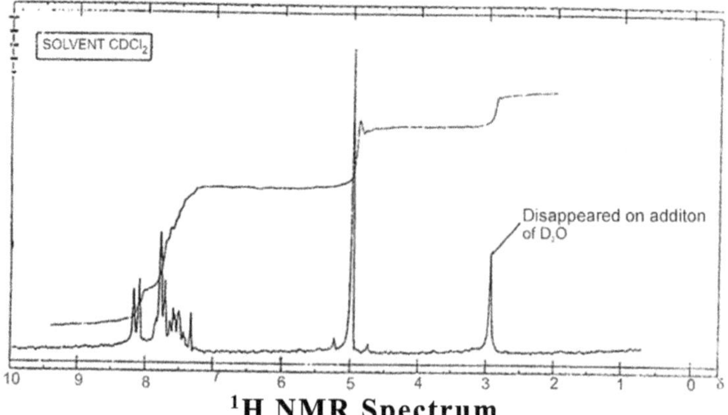

IR Spectrum

^{1}H NMR Spectrum

8.8.19 Solution

Molecular Formula $C_7H_7NO_3$

On the basis of the molecular formula, the carbon to hydrogen ratio and double bond equivalent indicates that the compound is an aromatic with conjugation.

Double bond equivalent $C_AH_BN_CO_D$ DBE = A - 1/2B + 1/2C + 1

Formula $C_7H_7NO_3$ $= 7 - 7/2 + 1/2 + 1 = 5$

Hence, one ring and three double bonds may be present.

Possibilities;

(a) 4-double bonds

(b) Aromatic ring and 1-double bond

Spectral data:

UV λ_{max} (nm)   :   260

$IR\ v_{max}$ (cm^{-1})   :   3500, 1560-1500,1350-1300,1230,1025,875

^{1}H NMR CDCl$_3$(δ)   :   3.0 (s), 5.0 (s), 7.3 to 8.2 (m)

^{13}C NMR (δ)   :   62, 124, 128, 129, 134, 137 & 147

MS (m/z)   :   150, 135, 121, 105, 78 & 77

(a) **UV Spectrum:** The intense K-band at 260 cm^{-1} is indicative of a chromophore conjugated with the aromatic ring (i.e., π - π^* transition)

(b) **Infrared spectrum:** The conspicuous feature of the infrared spectrum is the rather strong absorption at 3500 cm^{-1}. A peak at 1230 cm^{-1} is evidence for the confirmation of OH absorption. In the absence of carbonyl band in the IR spectrum and the presence of two intense absorption bands in the region 1560-1500 cm^{-1} and 1350-1300 cm^{-1} region of the spectrum due to asymmetric and symmetric stretching vibrations of the highly polar nitrogen –oxygen bonds. A strong absorption at 1025 cm^{-1} is characteristic of –C-O- stretching.

Ortho-substituted (like *p*-substituted benzenes) gives a single strong band at higher frequency around 875 cm^{-1} suggests a disubstituted benzene ring.

(c) **^{1}H NMR Spectrum:** The NMR spectrum is nicely in accord with the structure. The four protons appeared as a multiplet at δ 7.3 to 8.2 of course, are the four benzene ring protons. The singlet of two protons at δ 5.00 represents the methylene group highly deshielded by a phenyl and hydroxyl group.

(d) **^{13}C NMR Spectrum:** The peak at δ 32 is due to the benzyl alcohol carbon. The remaining ring carbons appeared in between δ 124-147 ppm.

(e) Mass Spectrum: We can obtain additional information by returning to the mass spectrum and considering the fragmentation pattern. The vicinity of very weak or missing molecular ion peak of an alcohol may be sometimes complicated by the presence of weak peaks corresponding to successive loss of H radical at M-1, M-2 and M-3 giving a peak at m/z 150. The peaks at 77, 78 and 79 are additional evidence for the benzene ring.

We can state with confidence that the compound represented by these spectra is *ortho*-nitrobenzylalcohol.

Fragmentation

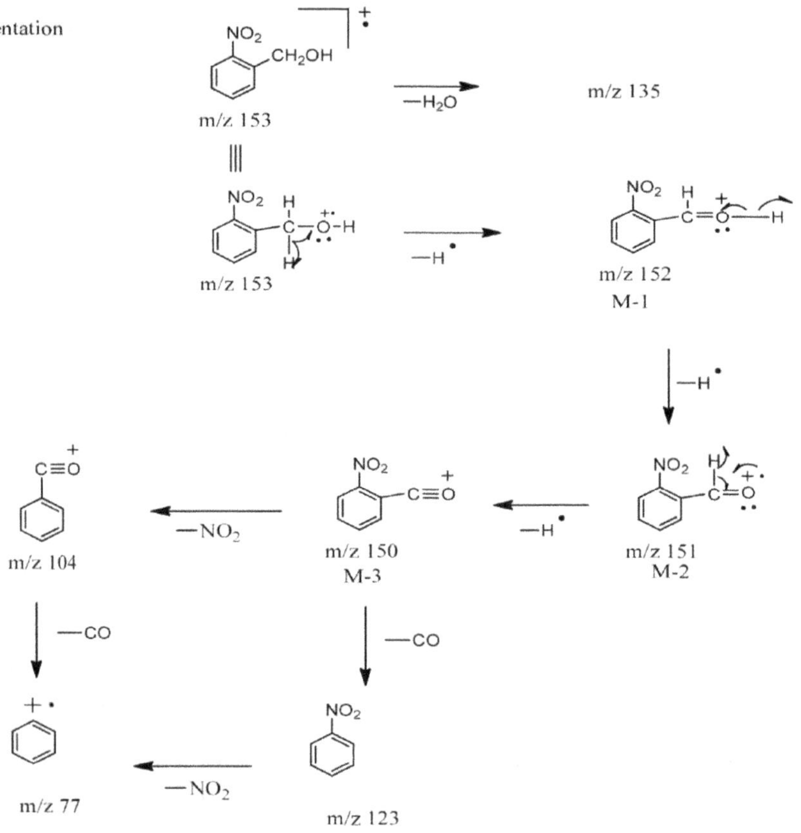

$-H_2O$ → m/z 135

m/z 153

m/z 153 → $-H^{\bullet}$ → m/z 152 M-1

m/z 152 M-1 → $-H^{\bullet}$ → m/z 151 M-2

m/z 151 M-2 → $-H^{\bullet}$ → m/z 150 M-3

m/z 150 M-3 → $-NO_2$ → m/z 104

m/z 104 → $-CO$ → m/z 77

m/z 150 M-3 → $-CO$ → m/z 123

m/z 123 → $-NO_2$ → m/z 77

8.8.20 Problem 20

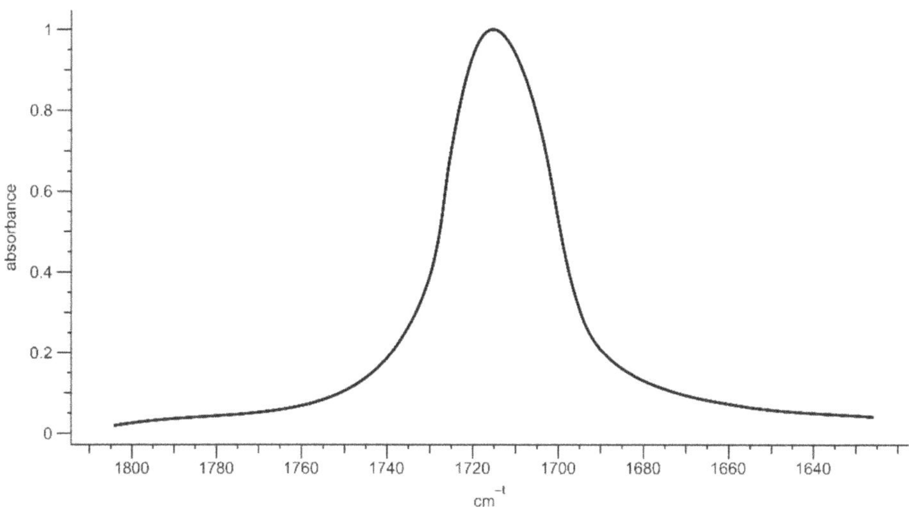

UV Spectrum

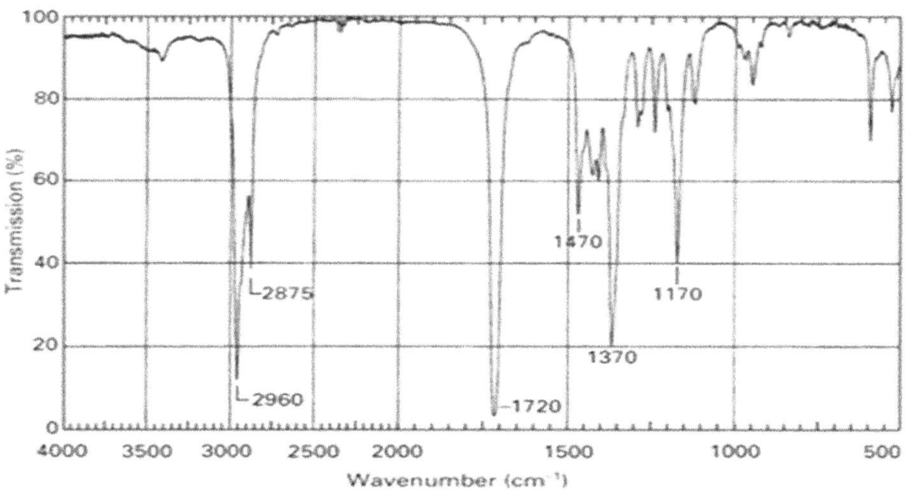

IR Spectrum

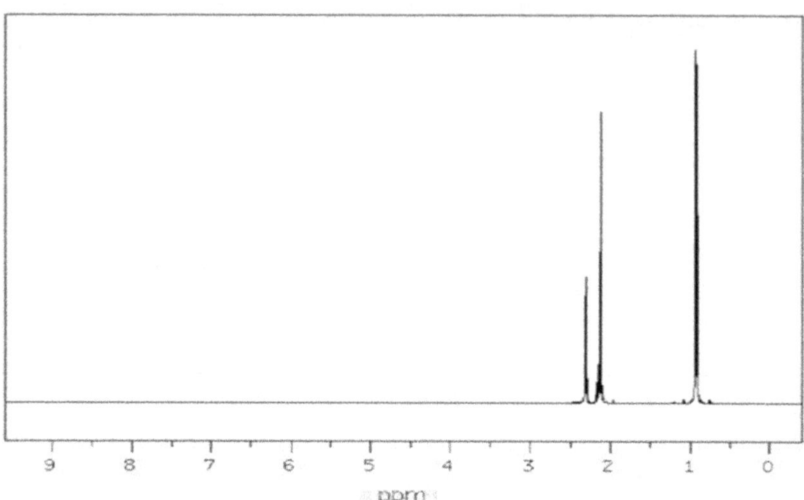

¹H NMR Spectrum

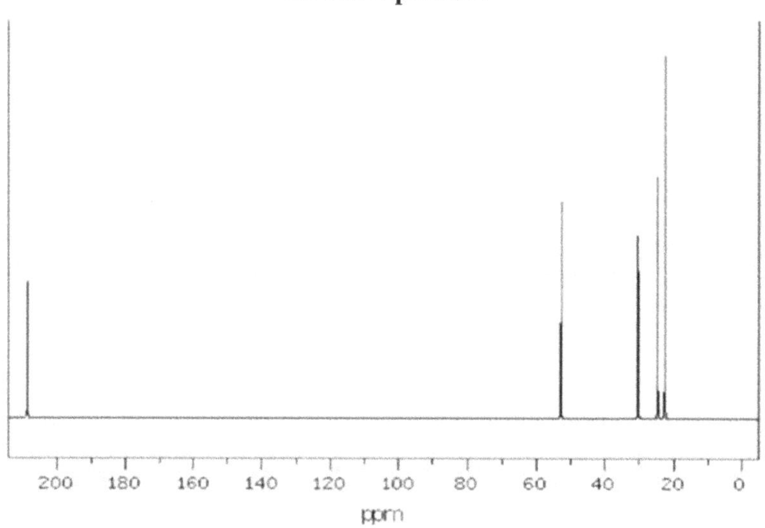

Carbon-13 Spectrum

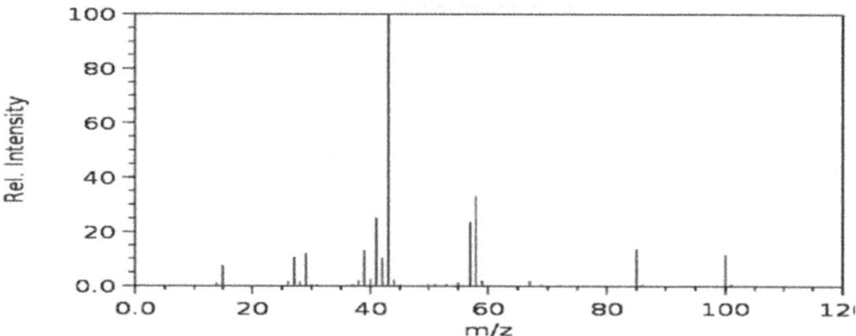

Mass Spectrum

8.8.20 Solution

Molecular Formula $C_6H_{12}O$

On the basis of the molecular formula, the carbon to hydrogen ratio indicates the compound is aliphatic in nature.

Double bond equivalence (DBE): $C_AH_BN_CO_D$ DBE = A - 1/2B + 1/2C + 1

Formula $C_6H_{12}O$ = 6 - 12/2 + 0 +1 = 1

It may contain one double bond. This compound has index of hydrogen deficiency of one.

Spectral data:

UV λ_{max} (nm)	:	237
IR vmax (cm^{-1})	:	2960, 2875, 1720, 1470, 1370, 1170
^1H NMR $CDCl_3(\delta)$	:	0.9, 1.1 (m), 2.4 (d)
^{13}C NMR (δ)	:	23, 25, 30, 53, 207.
MS (m/z)	:	100, 58, 57, 43,15

(a) **UV Spectrum**: The low intensity absorption at 237 nm suggests an aliphatic compound.

(b) **IR Spectrum**: The strong C=O band at 1720 cm^{-1} indicates the presence of an unconjugated ketone. The strong absorption band centred around 2960 cm^{-1} represents the C-H stretching of both methyl and methylene groups, the CH_2 and CH_3 bending vibrations at 1470 cm^{-1} and 1370 cm^{-1} suggest it is an aliphatic chain.

(c) **^1H NMR Spectrum**: The position of NMR signals show that all the 12 protons are of aliphatic type, confirms it is an aliphatic compound. The large singlet at $\delta 2.2$ must be the CH_3 group on the ketone carbonyl. We note another doublet at δ 2.45 represents two protons (possibly CH_2). The peak at δ 0.98 (triplet) (this must be a doublet which must be the methyls of the isopropyl part. The –CH proton which must be split by the two methyls and the CH_2 must be in the grass roots and difficult to observe and obscured by the 2.45 signal. Perhaps there is some restricted rotation in the molecule so there may be some nonequivalence in some of the protons.

(d) **^{13}C NMR Spectrum**: The peak at δ 207 is due to the carbonyl carbon which accounts for index of hydrogen deficiency. The mass spectrum shows a peak at m/z which may be due to resonance stabilized acylinium ion (CH_3CO+). The compound therefore, contains a methyl ketone group. Of the two carbons which appear farthest downfield (apart from C=O) if one is a methyl group, (δ 30) of the methyl ketone grouping (CH_3-CO), the other is CH_2 group (53) bonded to the electronegative carbonyl group. Three features accounts for the grouping CH_3-CO-CH_2- in the compound.

The remaining three carbons are therefore, in the form of an isopropyl group, of which the two methyl groups are chemically equivalent and thus both appear at the same position to give one peak at δ 23.

(e) Mass Spectrum: Mass spectrum shows a peak at m/z 43 (base peak) and another peak appeared at m/z 57. When a -hydrogen is available, the McLafferty rearrangement occurs, and this leads to the fission of the bond to give a major peak at m/z 58 depending on the -substitution. The peak at m/z 85 shows the loss of a methyl radical.

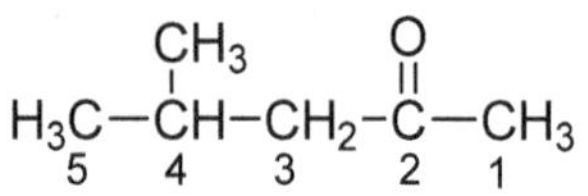

$$\underset{5}{H_3C}-\underset{4}{\overset{\overset{\displaystyle CH_3}{|}}{CH}}-\underset{3}{CH_2}-\underset{2}{\overset{\overset{\displaystyle O}{||}}{C}}-\underset{1}{CH_3}$$

4-Methyl-2-pentanone

We can write the complete structure as follows:

Fragmentation

(I)

$$CH_3 \ + \ :\overset{+}{O}{\equiv}C-CH_2-\overset{\overset{\displaystyle CH_3}{|}}{CH}-CH_3$$

m/z 85

$$H_3C-\overset{\overset{\displaystyle :\overset{+}{O}{\cdot}}{||}}{C}-CH_2-\overset{\overset{\displaystyle CH_3}{|}}{CH}-CH_3$$

m/z 100

(II)

$$CH_3-C{\equiv}O^+ \quad + \quad \overset{\overset{\displaystyle CH_3}{|}}{CH_2}-CH-CH_3$$

m/z 43   m/z 57

$$\underset{m/z\ 58}{\overset{\displaystyle :\overset{\cdot+}{O}-H}{\underset{\displaystyle H_3C}{}{C}{=}CH_2}} \quad + \quad \underset{m/z\ 42}{\overset{\displaystyle CH_2}{\underset{\displaystyle H}{}{C}{\diagdown}CH_3}}$$

MCLafferty rearrangement

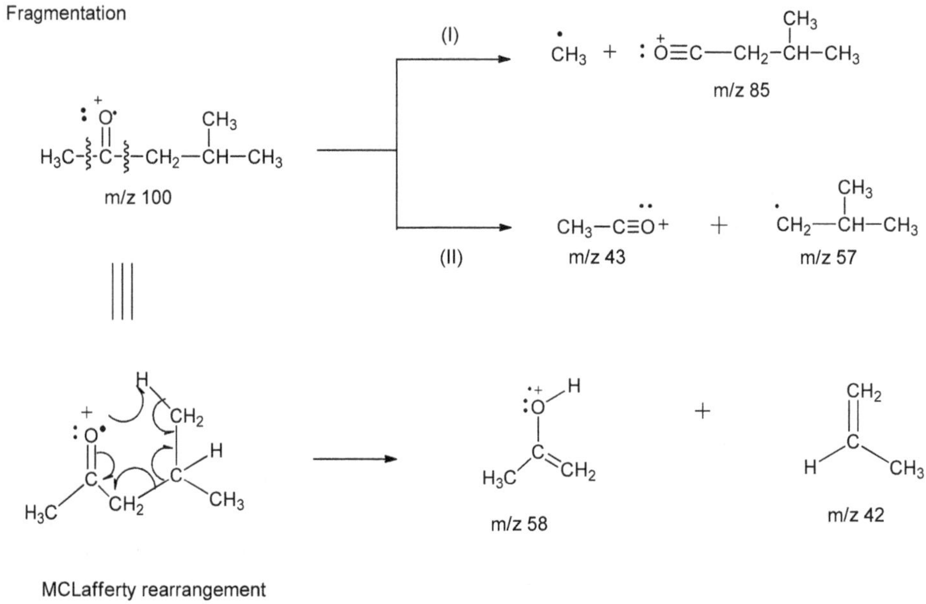

8.8.21 Problem 21

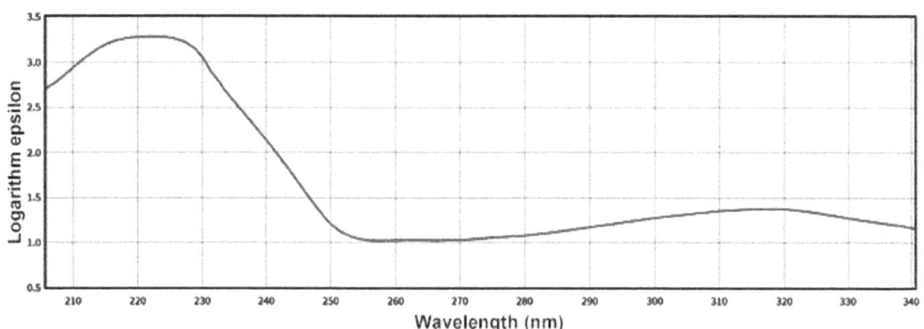

UV-Vis Spectra

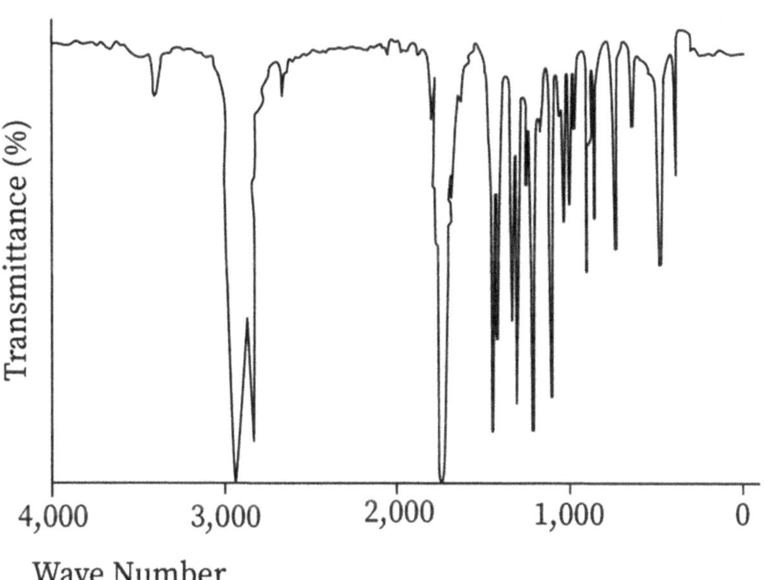

IR Spectra

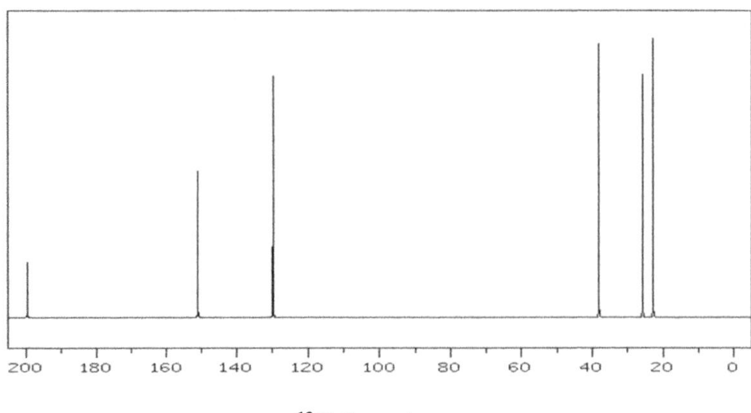

^{13}C Specctrum

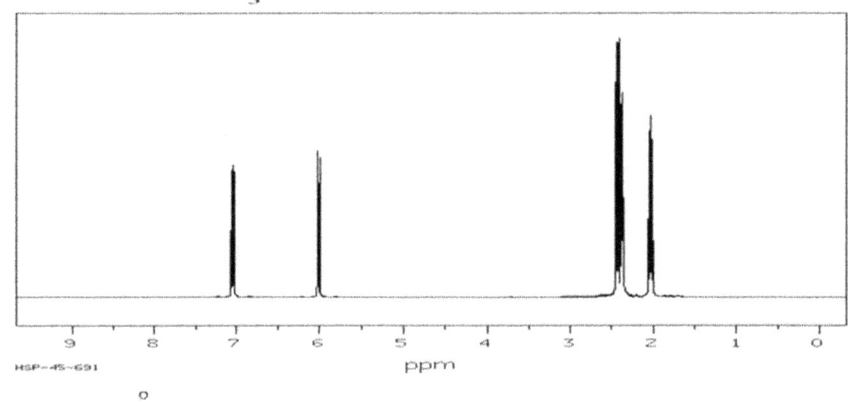

^{1}H-NMR Spectrum

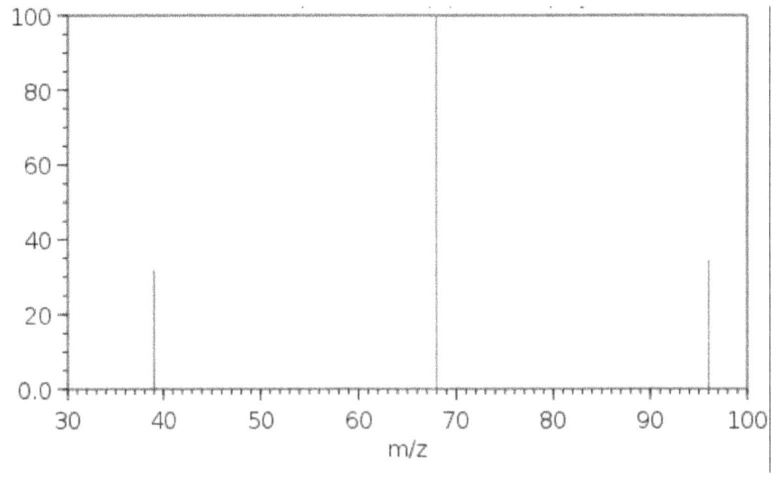

Mass spectrum

8.8.21 Solution

Molecular Formula C_6H_8O

On the basis of the molecular formula the carbon to hydrogen ratio indicates the compound is aliphatic in nature.

Double bond equivalent: $C_AH_BN_CO_D$ DBE = A - 1/2B + 1/2C + 1

Formula   C6H8O   = 6 - 8/2 + 0 + 1 = 3

Spectral data:

UV $\lambda_{max}^{(nm)}$	:	225, 318
IR vmax (cm^{-1})	:	2900 -2950, 1700 (strong),
^{1}H NMR CDCl$_3$ (δ)	:	5.9 (d), 7.0 (m), 2.4 (m) & 2.1 (m)
^{13}C NMR (δ)	:	22, 25, 39, 130,150 & 200
MS (m/z)	:	96, 68, 28

(a) **UV Spectrum:** A strong $\pi - \pi^*$ absorption in UV spectrum at 225 nm shows the presence of α, β - unsaturated ketone as it is not an aldehyde, since in the NMR the formyl proton is absent.

(b) **IR Spectrum:** The infrared spectrum confirms the presence of a conjugated carbonyl group stretching at 1690 cm^{-1} with the ring. The molecular formula has index of hydrogen deficiency of three, two of these elements of unsaturation are accounted by the enone system (-CH=CH - C=O), while the remaining may be a ring. This shows the presence of structural unit of mass 64 leaving 32 mass units (96-64=32).

(c) **^{1}H NMR Spectrum:** The NMR shows the presence of two alkenic protons on the double bond of the enone system (the hydrogen and the *beta* carbon being more deshielded). The doublet nature of the signal due to the proton on *alpha* carbon shows that the alkenic hydrogen is present on the beta carbon.

(d) **^{13}C NMR Spectrum:** The carbon-13 NMR spectrum shows six signals are non equivalent carbons. The carbon of the carbonyl group is Sp^2 hybridised and is directly bonded to an electro negative atom. It is shifted most downfield gives a peak atδ200 ppm. The two olefinic carbons appears at δ 130 and 150 respectively. The remaining ring carbons apperars at δ 22, 25 & 39 ppm respectively.

(e) **Mass Spectrum:** The base peak at 68 is due to "retro-Diels-Alder" reaction which is typical of cyclohexenone. The structure which fits into the data is 2- cyclohexenone.

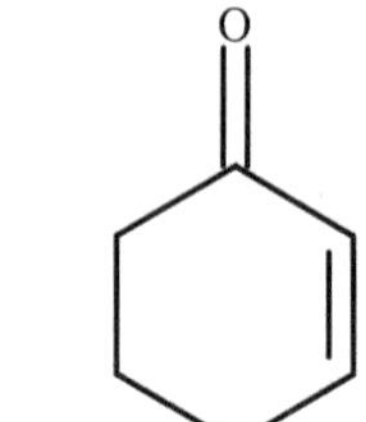

Fragmentation

PROBLEMS AND SOLUTIONS (UV, IR, NMR, MS)

Problems

9.1 Ultraviolet Spectroscopy

9.1.1　When *p*-nitro phenol is dissolved in water, the color is yellow, but when NaOH is added, the color deepens in intensity and moves to longer wavelength. Explain it.

9.1.2　When *p*-aminophenol is dissolved in water, λ_{max} is at longer wavelength than in acid solution. Explain it.

9.1.3　(a)　Which structure features may produce a bathochromic or a hypsochromic effect in an organic compound.

　　　　(a)　What is Red and Blue shifts in UV spectrum?

9.1.4.　The position of absorption of acetone shifts in different solvents: 279 nm (hexane); 272 nm (ethanol) and 265 (water). Explain.

9.2　Infrared Spectroscopy

9.2.1　How could IR spectroscopy be used to distinguish the following pairs? (Assume that you do not have access to any authenticated spectra).

　　　　(a)　Natural rubber (*cis*-polyisoprene) and butyl rubber (polyisobutene)

　　　　(b)　1-Hexyne and 3-Hexyne

　　　　(c)　The *cis* and *trans* isomers of 3-hexene

9.2.2　The hydrolysis of an ester, RCOOR, in dilute aqueous KOH produces the alcohol and the potassium salt of the carboxylic acid: the progress of the hydrolysis can be followed by IR spectroscopy. Suggest how this procedure might be carries out? Using a prominent infrared feature as a marker.

9.2.3   How might the following pairs of isomers be distinguished from their IR absorption spectra, assuming you have both spectra for each pair?

(a)   Propiophenone ($PhCOCH_2CH_3$) and Benzyl methyl ketone ($PhCH_2COCH_3$);

(b)   Propiophenone ($PhCOCH_2CH_3$); and 3-phenylpropanal ($PhCH_2CH_2CHO$);

(c)   2,5-Hexanedione ($CH_3COCH_2CH_2COCH_3$) and 2,4-hexanedione ($CH_3CH_2COCH_2COCH_3$);

(d)   Methyl benzoate (Ph $COOCH_3$) and phenyl acetate (Ph $OCOCH_3$).

9.2.4   Predict the frequency shift of the carbonyl absorption in the aldehyde (a-c).

(a)   Cyclohexane carboxaldehyde;

(b)   benzaldehyde;

(c)   salicylaldehyde

9.2.5   What is retro-Diels Alder process? Explain with an example.

9.3 NMR Spectroscopy

9.3.1   (a)   Define what is meant by shielded and deshielded protons

(b)   Tetramethylsilane (TMS) is used as internal standard substance in NMR spectra. Explain the advantage of using TMS.

9.3.2   Benzaldehyde (C_6H_5CHO) condenses in the presence of base with acetone (CH_3COCH_3) to give a compound (A) whose mass spectrum showed the molecular ion m/z 146, and two intense peaks at m/z 131 and 103. There was a strong absorption band in the infrared spectrum near 1650 cm^{-1}. The compound showed the following NMR absorptions. Proton NMR spectrum: 2.3 (3H, s), 6.7 (1H, d, J=16 Hz), 7.4 (5H, m), 7.5 (1H, d, J = 16 H z). 13C NMR spectrum: 27 (q), 127 (d), 128 (2C, d), 129 (2C, d), 130 (d), 135 (s), 144 (d), 198 (s). Deduce the structure and stereochemistry of compound A.

9.3.3   Predict the chemical shift positions for the protons in methyl acetate (CH_3COOCH_3).

9.3.4   Predict the chemical shift positions in methyl-phenoxyacetate ($C_6H_5OCH_2COOCH_3$).

9.3.5   Predict the chemical shift position for the proton in (a) benzyl alcohol ($PhCH_2OH$) and (b) acetic acid (CH_3COOH).

9.3.6   Predict the chemical shift positions for the protons in (a) succinic acid ($HOOCCH_2CH_2COOH$): and (b) paracetamol ($HOC_6H_4NHCOCH_3$)?

9.3.7   How many signals can be expected in the PMR spectrum in each of the following: (i) CH_3-$CH(Br)$-CH_2Br (ii) CH_3-$C(Cl) = CH_2$. Give reason for your answer.

9.3.8   A compound (C_9H_{10} O) shows a molecular ion at m/z 134 and a base peak at m/z 119. Its IR spectrum shows a strong band at 1680 cm^{-1}; Its NMR spectrum shows signals in three distinct regions at 2.3 (3H, s); 2.4 (3H, s) and at 6.4 – 7.5 (4H, a pair of doublets J=8Hz). Assign a structure.

9.3.9   How will you differentiate among *ortho, meta* and *para*–xylenes on the basis of their proton decoupled CMR spectra?

9.4 Mass Spectrometry

9.4.1   Write formulae for the fragment ions in the mass spectrum of

CH_3 with m/z values 83, 55, 41, 29 and 27

9.4.2   What is the mass of the 'McLafferty ion' for the following aldehydes?

(i) butanol   (ii)   pentanal   (iii) hexanal   (iv) heptanal

9.4.3   How do you distinguish the following compounds by mass spectrometry? Retro-Diels-Alder fragmentation (RDA).

i)   ii)

9.4.4   How can you differentiate these two compounds by mass spectral fragmentation?

H_3C–CH_2–CH_2–NH_2 and H_2N–CH–$(CH_3)_2$

9.4.5   In combined GC-MS analysis fatty acid methyl esters can be identified by the presence of a peak at m/z 74; why is this?

9.4.6   Two amines have respective relative molecular masses of 107 and 108 and in their mass spectra both show M-1 and M-27 peaks. What can be deduced from this, in relation to the number of nitrogen atoms in each?

9.4.7   The   mass   spectrum   of   1-(4-methylphenyl)-ethanol [p-$CH_3C_6H_4CH(OH)CH_3$] shows an abundant fragment ion at m/z 121, and a less abundant ion at m/z 119. What are feasible structures for these ions?

9.4.8   Predict the general appearance of the molecular ion peaks for 1-chlorobutane, considering the chlorine and carbon isotopes only.

9.4.9   How do you distinguish primary, secondary and tertiary alcohols by mass spectrometry?

9.4.10 The mass spectrum of n-butyl phenyl ketone ($C_6H_5COCH_2CH_2CH_2CH_3$) shows peaks at m/z 162, 120, 105 and 85: Interpret these.

9.4.11 When acetone is treated with base, a higher boiling liquid (b.p.130 oC) can be isolated from the reaction mixture. The spectroscopic properties of this liquid are: IR, 1620 (m), 1695 cm^{-1} (s); ^{1}HNMR, δ 1.9 (3H, singlet), δ 2.1 (6H, singlet), δ 6.15 (^{1}H, singlet); UV λ_{max} 700; mass m/z 55 (100), 83 (90), 43 (78), 98 (49), 29 (46), 27 (42); 13C NMR , δ 20, 27, 31, 124, 154 and 197. Make sketches of the NMR spectra and construct a mass spectrum bar diagram. Deduce the structure and account for all of the observed data.

Solutions to Problems

9.1 Ultraviolet Spectroscopy (UV)

9.1.1 In water the chromophore is the benzene ring with the phenolic oxygen nonbonding electrons; in alkali a full negative charge is on the oxygen – greater electron "availability" ; therefore, longer wavelength absorption.

9.1.2 Nitrogen nonbonding electrons are less 'available' as a chromophore after protonation by acid.

9.1.3 (a) A bathochromic shift (red shift) may occur by a change of medium or by the presence of an auxochrome. A hypsochromic shift (blue shift) may be caused by a change of the medium or by such structural changes like removal of conjugation

(b) A shift of absorption to a longer wavelength is called a bathochromic shift (red shift). A shift of absorption to a sorter wavelength is called a hypsochromic shift (blue shift).

9.1.4 This is the expected shift of the n to π^* transition of acetone to shorter wavelength (blue shift) by changing to solvents of increased polarity.

9.2 Infrared Spectroscopy (IR)

9.2.1 (a) Natural rubber still contains an alkene group, but butyl rubber does not. The presence of a weak/medium band for C = C stretching will identify the former with reasonable certainty.

(b) 1-Hexyne is a terminal alkyne and easily detected by the co-presence of C–H Stretching (strong band near 3300 cm⁻¹) and the weaker $C \equiv C$ stretching band near 2200 cm⁻¹ are easily assigned. Non-terminal alkynes are not easy to detect by IR, since they lack the valuable alkyne C–H *str* absorption, and *str* is extremely weak.

(c) *Trans* alkenes can often be distinguished from the *cis* isomer by the former's C–H *def* band around 970 cm⁻¹; The corresponding band for *cis* isomers is often appear near 700 cm⁻¹, but it is not so certainly identified as that of the *trans* isomer.

9.2.2 Follow the disappearance of the ester C=O stretching band near 1720 cm⁻¹; in aqueous alkali the reaction product shows the carboxylate absorption near 1600 cm⁻¹ and 1300 cm⁻¹. To avoid damage salt flats with aqueous medium, use calcium fluoride flats instead.

9.2.3 (a) Propiophenone is a **mono-aryl** ketone; in phenylacetone the carbonyl group has alkyl groups on both sides, and in infrared terms it is considered a dialkyl ketone. The C=O stretching band for the former will be at lower frequency than for the latter.

(b) Apart from the position of the C=O stretching bands (lower frequency in the former, since it is a monoaryl ketone), the aldehyde C-H stretching bands near 2800 cm⁻¹ will characterize the latter (3-phenylpropanal).

(c) 2,4-hexanedione is a β-diketone and will be highly enolic. It's spectrum will show broad O-H stretching absorption and a very–low frequency. C=O stretching band. 2,5-hexadione will show the IR spectral qualities of a simple dialkyl ketone.

(d) The C=O stretching bands of these two or near 1720 cm⁻¹ for the benzoate (conjugation on the C=O group of the ester)and near 1770 cm⁻¹ for the aryl acetate (conjugation on the – O - group of the ester).

9.2.4 Cyclohexancarboxaldehyde (a) being saturated will absorb around 1730 cm⁻¹. In the case of benzaldehyde; (b) the absorption will be shifted to lower frequency (1700 cm⁻¹) due to conjugation with the ring. In salicylaldehyde (c) the internal (chelated) hydrogen bonding causes a further large frequency shift to around 1634 - 1666 cm⁻¹.

9.2.5 A double bond in a cyclic system is not free to migrate, and if it is suitably positioned to produce energetically favorable fragments a characteristic fission process known as the retro-Diels Alder process. It gives a charged diene and an uncharged ethylene fragments respectively. This is represented as follows.

m/z 68

m/z 28

9.3 Nuclear Magnetic Responance (NMR) Spectroscopy

9.3.1 (a) A molecule giving rise to a signal further upfield than another nucleus is said to be more shielded.

   (b) TMS is a good reference substance for non-aqueous solutions since it is inert, readily volatile and gives rise to a single resonance line which is at a field higher than those of most organic substances. This gives rise to a sharp peak for the $(CH_3)_3Si$ (TMS) group at $\delta = 0.0$.

9.3.2 *trans*-4-Phenyl-3-buten-2-one ($C_6H_5\ CH = CHCOCH_3$)

9.3.3 Using the proton NMR, we can diagnose the presence of acetates or methyl esters in an organic molecule with some considerable certainty and these values are usually accurate to + or − 0.2 ppm, with rare exceptions.

   The CH_3 group attached to − COOR appears around δ 2.0, and that CH_3 attached to –OCOR appears around δ 3.6. Therefore, the NMR spectrum of methyl acetate shows two signals at δ 2.0 and δ 3.6, respectively.

9.3.4 Methyl esters of acids with an aliphatic residue on the carboxyl group show the methyl signal at around δ 3.6. The aromatic protons are all moved to lower frequency by the alkoxy group by approximately 0.2 ppm. The aromatic protons should come to resonance around δ 7.07 - 7.10.

   The methylene group in methyl phenoxyacetate, flanked by two electronegative groups is more difficult to deal with, but one can take base value of δ 1.2 and add 2.3 ppm for $–OC_6H_5$ and add 0.7 ppm for –COOR, giving δ 4.2. The error in predicting the δ values for methine protons frequently exceeds 0.5 ppm.

9.3.5 (a) It is a legitimate approximation to consider the substituent to be an alkyl group, R, and to ignore the more distant effect of the OH group. Thus, the ortho, meta and *para* positions are shifted respectively, by − 0.15 ppm, - 0.1 ppm and − 0.1 ppm: the ring protons is therefore all appear in between δ 7,17 - 7.3. The CH2 protons appear at δ4.2 ppm. The OH proton may appear anywhere between delta 0.5 and δ 4.5 dependent on hydrogen bonding.

(b) On the basis of the NMR the δ values for CH_3 attached to – COOH is δ 2.1. One can expect the carboxyl proton appears at δ 10-13.

9.3.6 (a) 2.5 (2.3 + 0.2; (CH_2 -CH_2), near δ 11 (COO<u>H</u>).

(b) Ring protons appeared at δ 6.27 (*ortho* to OH), 7.27. OH predicted near δ 4.5, but is observed at higher frequency, owing to H-bonding with C=O. N-H proton appeared around δ5.0 – 8.5, broad signal.

9.3.7 One can expect four NMR signals for 1,2-dibromopropane. The environment of the two protons on C^{-1} is not the same; the protons are not equivalent and will absorb at different field strengths. (ii) 2-chloropropene gives three signals in NMR spectrum. The environment of the two protons on C-1 is not the same; these protons are non-equivalent and we expect a NMR signal from each proton.

9.3.8 A strong C=O stretch at 1680 cm^{-1} shows the presence of a conjugated C=O group. That the conjugation with a phenyl group, and the presence of a 3H singlet at δ 2.3 for CH_3CO grouping shows the presence of grouping Ph-CO-CH_3 in the compound. The 3H singlet atδ 2.4 is due to –CH_3 group. The presence of 4H p-disubstituted benzene ring (a pair of doublets in the aromatic region) points to –CH_3 and CH_3-C=O as the two substituents. The compound thus comes out to be *p*-methoxyacetophenone. This is further confirmed by the presence of a base peak at m/z 119 in the mass spectrum is due to the loss of methyl radical leaving a resonance stabilized acylium ion.

9.3.9 *ortho*-Xylene gives 4 signals; *meta*-isomer gives 5 signals; *para*-isomer gives 3 signals on the basis of CMR spectra.

9.4 Mass Spectrometry (MS)

9.4.1 C_6H_{11}+ (cyclohexyl), C_4H_7+ (butenyl), C_3H_5+ (propenyl), C_2H_5+ (ethyl), C_2H_3+ (ethenyl, vinyl).

9.4.2 m/z 44 for all of them

9.4.3 Mass fragmentation:

i)

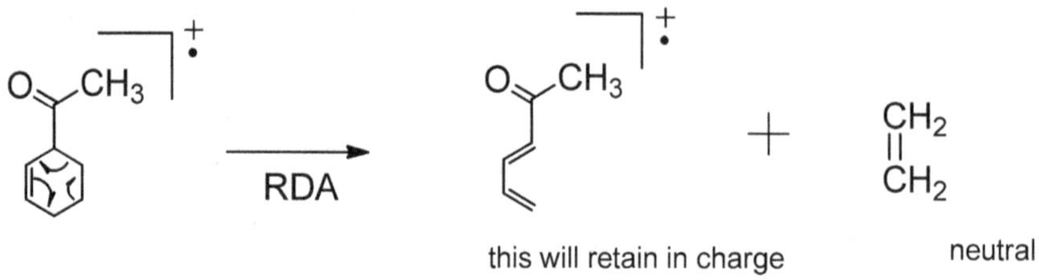

ii)

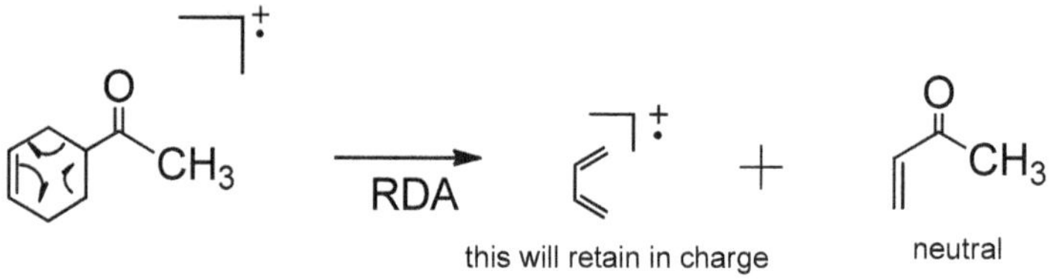

9.4.4 (i) $CH_3—CH_2—CH_2—NH_2\}+. \rightarrow [CH_2 = \overset{+}{N}H_2] + \overset{+}{C}_2H_5$

Iminium ion m/z 30 m/z 29

(ii) $(CH_3)_2 CH-NH_2\}+. \rightarrow \{CH_3- CH = \overset{+}{N}H_2\} + \overset{+}{C}H_3$

m/z 44m/z 15

9.4.5   m/z 74 is the McLafferty ion formed from the methyl esters of longer-chain fatty acids; it is highly characteristic.

9.4.6   The amine with odd number mass must contain 1, 3, 5 etc., nitrogen atoms; the other amine must contain 2, 4, 6, etc., nitrogens. They could be a toluidine ($CH_3 C_6H_4 NH_2$) and a phenylenediamine $\{C_6H_4(NH_2)_2\}$.

9.4.7   Benzylic alcohols readily fragment at the benzylic carbon, expelling a radical, and leaving an ion which rearranges to the hydroxytropylium structure. In this case the molecular ion will appear at m/z 136 and therefore the ion at m/z 121 corresponds simply to loss of CH_3+, and its structure is methylhydroxytropylium, C_8H_9O+.The ion at m/z 119 corresponds to loss of CH_5 (CH_3+,and two hydrogen atoms) and is the acyl ion $CH_3C_6H_4CO+$.

9.4.8   Two peaks in the abundance ratio 3:1 at m/z 92 and 94 (chlorine-35 and -37 isotopes), with two much smaller peaks at m/z 93 and 95 (13carbon isotope peaks).

9.4.9   We can distinguish 10, 20 & 30 alcohols on the basis of initial fragmentation of hydrogen ions.In primary alcohols, one can observe M-1, M-2 and M-3 are common fragments due to weak peaks corresponding to successive loss of hydrogen radicals from the molecular ion peak.

In secondary alcohols, one can observe M-1 & M-2 are common fragments.

In tertiary alcohols, M-1, M-2 & M-3 are not exhibited

9.4.10 With molecular formula $C_{11}H_{14}O$, the m/z 162 peak is the molecular ion. The m/z 120 fragment therefore corresponds to loss of 40 daltons from M+., which is accounted for by expulsion of propene, C_3H_6, in the McLafferty rearrangement. The ions m/z 105 and m/z 85 have formulae $C_6H_5CO^+$, and $C_4H_9CO^+$, respectively, and both arise from α fragmentation.

9.4.11 Structure of the isolated liquid on the basis of spectral data: $(CH_3)_2 - C = CH - CO-CH_3$ (mesityl oxide).

APPENDIX - I

Atom: The smallest particle of an element which can enter into a chemical combination.

Atomic number: The atomic number is the charge on the atomic nucleus i.e., the number of protons in the nucleus or the number of planetary electrons circulating about the nucleus.

Atomic mass: Atomic mass is the sum of the protons and the neutrons in the atom. Example, if there are 12 protons and 13 neutrons in one atom of a particular element it will have an atomic mass of 25.

Atomic weight: Atomic weight is the relative weight of the atom on the basis of 12C $\square\square$ 12. For a pure isotope, the atomic weight rounded off to the nearest integer gives the total number of neutrons and protons making up the atomic nucleus. If these weights are expressed in grams, they are called gram atomic weights.

Anode: The positive electrode

Allotropy: When an element can exist in two or more distinct physical forms it is said to exhibit allotropy. This arises when the atoms of the same element can be arranged in different ways giving rise to different physical properties. For example carbon is an allotropic element and exists as diamond and graphite.

Adsorption: The condensation of gases, liquid or dissolved substances on the surfaces of solids is called adsorption.

Activation energy: A reaction will need energy to start it unless it can proceed spontaneously. Minimum or initiation energy is called activation energy.

Alkali metals: Because they form Alkalis when dissolved in water. (e.g., group-1 of the periodic table).

Alkanes: A homologous series of hydrocarbons having the general formula CnH2n+2.

Alkenes: A homologous sereies of hydrocarbons having the genral formula CnH2n.

Alkynes: A homologous series of hydrocarbons having the general formula CnH2n-2.

Aliphatic: The derivatives of open-chain hydrocarbons are known as aliphatic cmpounds.

Aromatic: The derivatives of benzene or of any system containing a benzene nucleus are called aromatic compounds.

Acid: An acid can be defined, as a substance which when dissolved in water forms a positively charged hydroxonium ions (H_3^+O).

Anion: A negatively charged ion.

Base: A base is a substance which will react with an acid to form a salt and water only. A soluble base is called an alkali, in which hydroxyl ions (OH-) are the only negatively charged ions.

Catalyst: A catalyst is a substance which increases the rate of a chemical reaction by lowering the activation energy but takes no chemical part itself. i.e., it is unchanged in mass and chemical composition at the end of the reaction.

Chemical formula: It indicates the number of atoms of different elements present in a molecule.

Chemical change: In a chemical change a new substance is formed, and the change that has taken place is difficult to reverse. e.g., when iron is oxidised to iron oxide, the reverse process is quite complex.

Covalent bond: A covalent bond is formed by sharing of electrons.

Cracking: It is the name given to the process of splitting up higher aliphatic hydrocarbon molecule into much smaller substances. This process is carried out at high temperature, and a catalyst is used.

Compound: A compound is a substance made up of two or more elements which are chemically combined together, so that their individual properties have changed. Sodium is a highly reactive metal, chlorine a poisonous gas. When these two elements are combined together they form sodium chloride.

Chemiluminescence: Emission of light during chemical reaction.

Conservation of mass: In all ordinary chemical changes, the total of the reactants is always equal to the total mass of the products.

Chromatography: The analysis of a mixture by selective absorption techniques.

Chemical equation: A representation of a chemical change in terms of symbols.

Cathode: The negative electrode.

Drying agent: A drying agent will absorb water or water vapour from other substances. It is not to be confused with a dehydrating agent.

Dehydrating agent: A dehydrating agent will remove the elements of water from pure and perfectively dry compounds. e.g., conc. sulphuric acid converts blue copper(II) sulphate crystals to white anhydrous copper(II) sulpahte.

Deliquescence: When a substance absorbs moisture from the air and dissolves in it to form a solution, it is said to be deliquescence. e.g.,CaCl2 and NaOH etc.

Diffusion: This is a process whereby a gas passes through a porous material independent of gravity.

Density: Mass per the unit volume at specific temperature.

Distillation: The process of heating a liquid to above its boiling point in order to convert it to the gaseous state. It is employed in the purification and separation of liquids and liquid mixtures.

Electron: An electron is a particle found circulating around the nucleus of the atom and having a charge of -1 and negligible mass.

Endothermic: A reaction in which energy is given out.

Exothermic: A reaction in which energy is taken in.

Element: An element is a substance which cannot be split up chemically into another simpler substance.

Esterification: This is a process of producing an ester by heating an organic acid with alcohol in the presence of a few drops of conc. sulphuric acid.

Electrode: The conductor by which the electric current passes into or out of a solution during electrolysis.

Electrolyte: A solution or fused salt which conducts the electric current and is decomposed by it.

Formula: The formula for ethyl alcohol C2H5OH, which indicates that this substance is composed of two atoms of carbon combined with 6 atoms of hydrogen and one atom of oxygen.

Fluorescence: The property of emitting radiation as the result of absorption of radiation from some other source.

Gram mole, gram formula weight, gram equivalent: Mass in grams numerically equal to the molecular weight, formula weight or chemical equivalent respectively.

Gram atom or gram atomic weight: The mass in gram numerically equal to the atomic weight.

Gram molecular weight or gram molecule: A mass in gram of a substance numerically equal to its molecular weight. gram mole.

Hard water: Hard water does not readily form a lather with soap.

Hard water (Temporary): This type of hardness can be removed by boiling. It is due to the presence of the hydrogen carbonate of calcium and magnesium.

Hard water (Permanent): This type of hardness cannot be removed by boiling and is due to the presence of chlorides and / or sulphates of calcium and magnesium.

Ion: An ion is an electrically charged atom or radical formed by the transfer of electron(s).

Ionic equation: These are equations involving ions only. Reduction-oxidation (Redox) reactions are conveniently expressed in this way.

Indicators: They are chemically different compounds which indicate by changing their colour whether a solution is acidic or alkaline. They are usually vegetable dyes. e.g., litmus, methyl orange, phenolphthelein etc.

Isomerism: Existance of molecules having the same number and kinds of atoms but in different configurations.

Isotopes: Two or more nuclides having the same atomic number, hence constituting the same element, but differing in mass number.   Isotopes of a given element have the same number of nuclear protons but differing numbers of neutrons.

Mineral acids: These are sulphuric, hydrochloric and nitric acid etc.

Macromolecules: A macro molecule is a gaint molecule usually made up of smaller units. e.g., diamond is a macromolecule, being made up of carbon atoms.

Molar solution: A molar solution contains 1 mole of a substance in 1 liter of a solution e.g., a molar solution of sodium hydroxide would contain 40 g of sodium hydroxide in 1 liter.

Molecular weight: The sum of the atomic weights of the atoms in a molecule.

Mol volume: The volume occupied by a mol or a gram molecular weight of any gas measured at standard conditions is 22.414 liters.

A molal solution: It contains one mole/per 1000 grams of solvent.

A molar solution: It contains one mole or gram molecular weight of the solute in one liter of solution.

Mole: Mass numerically equal to the molecular weight i.e., as the weight of one mole expressed in grams.

Molecular volume: Volume occupied by one mole numerically equal to the molecular weight divided by the density.

Molecule: The smallest unit quantity of matter which can exist by itself and retains all the properties of the original substance.

Mixture: A mixture consists of two or more substances (elements or compounds) which are not combined together chemically.

Neutron: A neutron is the second particle found in the atomic nucleus. The neutron has no charge (neutral) but has a mass of $+1$.

Normal solution: It contains one gram molecular weight of the dissolved substance divided by the hydrogen equivalent of the substance (that is, one gram equivalent) per liter of solution.

Neutralisation: When an acid acts with a base to form a salt and water only, the acid is said to have been neutralised by the base.

Oxidation: It is a process in which one or more oxygen atoms are introduced or one or more hydrogens eliminated.

pH scale: The strength of various acids and bases can be measured according to this scale. The pH of a solution is the negative logarithm of the hydrogen ion concentration measured in mol/dm3. The pH = -log10 (H+). The scale ranges from 1-14. Strong acids having a pH of 1, while strong alkalies would register values of 14. Neutral solutions would obviously have pH of 7.

Pressure: Force applied to or distributed, over a surface; measured as force per unit area.

Photochemical reaction: Light increases the rate of some reactions. If the rate of a reaction is affected by light, it is called a photo chemical reaction.

Proton: A proton is one of the particles which is found in the nucleus of the atom. It has a charge +1 (equal and opposite to that of the electron) and a mass of +1.

Physical change: A physical change is one in which a substance undergoes without any major change in its properties. Processes such as melting point, boiling point and subliming are all examples of physical changes.

Reversible reaction: A reversible reaction is one which can work in both directions, by altering the condition under which the reaction is carried out.

Radiation: The emission and propagation of energy through space or through a material medium in the form of waves.

Resonance (Chemical): The moving of electrons from one atom of a molecule or ion to another atom of that molecule or ion. It is simply the oriented movement of the bonds between atoms.

Salt: Any substance which yields ions, other than hydrogen or hydroxyl ions. A salt is obtained by displacing the hydrogen of an acid by a metal.

Solid: A state of matter in which the relative motion of the molecules is restricted and they tend to retain a definite fixed position relative to each other, giving rise to crystalline structure. A solid may be said to have a definite shape and volume.

Specific gravity: The ratio of the mass of a body to the mass of an equal volume of water at 4°C or other specified temperature.

Specific heat: Specific heat of a substance is the ratio of its thermal capacity to that of water at 15°C.

Solute: The substance that dissolves when a solution is made. Example salt (solute) in solution (water).

Solvent: The substance that causes dissolving to take place when a solution is made. Example: salt (solute) dissolves in water (solvent).

Solution: A solution is formed when a solute (salt) dissolves in a solvent (water).

Saturated solution: A Saturated solution is formed when no more solute dissolves in a solvent at constant temperature.

Suspension: A mixture of a liquid and a finely divided insoluble solid 'held' by the liquid.

Supernatant liquid: The liquid found above a precipitate after the precipitate has settled.

Sublimation: A substance which sublimes from the solid state to the gaseous state without passing through the liquid state when heated upon cooling the vapour condenses to the solid. Examples, ammonium chloride, iodine and benzoic acid etc.

Vapor Pressure: The pressure exerted when a solid or liquid is in equilibrium with its own vapour. The vapour pressure is a function of the substance and of the temperature.

Valency: The valency of an element is the number of hydrogen atoms (or its equivalent) which will combine with or be displaced by one atom of another element, that is the combining capacity of an element.

APPENDIX – II

PHYSICAL CONSTANTS OF ORGANIC COMPOUNDS

The melting points, boiling points, and in appropriate instances the refractive indices and densities of some typical and common members of the functional groups are tabulated in this chapter, together with the melting points of some of their derivatives. The lists are far from comprehensive but should be adequate for introductory work. More comprehensive tables are to be found in the following books and elsewhere.

1. Text book of practical organic chemistry by A.I. Vogel (5th edition) pearson eduction (singapore) Pte Ltd.

2. Dictionary of Organic compounds. Eyre and Spottishwoode, London 1965, 5 Volumes.

3. The systematic Identification of organic compounds by R.L. Shriner, R.C. Fuson and D.Y. Curtin. Willy, New York, 1964.

4. Hand book of chemistry and physics 55th edition (1974-1975) CRC press.

The identification of derivatives prepared is not the sole use of these tables. Judicious use of them will indicate which derivatives should be prepared, so that the maximum of information may be gained from the minimum effort.

Transcription errors are almost inevitable in tables containing masses of numarical data. As much of the information as possible has been checked with original literature. The authors would appreciate notification of any remaining discrepancies.

ARYL SUBSTITUTED ALCOHOLS

Alcohols	B.P.	M.P.	3,5-Dinitro-phenylbenzoate	p-Nitro benzoate	Phenyl urethane	1-Naphthyl urethane
	°C	°C	°C (m.p)	°C (m.p)	°C (m.p)	°C (m.p)
Benzyl alcohol	205	--	113	86	76	134
1-Phenyl ethanol	203	20	94	43	92	106
2-Phenyl ethanol	220	--	108	63	80	119
Cinnamyl alcohol	257	33	121	78	91	114
Diphenylcarbinol	298	69	142	131	140	136
Triphenyl carbinol	--	165	--	--	--	--
o-Hydroxybenzyl alcohol	--	87	--	--	--	Benzoyl 51
m-Hydroxybenzyl alcohol	--	73	--	--	--	Acetyl 84
p-Hydroxybenzyl alcohol	--	125	--	--	--	--
o-Aminobenzyl alcohol	--	82	--	--	--	--
m-Aminobenzyl alcohol	--	97	--	--	--	N-acetyl 107
p-Aminobenzyl alcohol	--	65	--	--	--	--
o-Nitrobenzyl alcohol	270	74	--	--	--	Benzoyl 102
m-Nitrobenzyl alcohol	--	27	--	--	--	Benzoyl 72
p-Nitrobenzyl alcohol	185/12	93	--	--	--	Benzoyl 95
o-Chlorobenzyl alcohol	230	74	94	--	--	--
m-Chlorobenzyl alcohol	234	--	--	--	--	--
p-Chlorobenzyl alcohol	235	75	--	--	--	--
o-Bromobenzyl alcohol	--	80	--	--	--	--
m-Bromobenzyl alcohol	254	--	--	--	--	--
p-Bromobenzyl alcohol	--	77	--	--	--	--

ALIPHATIC ALCOHOLS

Name of the compound	B.P.	M.P.	nD20	3, 5-Dinitro benzoate	p-Nitro benzoate	Phenyl urethane	Alphna phthyl urethane
	(°C)	(°C)		Melting points			
Methyl alcohol	65	--	1.33	109	96	47	--
Ethyl alcohol	78	--	1.36	94	57	52	--
1-Propanol	97	--	1.38	75	35	57	--
2-Propanol	82	--	1.37	122	110	76	--
2-Methyl-1-propanol	108	--	1.39	88	68	86	--
2-Methyl-2-propanol	83	25	1.38	142	116	136	101
1-Pentanol	138	--	1.41	46	--	46	68
2-Pentanol	119		1.40	62	17	--	76
3-Pentanol	116	--	1.40	100	--	49	95
2-Methyl-1-butanol	129	--	1.41	70	--	--	82
3-Methyl-1-butanol	132		1.41	62	21	57	68
2-Methyl-2-butanol	102	--	1.40	118	85	42	72
3-Methyl-2-butanol	113	--	1.39	76	--	69	109
Cyclopentanol	141	--	1.41	115	62	132	118
Cyclohexanol	161	25	1.46	113	50	82	129
Ethylene glycol	197	--	1.43	169	141	157	176
Glycerol	290	--	1.47	--	188	180	192
Furfuryl alcohol	170	--	--	81	76	45	129
4-Methyl cyclohexanol	174	--	--	134	--	125	160
3-Methyl cyclohexanol	175	--	--	98	--	94	122
Trimethylene glycol	214	--	--	178	--	137	164
1-Butanol	118	--	1.39	64	36	61	--
2-Butanol	99	--	1.39	76	26	64	--

PHENOLS

Compond name	B.P. (°C)	M.P. (°C)	Bromo compd.	Acet-ate	Benzo ate	p-Nitro benzoate	3,5-Dinitro benzoate	Aryloxy acetic acid	p-Toluene sulfonate
						Melting points (°C)			
Phenol	182	43	95	liq.	69	126	146	99	96
o-Cresol	191	30	56	liq.	liq.	94	138	152	55
m-Cresol	202	--	84	liq.	55	90	165	103	56
p-Cresol	202	36	49	liq.	71	98	181	136	70
o-Chloro phenol	176	--	--	liq.	liq.	115	143	145	74
m-Chloro phenol	214	33	--	liq.	71	99	156	110	--
p-Chloro phenol	217	43	--	liq.	90	168	186	156	71
o-Bromo phenol	194	05	95	liq.	--	--	--	143	78
m-Bromo phenol	236	33	--	liq.	86	--	--	108	53
p-Bromo phenol	235	64	95	21	102	180	191	159	94
o-Iodo phenol	--	43	--	--	35	--	--	135	80
m-Iodo phenol	--	40	--	38	73	133	183	115	61
p-Iodo phenol	--	94	--	32	119	--	--	156	99
o-Nitro phenol	216	45	117*	41	59	141	155	158	83
m-Nitro phenol	--	97	91*	56	95	174	159	156	113
p-Nitro phenol	--	114	142*	83	142	159	186	187	97
Catechol	240	105	192	65	84	169	152	--	--
Resorcinol	280	110	112*	liq.	117	182	201	195	81
Hydroquinone	286	170	186*	124	205	250	317	250	159
2,3-Dimethylphenol	218	75	--	--	--	126	--	187	--
3,4-Dimethylphenol	225	62	171	--	58	--	181	163	--

PHENOLS Contd.,

Compond name	B.P. (°C)	M.P. (°C)	Bromo compd.	Acet- ate	Benzo ate	p-Nitro benzoate	3,5-Dini troben zoate	Aryloxy acetic acid	p-Toluene sulfonate
						Melting points (°C)			
2,4-Dimethylphenol	211	28	--	--	38	105	164	142	--
3,5-Dimethylphenol	219	68	166	--	--	109	195	86	83
2,5-Dimethylphenol	211	75	108	liq.	61	87	137	118	--
2,6-Dimethylphenol	203	49	79	--	--	--	159	140	--
2,4-Dichlorophenol	210	45	68	--	96	--	--	140	125
2,4-Dibromophenol	239	40	--	36	98	184	--	153	120
2,4-Diiodophenol	--	72	--	71	98	--	--	166	--
2,4,6-Trichloro phenol	246	96	--	--	75	106	--	182	--
2,4,6-Tribromo phenol	--	94	120*	87	81	153	174	200	113
2,4,6-Triiodo phenol	--	159	--	156	--	--	181	224	--
Salicylic acid	--	159	--	135	132	205	--	191	155
2,4-Dinitro phenol	--	113	118	72	132	139	--	--	121
Picric acid	--	122	--	76	--	143	--	--	--
α-Naphthol	279	94	105*	49	56	143	217	192	88
β-Naphthol	285	123	84	107	72	169	210	154	125
Vanillin	--	81	160	102	78	--	--	189	115
Pyrogallol	309	133	158*	173	90	230	205	198	--
o-Hydroxybiphenyl	275	58	--	--	76	--	--	--	65
p-Hydroxydiphenyl	306	165	--	--	151	--	--	--	177

CARBOXYLIC ACIDS

Aliphatic Carboxylic acids	B.P. (°C)	M.P. (°C)	Ani-lide	p-Tolu idide	Amide	p-Bromo phenacyl ester	p-Nitro benzyl ester	p-Phenyl phenacyl ester	p-Bromo anilide
						Melting points (°C)			
Formic	101	8	50	53	03	140	31	74	119
Acetic	118	16	114	153	82	86	78	111	166
Propionic	141	--	106	126	79	59	31	102	--
n-Butyric	163	--	96	75	116	63	35	82	111
Iso-Butyric	154	--	105	109	129	77	--	89	151
n-Valeric	187	--	63	74	106	75	--	63	--
Iso-Valeric	174	--	113	109	136	68	--	78	129
n-Hexanoic	205	--	95	74	100	72	--	70	105
n-Heptanoic	223	--	65	80	95	72	--	62	95
Palmitic	--	63	91	98	106	86	43	94	--
Stearic	--	70	94	102	109	90	--	97	--
Chloroacetic	189	63	137	162	120	105	--	116	--
Dichloro acetic	194	10	119	153	97	99	--	--	--
Trichloro acetic	196	58	95	113	141	--	80	--	--
Bromo acetic	208	50	130	91	91	--	89	--	--
Iodoacetic	--	83	144	--	95	--	--	--	--
Acrylic	140	--	105	141	85	--	--	--	--
Crotonic	189	72	118	132	150	95	67	--	--
Oxalic	--	100	246	268	--	242	204	165	--
Malonic	--	135	225	253	170	--	86	175	--
Succinic	--	185	229	255	260	211	88	208	--
Glutaric	--	98	224	218	175	137	69	152	--

CARBOXYLIC ACIDS *contd...*

Compond name	B.P. (°C)	M.P. (°C)	Bromo compd.	Acetate	Benzoate	p-Nitro benzoate	3,5-Dinitro benzoate	Aryloxy acetic acid	p-Toluene sulfonate
						Melting points (°C)			
Adipic	--	150	239	241	220	155	106	148	--
Pimelic	--	105	156	206	--	137	--	146	--
Azelaic	--	106	187	202	172	131	44	141	--
Malic	--	101	197	207	157	179	124	--	--
Maleic	--	135	187	142	181	--	89	--	--
Fumaric	--	286	314	--	266	--	151	--	--

AROMATIC CARBOXYLIC ACIDS

Aromatic Carboxylic acids	B.P. (°C)	M.P. (°C)	Ani-lide	p-Tolu idide	Amide	p-Bromo phenacyl ester	p-Nitro benzyl ester	p-Phenyl phenacyl ester	p-Bromo anilide
Benzoic	--	122	162	158	129	119	89	167	106
o-Toluic	--	105	125	144	143	57	91	95	--
m-Toluic	--	112	125	118	95	108	87	136	75
p-Toluic	--	181	146	160	159	153	104	165	133
o-Chlorobenzoic	--	141	118	131	141	107	106	123	--
m-Chlorobenzoic	--	158	124	--	134	117	107	154	--
p-Chlorobenzoic	--	240	194	--	179	126	130	160	--
o-Bromobenzoic	--	150	141	--	155	102	110	98	--
m-Bromobenzoic	--	155	146	--	155	126	105	155	--
p-Bromobenzoic	--	252	197	--	189	134	141	160	--
o-Iodobenzoic	--	162	141	--	184	110	111	143	110
m-Iodobenzoic	--	187	--	--	186	128	121	--	--
p-Iodobenzoic	--	270	210	--	218	146	141	171	--
o-Nitrobenzoic	--	147	155	--	175	107	112	140	--
m-Nitrobenzoic	--	141	154	162	142	137	142	153	101
p-Nitrobenzoic	--	239	211	203	200	136	169	182	142
Phenyl acetic	265	76	218	136	157	89	65	88	122
Diphenyl acetic	--	148	180	173	168	--	104	111	--
Salicylic	--	159	135	156	139	140	98	148	136
m-Hydroxybenzoic	--	201	157	163	167	176	108	--	142
p-Hydroxybenzoic	--	215	197	204	162	184	192	240	--

AROMATIC CARBOXYLIC ACIDS contd….

Aromatic Carboxylic acids	B.P. (°C)	M.P. (°C)	Ani-lide	p-Tolu idide	Amide	p-Bromo phenacyl ester	p-Nitro benzyl ester	p-Phenyl phenacyl ester	p-Bromo anilide
					Melting points (°C)				
o-Methoxybenzoic	--	101	131	--	129	--	113	131	--
Anisic	--	184	171	186	163	152	132	160	--
Phthalic	--	208	254	--	220	153	155	167	179
Iso-Phthalic	--	347	--	--	280	179	203	--	--
Terephthalic	--	>300	337	--	--	225	264	--	266
Cinnamic	--	133	153	168	147	146	117	182	225
3-Nitrophthalic	--	219	234	223	201	--	190	--	--
4-Nitrophthalic	--	165	192	--	200	--	--	120	--
o-Nitrocinnamic	--	240	--	--	185	142	132	146	--
m-Nitrocinnamic	--	205	--	--	196	178	174	--	--
1-Naphthoic	--	161	163	--	202	135	--	--	--
2-Naphthoic	--	185	170	191	192	--	--	--	--
Biphenyl acetic	--	148	180	173	168	--	104	111	--
Acetyl salicylic	--	135	136	--	138	--	90	--	--
p-Nitrocinnamic	--	287	--	--	217	191	187	192	--
Hippuric	--	187	208	--	183	151	136	163	--

ALDEHYDES AND KETONES

Aliphatic Aldehydes	B.P. (°C)	2,4-Dinitrophenyl hydrazone	Oxime	Semi carbazone	Phenyl hydrozone	p-Nitro phenyl hydrazone	Dimethone
				Melting points (°C)			
Formaldehyde	-21	166	--	--	--	182	189
Acetaldehyde	20	168	47	162	--	129	141
Propionaldehyde	49	155	40	89	--	124	155
n-Butyraldehyde	75	123	--	106	--	91	134
Iso-Butyraldehyde	64	187	--	126	--	131	154
n-Valeraldehyde	104	108	52	--	--	--	105
Iso-Valeraldehyde	92	123	48	132	--	110	115
Glyoxal	50	328	178	270	180	--	--
Acrolein	52	165	--	171	--	151	--
n-Hexaldehyde	128	107	50	106	--	80	109
Diethylacetaldehyde	117	130	--	99	--	--	102
n-Heptaldehyde	154	108	57	109	--	73	103
n-Octaldehyde	170	106	60	101	--	80	90
n-Nonaldehdye	190	100	64	100	--	--	86
Crotonaldehyde	102	190	119	201	--	--	92
Aliphatic Ketones							
Acetone	56	128	59	190	42	149	--
Diethyl ketone	102	156	--	139	--	144	--

Aliphatic ketones	B.P. (°C)	2,4-Dinitrophenyl hydrazone	Oxime	Semi carbazone	Phenyl hydrozone	p-Nitro phenyl hydrazone	Dimethone
				Melting points (°C)			
Di-n-Propyl ketone	144	75	--	133	--	--	--
*Di-iso-P*ropyl ketone	124	88	34	160	--	--	--
Di-n-Butyl ketone	185	--	--	90	--	--	--
Methylethyl ketone	80	115	--	146	--	129	--
Methyl-n-propyl ketone	102	144	--	106	--	117	--
Methyl-iso-propyl ketone	94	120	--	114	--	109	--
Methyl-iso-butyl ketone	117	95	58	132	--	79	--
Methyl-n-butyl ketone	128	107	49	125	--	88	--
Cyclopentanone	131	146	57	210	55	154	--
Cyclohexanone	156	--	162	91	167	81	147
Cycloheptanone	180	--	148	--	162	--	--
2-Methylcyclopentanone	139	--	--	--	184	--	--
2-Methylcyclohexanone	165	--	137	43	197	--	132
3-Methylcyclohexanone	170	--	155	--	191	94	119
4-Methylcyclohexanone	171	--	134	39	200	110	128
Camphor	209	179	177	119	238	233	217

Aliphatic ketones	B.P. (°C)	2,4-Dinitrophenyl hydrazone	Oxime	Semi carbazone	Phenyl hydrozone	p-Nitro phenyl hydrazone	Dimethone
				Melting points (°C)			
Aromatic Ketones							
Acetophenone	202	20	249	60	200	106	185
Propiophenone	218	20	191	54	182	--	--
Butyrophenone	229	--	190	50	188	--	--
o-Methylacetophenone	214	--	159	61	212	--	--
m-Methylacetophenone	220	--	207	55	205	--	--
p-Methylacetophenone	226	28	248	88	208	96	198
Benzophenone	306	49	238	144	167	137	155
Methylbenzyl ketone	216	27	156	69	190	87	145
1-Tetralone	129/12	--	257	--	226	84	231
Methyl-1-napthyl-ketone	--	34	--	139	229	146	--
m-Nitroacetophenone	--	81	228	132	257	135	--
o-Chloroacetophenone	229	--	--	113	160	--	--
p-Chloroacetophenone	236	20	231	94	203	114	239
o-Bromoacetophenone	112/10	--	--	--	177	--	--
m-Bromoacetophenone	131/16	--	--	--	238	--	--
p-Bromoacetophenone	256	51	230	129	208	126	248
p-Hydroxybenzophenone	--	135	242	--	194	144	--

AROMATIC ALDEHYDES

AROMATIC ALDEHYDES	B.P (°C)	M.P. (°C)	2,4-Dinitrophenyl hydrazine	oxime	Semi-carbazone	Phenyl hydrazone	p-Nitro phenyl hydrazone	Dimethone
				Melting points (°C)				
Benzaldehyde	179	--	237	35	224	158	192	195
o-Tolualdehyde	200	--	164	49	212	106	222	167
m-Tolualdehyde	199	--	194	60	223	91	157	172
p-Tolualdehyde	204	--	233	80	234	112	201	--
o-Chlorobenz aldehyde	213	11	209	76	226	86	249	205
m-Chlorobenz aldehyde	214	18	248	71	229	134	216	--
p-Chlorobenz aldehyde	214	47	265	110	232	127	220	--
m-Bromo benzldehyde	234	--	257	72	205	141	220	--
p-Bromo benzldehyde	--	67	266d	111	228	113	208	--
o-Nitro benzldehyde	--	44	265	103	256	156	263	--
m-Nitro benzldehyde	--	58	292	122	246	121	247	198
p-Nitro benzldehyde	--	106	320	133	221	159	249	190
Salicylaldehyde	197	--	252	63	231	143	228	211
p-Hydroxy benzldehyde	--	116	280	72	224	177	266	189
Anisaldehyde	248	--	254	132	209	121	161	145
Phenylacetaldehyde	195	34	121	99	156	63	151	165
Cinnamaldehyde	252	--	255d	139	215	168	195	213
Vanillin	--	81	269	117	239	105	226	197
Veratraldehde	285	58	264	95	177	121	--	--
Piperonal	263	37	266d	110	234	106	200	178
1-Naphthaldehyde	292	34	--	98	221	80	234	--
2-Naphthaldehyde	--	61	270	156	245	206	230	--

ALIPHATIC AMINES

ALIPHATIC PRIMARY AMINES	B.P (°C)	p-Toluene sulfon amide	Benzene sulfon amide	Benz amide	Acet- amide	Phtha- lamide	Picrate
			Melting points (°C)				
Methyl amine	-7	75	30	80	--	134	215
Ethyl amine	17	63	58	71	--	78	165
n-Propylamine	49	52	36	84	--	66	135
Iso-Propylamine	35	51	26	100	--	86	140
Allylamine	55	64	39	--	--	70	100
n-Butylamine	77	--	--	42	--	34	151
Iso-Butylamine	68	78	53	57	--	93	151
sec. Butylamine	64	55	70	76	--	--	140
n-Hexylamine	129	--	96	40	--	--	127
Cyclohexylamine	134	87	89	149	104	168	158
n-Heptylamine	155	--	--	--	--	40	121
Benzylamine	185	116	88	105	60	115	198
2-Phenylethylamine	198	66	69	116	114	130	174
Ethanolamine	171	--	--	--	--	127	160
Ethylenediamine	117	160	168	244	172	--	233
Propylenediamine	120	--	--	193	139	--	137
Aliphatic secondary and tertiary amines							
Dimethylamine	07	79	47	41	--	--	158
Diethylamine	55	60	42	42	--	--	155
di-n-Propylamine	110	--	51	--	--	--	75
di-iso-Propyl amine	84	--	94	--	--	--	140
di-n-Butyl amine	159	--	--	--	--	--	59
di-iso-Butyl amine	137	--	55	--	--	--	119
Piperidine	106	96	94	--	--	--	152
Pyrrolidine	89	123	--	--	--	--	112
Morpholine	130	147	119	--	--	--	148
Piperazine	140	173	282	--	--	--	280
tri-Methyl amine	03	--	--	--	--	--	216
tri-Ethyl amine	89	--	--	--	--	--	173
tri-n-Propyl amine	156	--	--	--	--	--	117
tri-n-Butyl amine	212	--	--	--	--	--	106

AROMATIC AMINES

Aromatic Primary amines	B.P. (°C)	M.P. (°C)	p-Toluene sulfon- amide	Benzene sulfon amide	Benz amide	Acet amide	Form amide	Picrate
					Melting points (°C)			
Aniline	183	--	103	112	163	114	47	165
o-Toluidine	200	--	110	124	144	112	59	213
m-Toluidine	203	--	114	95	125	66	--	200
p-Toluidine	200	44	118	120	158	153	53	181
o-Toluidine	225	--	127	89	60	88	83	200
m-Anisidine	251	--	68	--	--	80	57	169
p-Anisidine	246	57	114	96	154	127	81	--
o-Phenetidine	228	--	164	102	104	79	62	--
m-Phenetidine	248	--	157	--	103	96	52	158
p-Phenetidine	254	--	107	143	173	135	76	--
o-Chloro aniline	209	--	105	130	99	88	78	134
m-Chloro aniline	230	--	138	121	122	73	58	177
p-Chloro aniline	232	71	95	122	193	179	102	178
o-Bromo aniline	229	32	90	--	116	99	89	129
m-Bromo aniline	251	18	--	--	120	88	63	180
p-Bromo aniline	--	66	101	134	204	167	119	180
o-Iodo aniline	--	58	--	--	139	110	--	112
m-Iodo aniline	--	25	128	--	151	119	--	--
p-Iodo aniline	--	63	--	--	222	184	109	--
AromaticSecondaryand Tertiaryamines								
N-methyl Aniline	194	--	95	79	63	103	--	145
N-ethyl Aniline	205	--	88	--	60	55	--	138
N-n-propyl Amine	222	--	56	54	--	47	--	--
N-n-butyl Amine	240	--	56	--	56	--	--	--
N-methyl- *o*-Toluidine	208	--	120	--	66	56	--	90
N-methyl-	206	--	--	--	--	66	--	--

AROMATIC AMINES *contd....*

Aromatic Primary amines	B.P. (°C)	M.P. (°C)	p-Toluene sulfonamide	Benzene sulfonamide	Benz amide	Acet amide	Form amide	Picrate
				Melting points (°C)				
m-Toluidine								
N-methyl-p-Toluidine	210	--	60	64	53	83	--	131
N-ethyl-o-Toluidine	214	--	75	62	73	--	--	--
N-ethyl-m-Toluidine	221	--	--	--	72	--	--	--
N-ethyl-p-toluidine	217	--	71	66	39	--	--	--
o-Nitro-N-methyl aniline	--	37	--	--	--	70	--	--
m-Nitro-N-Methyl aniline	--	68	--	83	155	95	--	--
p-nitro-N-Methyl aniline	--	152	177	120	111	152	119	--
N-Benzyl aniline	306	38	140	119	107	58	48	--
Diphenyl amine	302	54	142	123	180	103	74	182
Dibenzyl amine	300	--	81	68	112	93	52	--
N-ethyl-o-Toludine	214	--	75	62	73	--	--	--
*N-methyl-*1-Naphthylamine	294	--	164	--	121	94	--	--
*N-methyl-*2-Naphthylamine	317	--	78	107	84	51	--	145
Dimethyl aniline		193		--	87	164		228d
Diethyl aniline		218		--	84	142		102
Methyl ethyl aniline		201		--	66	134		125

AROMATIC AMINES *contd....*

Aromatic Primary amines	B.P. (°C)	M.P. (°C)	p-Toluene sulfon-amide	Benzene sulfon amide	Benz amide	Acet amide	Form amide	Picrate
				Melting points (°C)				
Methyl benzyl aniline	306	--		44		127		164
Dibenzyl aniline	300	70		91		132		135
Dimethyl-o-toluidine	185	--		--		122		210
Dimethyl-m-toludine	212	--		--		131		177
Dimethyl-p-toludine	211	--		--		130		220
Dimethyl-1-Naphthyl amine	273	--		--		145		164
Dimethyl-2-Naphthyl amine	305	47		--		206		193d
p-Bromodimethyl aniline	264	55		--		--		185d
Triphenyl amine	365	127		--		--		--
Tribenzyl amine	380	92		--		190		184
Hetero cyclic amines								
Pyridine	115	--		--		167		118
2-Picoline	129	--		--		169		227
3-Picoline	144	--		--		150		92
4-Picoline	143	--		--		167		152
Piperidine	106	--		--		--		--
Quinoline	238	--		--		203		134
Trimethyl amine	3.5	--		--		216		230
Triethyl amine	89	--		--		--		173
Isoquinoline	242	24		--		223		159

ALIPHATIC ESTERS

ALIPHATIC ESTERS	B.P.	D_4^{20}
Methyl formate	32	0.974
Ethyl formate	53	0.923
Propyl formate	81	0.904
Isopropyl formate	71	0.873
n-Butyl formate	106	0.892
Iso-Butyl formate	98	0.876
s-Butyl formate	97	0.884
t-Butyl formate	83	--
Methyl acetate	56	0.939
Ethyl acetate	77	0.901
Propyl acetate	101	0.887
Iso propyl acetate	88	0.872
n-Butyl acetate	124	0.881
Iso-Butyl acetate	116	0.871
Sec-butyl acetate	112	0.872
Tert-butyl acetate	97	0.867
Methyl propionate	79	0.967
Ethyl propionate	98	0.892
Propyl propionate	122	0.882
Iso-propyl propionate	111	--
Butyl propionate	145	0.875
Methyl-n-butyrate	102	0.898
Ethyl-n-butyrate	120	0.879
Propyl butyrate	142	0.872
Iso-propyl-n-butyrate	128	--
Methyl isobutyrate	91	0.888
Ethyl isobutyrate	110	0.869
Methyl valerate	127	0.890
Ethyl valerate	144	0.874

ESTERS *contd...*

AROMATIC ESTERS	B.P. °C	M.P. °C	D_4^{20}
Methyl benzoate	199		1.089
Ethyl benzoate	212		1.047
Propyl benzoate	230		1.023
Isopropyl benzoate	218		1.015
Butyl benzoate	248		1.005
Iso butyl benzoate	242		0.997
Methyl phenylacetate	215		1.068
Ethyl phenyl acetate	228	--	1.033
Methyl-o-toluate	213		1.068
Ethyl-o-toluate	227		1.034
Methyl-m-toluate	215		1.061
Ethyl-m-toluate	227		1.028
Methyl-p-toluate	217	34	--
Ethyl-p-toluate	228		1.025
Methyl cinnamate	261	36	--
Ethyl cinnamate	273		1.049
Methyl salicylate	223		1.184
Ethyl salicylate	234		1.125
n-Butyl salicylate	260		1.073
Methyl-m-hydroxy benzoate		70	
Ethyl-m-hydroxy benzoate	295	73	--
Methyl-p-hydroxy benzoate	--	131	--
Ethyl-p-hydroxy benzoate	297	116	--
Methyl-o-methoxy benzoate	246		1.156
Ethyl-o-methoxy benzoate	261		1.104
Methyl-m-methoxy benzoate	237		1.131
Ethyl-m-methoxy benzoate	251		1.100
Methyl anisate	255	49	--
Ethyl anisate	263	7	1.103
Phenyl acetate	196		1.078
Phenyl benzoate	299	68	--

AROMATIC ETHERS

Aromatic Ethers	B.P. (°C)	M.P. (°C)	d420	Nitro	Bromo	Sulphon amide	Picrate	Acid*
						m.p (°C)		
Anisole	153		0.996	87	61	111	81	--
Phenetol	170		0.965	58	--	150	92	--
Furan	32	--	0.937	--	--	--	--	--
Benzyl methyl ether	171		0.965	--	--	--	--	--
Methyl-o-tolyl ether	171		0.980	69	64	137	119	101
Methyl-p-tolyl ether	175		0.970	122	--	182	89	184
Methyl-m-tolyl ether	177		0.972	54	--	130	114	110
Ethyl-o-tolyl ether	184		0.953	--	--	--	118	--
Ethyl-p-tolyl ether	189		0.949			138	111	196
Benzyl ethyl ether	186		0.948	--	--	--	--	--
Ethyl-m-tolyl ether	191		0.949			111	115	137
o-Chloroanisole	195		1.191	95	--	131	--	--
m-Chloro anisole	194							
p-Chloro anisole	198			98	--	151	--	--
o-Bromo anisole	210			106	--	140	--	--
m-Bromo anisole	211							
p-Bromo anisole	215	11	--	88	--	148	--	--
1-Methoxy naphthalene	271	--	1.092	--	55	157	129	--
2-Methoxy naphthalene	274	72	--	--	63	151	117	--
1,4-Dimethoxy benzene	212	56	--	72	142	148	48	--
Dibenzyl ether	296	--	1.034	--	--	--	--	--
Diphenyl ether	259	28	--	144	55	159	--	--
Veratrole	206	22	--	95	93	136	--	--
Butyl phenyl ether	208	--	0.934	--	--	--	--	--
Guaiacol	205	28	1.129	--	116	--	--	--

* Alkoxybenzoic acid

AROMATIC HYDROCARBONS

	B.P. (°C)	M.P. (°C)	d420	Sulfon amide	Aroyl benzoic acid	Nitro Compd	Picr- ate	Bromo
						m.p. (°C)		
Benzene	80	--	0.879	153	128	90	--	89
Toluene	110	--	0.867	137	138	71	--	--
Ethyl benzene	135	--	0.868	109	122	37		
o-Xylene	144	--	0.880	147	129	71	--	88
m-Xylene	139	--	0.864	137	126	182	--	72
p-Xylene	138	--	0.861	144	132	139	--	75
n-Propyl benzene	159	--	0.864	--	126	--	--	--
i-Propyl benzene	153	--	0.862	--	134	109	--	--
Mesitylene	164	--	0.865	--	212	235	97	224
n-Butyl benzene	182	--	0.860	--	98	--	--	--
t-Butyl benzene	169	--	0.867	--	--	62	--	--
Styrene	146	--	0.909	--	--	--	--	73
Naphthalene	218	80	--	--	173	61	150	--
Anthracene	340	216	--	--	--	226	138	--
Phenanthrene	340	100	--	--	--	63	143	--
Diphenyl	254	70	--	--	220	--	--	164
Diphenyl methane	262	25	--	--	--	172	--	64
Triphenyl methane	358	92	--	--	--	206	--	--

ARYL HALIDES

	B.P. (°C)	M.P. (°C)	d420	Sulfon amide	Nitro Compd	Acid*
					(°C)	
Chlorobenzene	132		1.107	143	52	
Bromobenzene	156		1.494	160	75	
Iodobenzene	188		1.831		174	
o-Chlorotoluene	159		1.082	126	45	141
m-Chlorotoluene	162		1.072	185	91	158
p-Chlorotoluene	162		1.071	143	38	242
o-Bromo toluene	181		1.425	146	82	148
m-Bromo toluene	184		1.410	168	103	155
p-Bromo toluene	185	28	1.390	165	47	251
o-Iodo toluene	207		1.698	--	103	162
m-Iodotoluene	204		1.698	--	108	186
p-Iodotoluene	211	35	--	--	--	269
o-Dichlorobenzene	180		1.305	135	110	
p-Dichlorobenzene	174	53		180	54	
o-Dibromobenzene	224		1.956	176	114	
p-Dibromobenzene	219	89		195	84	
p-Bromochloro benzene	195	67			72	
1-Chloronaphthalene	263		1.191	186	180	
2-Chloronaphthalene	265	61	--	232	175	
1-Bromonaphthalene	281		1.484	193	85	
2-Bromonaphthalene	282	59	--	208	--	
1,2-Dichloro naphthalene	296	35	--	--	--	
1,2,4-Tribromobenzene	275	44	--	--	--	
1,2,3-Trichlorobenzene	218	53	--	230	56	

* Halogeno acid from side chain oxidation

AMIDES

Primary aliphatic amides	M.P. (°C)	Xanthylamide
Formamide	2	184
Propionamide	79	214
Acetamide	82	245
Acrylamide	86	--
Heptamide	96	154
Hexanamide	101	160
Butyramide	115	187
N-Allyl urea	80	--
N-Methyl urea	102	230
Urea	132	274
N,N-Dimethyl urea	182	250
N-Acetyl urea	218	--
Thio urea	182	--
Ethyl carbonate	49	169
Methyl carbonate	54	193
Butyl carbamate	54	--
Isobutyl carbamate	55	--
Propyl carbamate	61	--
Malonamide	170	270
Succinamide	260d	--
Glutaramide	175	--
Adipamide	220	--
Aromatic amides		
Anisamide	162	--
Benzamide	129	224
N-Benzyl urea	149	--
o-Bromobenzamide	155	--
p-Bromobenzamide	189	--
m-Chlorobenzamide	134	--
o-Chlorobenzamide	141	--
Cinnamamide	148	--

AMIDES *contd...*

Primary aliphatic amides	M.P. (°C)	Xanthylamide
Aromatic amides	M.P. (°C)	Xanthylamide
p-Chlorobenzamide	179	--
Diphenylacetamide	167	--
3,4-Dinitrobenzamide	183	--
N,N-Diphenyl urea	189	180
p-Ethoxy benzamide	202	--
p-Hydroxy benzamide	162	--
p-Iodobenzamide	218	--
m-Iodobenzamide	186	--
o-Iodobenzamide	184	--
o-Methoxybenzamide	129	--
m-Nitrobenzamide	142	--
o-Nitrobenzamide	175	--
1-Naphthamide	202	--
2-Naphthamide	192	--
Phthalamide	219d	--
N-Phenyl urea	147	225
Phenyl acetamide	157	196
Salicylamide	139	--
m-Toluamide	95	--
o-Toluamide	143	200
p-Toluamide	159	225

AROMATIC NITRO COMPOUNDS

Nitro compd	B.P. °C	M.P. °C
Nitrobenzene	221	9
m-Dinitro benzene	--	90
o-Dinitro benzene	--	118
p-Dinitro benzene	--	173
o-Nitro toluene	222	
m-Nitro toluene	229	16
p-Nitro toluene	238	54
2,4-Dinitro toluene	--	71
2,4,6-Trinitro toluene	--	82
o-Nitro phenol	216	46
m-Nitro phenol	--	97
p-Nitro phenol	--	114
2,4-Dinitro phenol	--	122
o-Nitro aniline	--	71
m-Nitro aniline	--	114
p-Nitro aniline	--	148
1-Nitronaphthalene	304	61
2-Nitro naphthalene	--	79
o-Nitro anisole	265	10
m-nitro anisole	258	39
p-Nitro anisole	259	54
o-Nitro chlorobenzene	254	33
m-Nitroiodo benzene	--	38
o-Nitrobromo benzene	261	42
o-Nitro benzaldehyde	--	44
m-Nitrobenzaldehyde	--	58
p-Nitrobenzaldehyde	--	106
o-Nitrobenzyl chloride	--	48
m-Nitrobenzyl chloride	--	46

AROMATIC NITRO COMPOUNDS *contd...*

Nitro compd	B.P. °C	M.P. °C
p-Nitrobenzyl chloride	--	71
o-Nitrobenzoic acid	--	148
m-Nitrobenzoic acid	--	141
p-Nitrobenzoic acid	--	241
2-Nitro-p-xylene	237	--
3-Nitro-o-xylene	240	--
4-Nitro-m-xylene	244	--
p-Nitrobenzyl alcohol	--	93
o-Nitro benzyl alcohol	270	74
m-Nitro benzyl alcohol	--	27
2,4-Dinitro chlorobenzene	315	51
2,4-Dinitro bromobenzene	--	75
2,4-Dinitro Iodo benzene	--	88
Ethyl-m-nitro benzoate	297	47
Ethyl-p-nitro benzoate	--	57
Methyl-p-Nitrobenzoate	--	96
Methyl-o-nitro benzoate	275	--
Methyl-m-nitro benzoate	--	78
Aliphatic nitro compounds		
Nitro methane	101	--
Nitro ethane	114	--
2-Nitropropane	120	--
1-Nitro propane	131	--
1-Nitrobutane	152	--

AMINO ACIDS

AMINO ACIDS	M.P. (°C)	Acetyl	Bezo-ate	3,5-Dinitro-benzoate	Toluene p-sulph-onate	2,4-Dinitro phenyl
				(°C)		
N-Phenyl glycine	126	194	63	--	--	--
Glycine	232	206	187	179	150	204
Anthranilic acid	145	185	182	278	217	--
Alanine (dl)	295	138	166	177	139	--
(±)Tyrosine	318	172	197	254	--	--
3-Aminobenzoic acid	174	250	248	270	--	--
4-Aminobenzoic acid	186	252	278	290	223	--
Phenyl alanine (dl)	274	--	187	93	135	186
Sarcosine	210	135	104	153	102	--
(±)Proline	203	--	--	217	--	181
Cystine (±)	260	--	181	180	205	109
(±) Tryptophan	275	--	188	240	176	--
3-Aminopropanoic acid	196	--	165	202	--	146
(±) Glutamic acid	227	--	156	--	213	149
p-Aminophenyl acetic acid	200	--	206	--	--	--
(+) or (-) Phenyl alanine	320	--	146	93	165	189
Valine (dl)	298	--	132	158	110	184
(±) Aspartic acid	280	--	165	--	--	196
(±) Asparagine	>300	--	147	--	--	143
(±) Leucine	332	--	141	--	--	--
(±) 2-Aminobutanoic acid	307	--	147	--	--	--

NITROSO, AZO AND AZOXY COMPOUNDS

COMPOUND	M.P. (°C)	B.P.(°C)
Methyl phenyl nitrosoamine		120 °C / 13 mmHg
Ethyl phenyl nitroso amine		134 °C / 16 mmHg
Nitroso benzene	68	
p-Nitroso phenol	125d	
p-Nitroso toluene	48	
p-Nitrosodimethyl aniline	87	
p-Nitrosodiethyl aniline	84	
p-Nitroso-N-ethylaniline	78	
1-Nitroso-2-naphthol	109	
p-Nitroso-N-methyl aniline	118	
2-Nitroso-1-naphthol	152d	
1-Nitroso-naphthalene	98	
1-Nitroso-diphenylamine	144	
Azo Compounds		
Azobenzene	68	
p-Amine azobenzene	126	
2,2'-Dimethyl azobenzene	55	
4-Anilino azobenzene	82	
1-Phenylazo-2-naphthol	134	
Azo-p-toluene	144	
Azo-o-Toluene	55	
2-Phenylazo-1-naphthol	138	
p-Hydroxy azobenzene	152	
Azoxy Compounds		
Azoxy benzene	36	
o-Azoxy toluene	59	
p-Azoxy toluene	75	

SULPHONIC ACIDS* (M.P, °C)

COMPOUND	Sulpho-namide	Sulpho-nanilide	Sulphonyl chloride	Benzyl isothiau ronium salt
Benzene sulphonic	153	110	--	150
Toluene-o-Sulphonic	156	136	68	170
Toluene-m-sulphonic	108	96	12	--
Toluene-p-sulphonic	137	103	71	182
o-Chlorobenzene sulphonic	188	--	27	--
m-Chlorobenzene sulphonic	148	--	--	--
p-Chlorobenzene sulphonic	144	104	53	175
o-Bromobenzene sulphonic	186	--	51	--
m-Bromobenzene sulphonic	154	--	--	--
p-Bromo benzene sulphonic	166	119	75	170
o-Nitrobenzene sulphonic	193	115	69	--
m-Nitrobenzene sulphonic	168	126	64	146
p-Nitrobenzene sulphonic	179	136	80	--
Sulphanilic	164	200	--	187
Orthanilic	153	--	--	132

** The aromatic sulphonic acids do not genrally have sharp melting points; they are supplied in the form of their salts.*

Metanilic	142	--	--	148
Naphthalene-1-sulphonic	150	112	68	137
Naphthalene-2-sulphonic	217	132	79	191
1-Naphthol-2-sulphonic	--	--	--	170
1-Naphthol-4-sulphonic	--	200	--	104
2-Naphthol-1-sulphonic	--	--	124	136
2-Naphthol-6-sulphonic	238	161	--	217
Phenol-p-sulphonic	177	--	--	169
Anthraquinone-1-sulphonic	--	216	217	191
Anthraquinone-2-sulphonic	261	193	197	211
2-Naphthol-8-sulphonic	--	195	--	218

QUINONES (M.P., °C)

COMPOUND	M.P	Oxime	Semicar-bazone	Diacetate of hydro quinone
Benzoquinone	116	240d	243d	123
Toluquinone	69	234d	240d	--
1,4-Naphthaquinone	125	198	247	128
1,2-Naphthaquinone	116-120	162	184	105
Anthraquinone	286	224	--	260
9,10-Phenanthraquinone	206	162	220d	202
Alizarin	290	--	--	182
1-Methyl anthraquinone	172	--	--	--
2-Methylanthraquinone	177	--	--	217
Camphorquinone	199	170	236	--
2-Methyl-1,4-benzoquinone	69	135	179	52
2-Methyl-1,4-naphthaquinone	106	167	247	--

CARBOHYDRATES

COMPOUND	M.P (°C)	Osazone		Acetate	
		M.P (°C)	Time of formation (minutes)	α	β
D-Glucose (anhydrous)	146	205	5	--	--
D-Ribose	95	166	--	--	--
D-Fructose	104	205	2	70	109
D-Lactose (anhydrous)	233	200	--	--	--
Sucrose	185	205	30	70	--
D-Mannose	132	205	05	74	115
D-Galactose (anhydrous)	170	201	19	95	142
Maltose (anhydrous)	165	206	--	125	160
L-Arabinose	160	166	9	94	86
L-Sorbose	161	162	4	--	--
Cellobiose	225	198	--	230	192
Gentiobiose	190	162	--	188	193

GENERAL SOLVENTS PROPERTIES

Solvent	B.P. (°C)	M.P. (°C)	M.Wt.	Density at 20°C
Acetone	56	--	58	0.79
Acetonitrile	82	--	41	0.78
Acetic acid	118	--	60	0.96
Acetic anhydride	140	--	102	1.08
Anisole	154	--	108	0.99
Benzene	80	--	78	0.88
t-Butanol	82	--	74	0.78
Butane-1-ol	118	--	74	0.81
Chloroform	61	--	119	1.49
Carbon disulphide	46	--	76	1.26
Carbon tetrachloride	77	--	154	1.59
Cyclohexane	81	--	84	0.78
Chlorobenzene	132	--	113	1.11
Dichloromethane	40	--	85	1.33
1,4-Dioxane	101	--	88	1.03
Dibutyl ether	142	--	130	0.77
Dimethylformamide	153	--	73	0.95
Diethylene glycol dimethyl ether	160	--	134	0.94
Dimethylsulfoxide	189	--	78	1.10
Diethylene glycol monomethyl ether	194	--	120	1.02
Diethylene glycol	245	--	106	1.11
Diphenyl ether	258	--	170	1.07
Dibutyl phthalate	340	--	278	1.05
Ether	35	--	74	0.71
Ethyl acetate	77	--	88	0.90
Ethanol	78	--	46	0.79
Ethylene glycol dimethyl ether	83	--	90	0.79
Ethylene glycol monomethyl ether	125	--	76	0.96
Ethylene glycol (Ethane-1,2-diol)	197	--	62	1.11

APPENDIX - III

Polarity List of Solvents

The solvents are grouped into non-polar, polar aprotic, and polar protic solvents and ordered increasing polarity. The polarity is given as the dielectric constant. The properties of solvents that exceed those of water are bolded.

NON POLAR SOLVENT

Solvent	Chemical formula	BP	Density constant	Dielectric constant	Dipole
1	2	3	4	5	6
Pentane	C5 H12	36 ^{O}C	0.626g/ml	1.84	0.00 D
Cyclopentane	C5 H10	40 ^{O}C	0.751g/ml	1.97	0.00 D
Hexane	C6 H14	69 ^{O}C	0.655g/ml	1.88	0.00 D
Cyclohexane	C6 H12	81 ^{O}C	0.779g/ml	2.02	0.00 D
Benzene	C6 H6	80 ^{O}C	0.879g/ml	2.3	0.00 D
Toluene	C6H5-CH3	111 ^{O}C	0.867g/ml	2.38	0.36 D
1,4-Dioxane	-C4 H8 O2-	101 ^{O}C	**1.033g/ml**	2.3	**0.45 D**
Chloroform	CHCl3	61 ^{O}C	**1.498g/ml**	4.81	**1.04 D**
Diethyl ether	C2H5OC2H5	35 ^{O}C	0.713g/ml	4.3	1.15 D

Polar aprotic solvents

1	2	3	4	5	6
Dichloromethane	CH2 Cl2	40 ^{O}C	**1.326g/ml**	9.1	1.60 D
Tetrahydrofuran	/-C2H4-0-C2H4-/	66 ^{O}C	0.886g/ml	7.5	1.75 D
Ethyl acetate	C4 H8O2	77 ^{O}C	0.894g/ml	6.02	1.78 D
Acetone	CH3COCH3	56 ^{O}C	0.786g/ml	21	2.88 D
DMF	C3H7NO	153 ^{O}C	0.944g/ml	38	3.82 D
Acetonitrile	CH3CN	82 ^{O}C	0.786g/ml	37.5	3.92 D
DMSO	CH3SOCH3	**189 ^{O}C**	**1.092g/ml**	46.7	3.96 D

Polar protic solvents

1	2	3	4	5	6
Formic acid	HCOOH	**101 ^{O}C**	**1.2g/ml**	58	1.41 D
n-Butanol	C4H10O	118 ^{O}C	0.810g/ml	18	1.63 D
Isopropanol	C3H8O	82 ^{O}C	0.785g/ml	18	1.66 D
n-Propanol	C3H8O	97 ^{O}C	0.803g/ml	20	1.68 D
Ethanol	C2H5O	79 ^{O}C	0.789g/ml	24.55	1.69 D
Methanol	CH3OH	65 ^{O}C	0.791g/ml	33	1.70 D
Acetic acid	CH3COOH	118 ^{O}C	**1.049g/ml**	6.2	1.74 D
Water	H2O	100 ^{O}C	1.0g/ml	80	1.85 D

APPENDIX - IV

Reducing Agents

Table: Reactivity of various functional groups wth some metal hydrides and toward catalytic hydrogenation, +/- indicates a borderlines case														
Reaction	$NaBH_4$ in EtOH	$NaBH_4$ + LiCl in diglyme	$NaBH4$ + $AlCl_3$ in diglyme	BH_3–THF	Bis-3-methyl-2-butyl borane in THF	9-BBN	$LiAlH$ (O-t-Bu)$_3$ in THF	$LiAlH$ (OMe)$_3$ in THF	$LiAl$ H_4 in ether	AlH_3 in THF	$LiBEt_3H$	(i-Bu)$_2$AlH (DIBALH)	$NaAlEt_2H_2$	Catalytic hydrogenation
S.No														
1 $RCHO \rightarrow RCH_2OH$	+	+	+	+	+	+	+	+	+	+	+	+	+	+
2 $RCOR \rightarrow RCHOHR$	+	+	+	+	+	+	+	+	+	+	+	+	+	+
3 $RCOCl \rightarrow$ RCHO / RCH_2OH	+	+	+	-	-	+	+	+	+	+	+	+	+	+
4 Lactone $\rightarrow$ diol	-	+	+	+	+	+	±	+	+	+	+	+	+	+
5 Epoxide $\rightarrow$ alcohol	-	+	+	+	±	±	±	+	+	+	+	+	+	+
6 $RCOOR' \rightarrow RCH_2OH + R'OH$	-	+	+	±	-	±	±	+	+	+	+	+	+	+
7 $RCOOH \rightarrow RCH_2OH$	-	-	+	+	-	±	-	+	+	+	-	+	+	-
8 $RCOO^- \rightarrow RCH_2OH$	-	-	+	-	-	-	-	+	+	+	-	-	+	-
9 $RCONR'_2 \rightarrow$ RCHO / $RCH_2NR'_2$	-	-	-	+	+	+	-	+	+	+	+	+	+	+
10 $RCN \rightarrow RCH2NH2$	-	-	-	+	-	±	-	+	+	+	±	+	+	+
11 $RNO_2 \rightarrow$ RNH_2 / RN=NR	-	-	-	-	-	-	-	+	+	-	-	+	+	+
12 $RCH=CHR \rightarrow RCH_2CH_2R$	-	-	-	+	+	+	-	-	-	-	+	-	-	+

APPENDIX - V

CALCULATION OF EXPERIMENTAL UNCERTAINTIES (ERRORS) & LABELLING OF GRAPHS

Introduction

Experimental error is the uncertainty associated with the accuracy of the measurements which are used to calculate the final results. It should ne distinguished from *systematic* error which arises from intrinsic faults in the equipment (which will be unknown in most work, being revealed only by calibration against international standards) or application of wrong interpretation to the results.

Experimental errors requires to be calculated because the measured data of the experiment are usually processed through some calculation to arrive at the final result. It is therefore essential to know how the uncertainties in the original data affect this final result.

Calculation of Uncertainties

All data processing calculations consist of two types of operation. These are: add or subtract and multiply or divide. The calculation of uncertainties is different for each of these two operation types.

Add or Subtract

Let the final result be R and its uncertainty be ΔR. Let the result be calculated from the expression.

$$R = a + b - c,$$

and the uncertainties id the measured quantities a, b, and c be and . Theoveralluncertainty in the result, ÄRis given by

$$\Delta R = \delta a + \delta b + \delta c$$

N.B. Again irrespective of the presence of the minus sign in the expression for R, all data uncertainties are *added* (i.e. $+\delta c$ in the above equation *not* $-\delta c$).

ΔR is the same units as a, b, and c and their uncertainties.

Multiply or Divide

Let the final result be R and its uncertainty be ΔR. Let the result be calculated from the expression.

$$R = a \times b/c$$

And the uncertainties in the measured quantities a, b and c be $\delta a, \delta b$ and δc. The overall uncertainty in the result, is given by

$$(\Delta R / R) = (\delta a / a) + (\delta b + b) + (\delta c / c)$$

N.B. Again irrespective of the presence of the division sign in the calculations, all fractional uncertainties are added (i.e $\delta c / c$ in the above equation, not - $\delta c/c$). Since fractions are dealt with through out these calculations, each term in the equation is dimensionless (has no units).

Unlike the case for Add or Subtract, the overall uncertainty is calculated as a *fraction* of the final result from the uncertainties in the original data also expressed as *fractions* of each piece of data.

Percentage uncertainties can be used instead of fractional of uncertainties for the data in which case the percentage uncertainty in R is calculated. This is readily verifies by multiplying by above equation by 100 throughout.

The above method gave a Maximum Possible Error (MPE). As for an

Engineer building a bridge to carry a given maximum load, the error is always best

Calculated as such a maximum, since, in presenting a result and its associated uncertainty, the experimenter is stating that he/she is 99% certain that the true result lies between the limits $R - \Delta R$ to $R + \Delta R$. More sophisticated statistical analysis are possible, but in everyday experiments the low number of measurements rarely justifies these. Such methods have a place in the biosciences experiments where large numbers of data points are more usual than in the physical sciences.

The two operation types discussed above (Add or Subtract and Multiply or Divide) allow uncertainties to be calculated for more complicated expressions where such operations are combined.

For uncertainties in more complicated functions e.g. logarithms or trigonometrical functions it is simplest to evaluate the maximum value of R using data with errors in $+ \delta a$ sense and a maximum value with errors in $- \delta a$ sense. R is then quoted as the R value $\pm$ half the difference between the maximum and minimum value of R. Some care may be needed to decide whether the maximum value of R does indeed need the data error in the $+ \delta a$ sense - sometimes it has to be in the δa sense. Consult the example below for such rare, complicated cases.

Examples – N.B. numbers 1) and 2) should be studied carefully.

1. { simple Add or Subtract} In titration a volume of liquid, VB, is measured by subtracting from a burette reading at the end point, 25.37 cm3, an initial reading, 0.45 cm3. If the uncertainty in reading the burette is ± 0.03 cm3, calculate the volume dispensed from the burette and its associated error or uncertainty.

$$VB \pm \, VB \; = \; (25.37 \pm 0.03 - 0.45 \pm \, 0.03) \, cm3$$

$$= \, 24.93 \pm 0.06 \, cm3$$

(Here a = 25.37 cm3 and b = 0.45 cm3. δa= 0.03 cm3 and δb = 0.03 cm3. The overall uncertainty is $\delta a + \delta b = 0.06 \, cm^3$).

2. [simple Multiply or Divide}   The molarity of the solution titrated, MA, is calculated from the equation

$$MA \; VA = MB \; VB$$

Where VA, 25.00 cm3, is the volume of the pipette used to dispense the unknown solution and MB, 1.016M, is the molarity of the liquid in the burette. Calculate the molarity of the unknown solution and its associated uncertainty or error. Use the data from example 1.

We note that VA is quoted as 25.00 cm3 and, in the absence of other information, we may assume that this method of quoting VA is indicative of the uncertainty. In VA, i.e. 5VA = ± 0.01 cm3. Similarly the uncertainty in MB is assumed to be ± 0.001. Rearranging the given equation for MA we have

$$MA = MB \; VB/VA:$$

We immediately express the errors in MB, VB and VA as fractions, since this equation is in the multiply or Divide format. Thus δMB/MB = 0.001/1.016 = 9.8425197E -4 and δVA/VA = 0.01/25.00 = 4.0000000E -4. From example 1) we already have δ VB/VB = 0.06/24.92 = 2.407705E -3 {E-4 etc. is calculator display for 10-4 etc.}

Thus δMA/MA = 9.842519E-4 + 4.0000E-4 + 2.407705E-3 = 3.791957E-3

Now MA = 1.016 x 24.92/25.00 = 1.0127488

Whence δ MA = 3.8403E-3 = 4E-3, since uncertainties are only ever known to 1 significant figure.

Thus MA = 1.01 ± 0.004 M   '

Be careful to keep the accuracy of all intermediate figures up to the final result for higher than the finally estimated uncertainty.

Now one important feature of uncertainty calculation is that it allows "rounding off" of results such as these, commensurate with their uncertainty. This must be performed on all calculated results and failure to do so *constitute a mistake which will be penalised..*

Applying this philosophy to the above, we have MA = 1.01 ± 0.004 M or just quote 1.01 M which implies an intrinsic error of ± 0.01 rather than 0.004 i.e. overestimates the error by a factor of 2.5. It would be quite wrong to quote 1.0127 M (implying ± 0.0001, 40 times better than the 0.004 calculated). Likewise 1.013 (which implies ± 0.001 4 times better than calculated) would be less acceptable.

Note that had MA, been calculated as 0.98914365 M, the same philosophy as above would lead to a final quote of 0.99 M implying an error of ± 0.01 as above. Here 2 significant figures are used whereas 3 were used in the 1.01 M result. To discuss uncertainties in terms of significant figures is thus seen to be confusing and misleading. Had a stronger solution been analysed and MA been calculated to be 10.142456 M, retaining the fractional error of 0.004 as above, gives a final quote of 10.1 M commensurate with the error, showing that to discuss errors in terms of decimal places is also confusing and misleading, since in this case 1 decimal place is used and in example 2) above 2 decimal places were used.

Note also that if the fractional error had been calculated to be 0.002 instead of 0.004, the final quote would be 1.013 M in example 2). This underestimates the error by a factor of 2 and this is more acceptable than a quote of 1.01 M which overestimates the error by a factor of 5. It is arguable that, in the case of fractional error of 0.003, 1.01 M overestimates the error by a factor of 3 whereas a quote of 1.013 M underestimates the error by a factor of 3, so that either quote is acceptable. Generally, however, overestimation of the error is preferable.

3. {Combination of Add – Subtract and Multiply – Divide}

Let the quantity to be calculated by J and its uncertainty by δJ. Let J be given by:

$$J = P + Q/(R+S)$$

To calculate J, begin with the fraction and focus on its denominator as the most complicated area.

Following the procedure in 1) above, putting A = R + S, we have

$$\delta A = \delta(R+S) = \delta R + \delta S$$

Following the procedure in 2) above and expressing uncertainties in fractional form for the quotient, let B = Q/(R +S), so that

$$\delta B/B = \delta Q/Q + A/A$$

Finally, treating the remaining expression as a sum,

$$\delta J = \delta P + \delta B$$

In all cases where a value of an expression is required to evaluate an uncertainty e.g. B is required to evaluate B, use the average value from the data set (is more than one value is available).

Graph Labelling

In labeling axes of graphs, if the quantities have units, write the label name *divided* by the units. Powers of ten used to adjust the figures on the axes, should appear *in front of the axis labels* e.g. if a number to be plotted is a mass of 0.00375 kg and the axis is to be labeled 1,2,3,4 etc., the label should read 103 x mass/kg. Other versions such as mass (kg x 103) are ambiguous in that it is unclear whether the number has been multiplied by 103 or the unit has been multiplied by 103 and these quite different.

Error bars must always appear on a graph. These indicate the uncertainty in the coordinates and are displayed parallel with the approximate axis. They are calculated as in the above treatment of errors.

$$Y = +1.00E\text{-}4x2 + 3.61E\text{-}10x1 - 5.41E\text{-}9, \ R: 7.60E\text{-}10, \ \text{max dev: } 1.86E\text{-}9$$

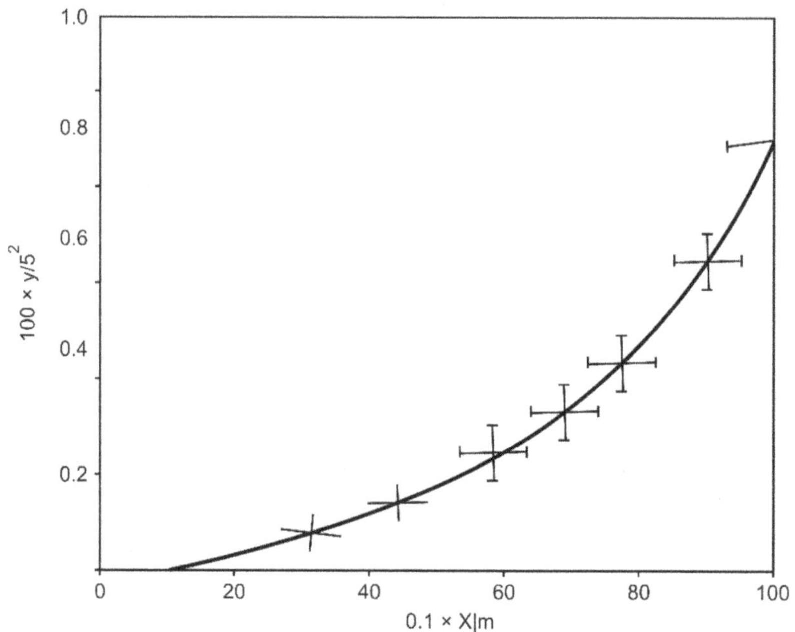

Data Set

xy

x	y
100 m	.0001 s2
200 m	.0004 s2
300 m	.0009 s2
400 m	.0016 s2
500 m	.0025 s2
600 m	.0036 s2
700 m	.0049 s2
800 m	.0064 s2
900 m	.0081 s2

In the above example the errors in X were ± 10% and in Y were ± 20%.

APPENDIX - VI

PROBLEMS (to be Solved)

Model Problems

(i) State the Lewis concepts of acids and bases.

(ii) Classify the following into Lewis acids, bases or neither:

HF, BF_3, $HC \equiv CH$, NO_2+, CH_3CN, H_2O, CH_3OH, C_2H_6, $Ag+$.

Solution:

(i) Lewis acids are species that can accept a pair of electrons, and those which donate are Lewis basis.

(ii) Acids: BF_3, NO_2+, $Ag+$

Bases: CH_3CN, H_2O, CH_3OH

Neither: $HC \equiv CH$, C_2H_6, HF

Work-out answers for the following:

Q1. Identify the class or classes of compounds indicated in each of the following:

a) A compound reacts with both acetic anhydride and methanol.

b) A compound does not react with 2, 4- dinitrophenylhydrazine but gives a positive test with Tollen's reagent.

c) A compound reacts in aqueous base to form two organic products.

d) A compound reacts with aqueous sodium hydroxide but not with aqueous sodium bicarbonate.

e) A compound reacts with aqueous sodium hydroxide and also with aqueous sodium bicarbonate.

f) A nitrogen-containing compound does not react with benzoyl chloride

g) Refluxing a compound in aqueous sodium hydroxide gives rise to ammonia.

Q 2. Offer explanation for the following observations:

a) Bezaldehyde does not undergo Aldol condensation

b) 2,4- Pentanedione is highly acidic.

c) Hydroxylamine is employed in the form of its hydrochloride salt.

d) Acetophenone but not benzophenone forms an adduct with $NaHSO_3$

Q3. Identify all compounds in each of the following and write equations for reactions involved.

a) Compound **A**, $C_7 H_7 O_2 N$, undergoes the Schotten Baumann reaction with benzoyl chloride to give **B**, $C_{14} H_{11} O_3 N$. Compound **A** dissolves in an aqueous solution of sodium bicarbonate. The reaction of **A** with sodium nitrite in the presence of an excess of hydrochloric acid at 0 °C. gives **C**. The solution in which **C** is formed is poured into boiling water. A gas is evolved and an acidic substance **D** is isolated from the solution. **D,** which also reacts with benzoyl chloride, has a melting point of 210- 2120 C.

b) **A** and **B** ($C_4 H_6 O_2$) are isomeric. Treatment of either with hydrogen bromide yields the same compound **C** ($C_4 H_7 O_2 Br$).

c) When compound **A**, $C_6 H_{14} O_2$, is treated with periodic acid, a single organic product (**B**) is obtained. **B** reacts with Fehling's solution.

d) Compound **A**, $C_4 H_8 O_2$, gives a positive test with Fehling's solution. Compound **A** reacts with acetic anhydride and yields iodoform when reacted with sodium hydroxide and iodine. **A** and an excess of phenylhydrazine produces **B** ($C_{16} H_{18} N_4$)

e) The careful oxidation of **A**, $C_4 H_8 O_3$, forms **B**. When **B** is heated above its melting point, **C** is formed with carbon dioxide evolution. Heating **A** gives **D**, which reacts with hydrogen bromide to form **E** ($C_4 H_7 O_2 Br$).

f) Compound **A**, $C_8 H_8 O$ reacts with Tollen's reagent. The vigorous oxidation of **A** yields **B**, which forms but a single mononitro derivative (**C**). The reaction of **A** with acetic anhydride and sodium acetate provides **D**, which is obtainable in *cis* and *trans* – isomeric forms.

Q4. Select a reagent which permits a simple and rapid differentiation of each of the following pairs without resorting to physical constants. Write equations for reactions concerned.

a) 2-pentanone   ($CH_3COCH_2CH_2CH_3$)   and   3-pentanone ($CH_3CH_2COCH_2CH_3$)

b) phenol (PhOH) and n-hexyl alcohol ($C_6H_{13}OH$)

c) propionaldehyde (CH_3CH_2CHO) and acetone (CH_3COCH_3)

d) benzoyl chloride (C_6H_5COCl) and chlorobenzene (C_6H_5Cl)

e) acetanilide ($C_6H_5NHCOCH_3$) and p-aminoacetophenone ($NH_2C_6H_4COCH_3$)

f) 1, 2 - dihydroxypropane ($CH_3\ CHOH\ CH_2OH$) and 1, 3 – dihydroxypropane ($OHCH_2\ CH_2\ CH_2OH$)

g) diethyl malonate $[CH_2\text{-}(CO_2C_2H_5)_2]$ and diethyl succinate ($CH_2CO_2C_2H_5)_2$.

Q5. What other oxidising agents could be used, other than MnO_2, for the oxidation of benzhydrol to benzophenone?

Q6. What products would you expect to obtain from the oxidation of 4,4' – dimethylbenzhydrol with potassium permanganate ?

Q7. How may be following selective reductions be carried out in the laboratory ? Give adequate reasons for your choice of reagents.

a) $CH_3CH = CH\ CO\ CH_3 \rightarrow CH_3\ CH_2CH_2\ CO\ CH_3$

b) $CH_3CH\ CH_2\ CO_2\ C_2H_5 \rightarrow CH_3\ CH\ (OH)CO_2C_2H_5$

c) $Ph\ COO\ Et \rightarrow Ph\ CH_2OH$

d)
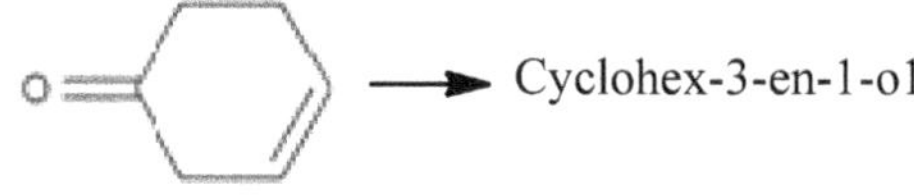

Q8. How would you account for the relative Rf values of the product and starting material ?

Q9. Chromatographic methods have been considered as the most viable techniques for the separation and purification of compounds. List the different chromatographic technique.

APPENDIX - VII

GLASS APPARATUS

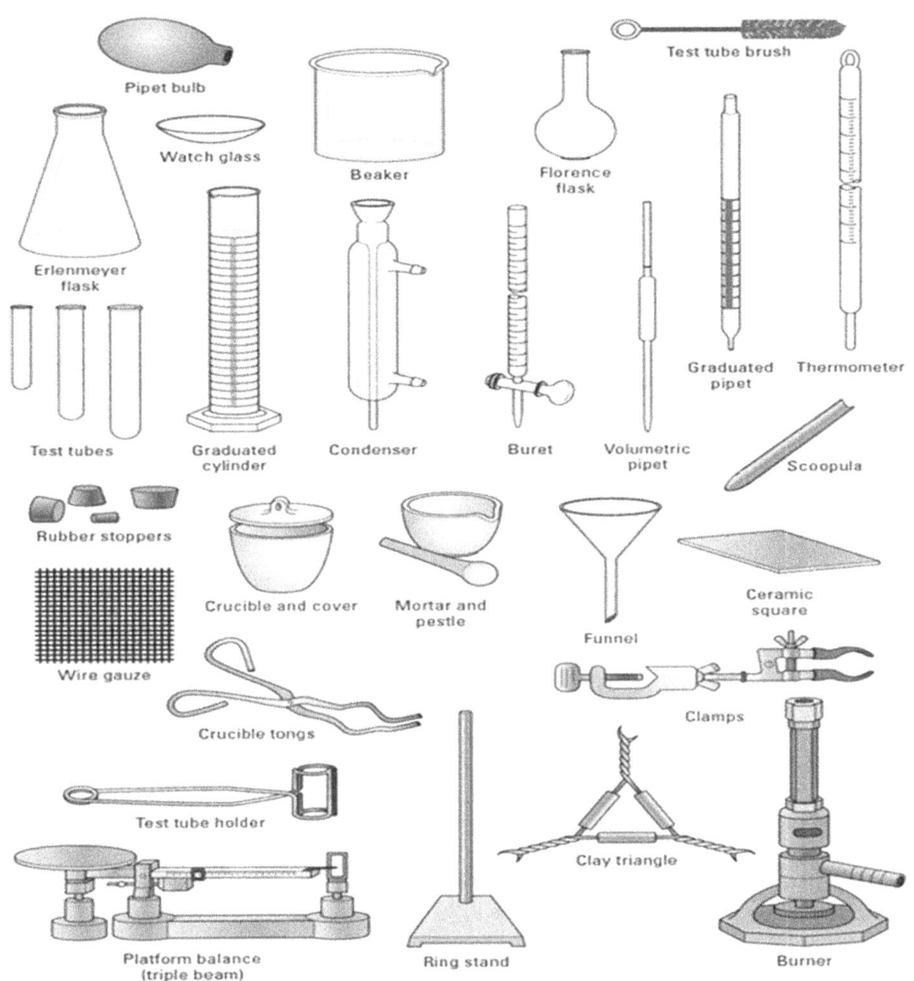

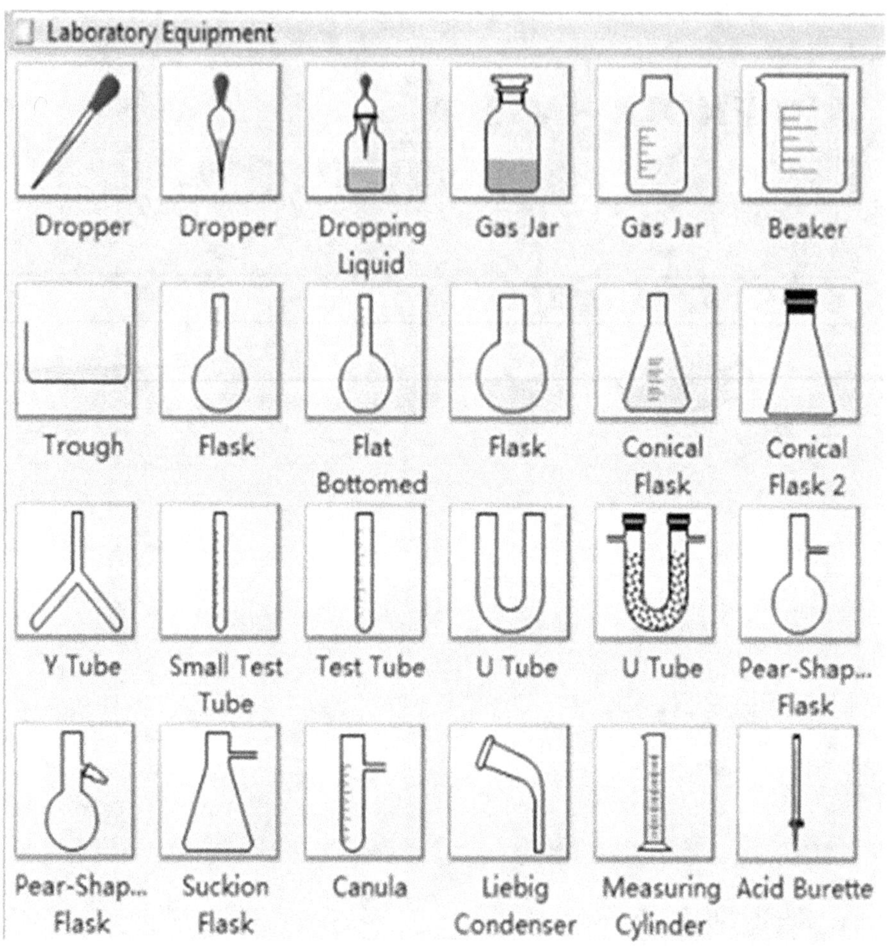
Laboratory Equipment
Dropper
Dropper
Dropping Liquid
Gas Jar
Gas Jar
Beaker
Trough
Flask
Flat Bottomed
Flask
Conical Flask
Conical Flask 2
Y Tube
Small Test Tube
Test Tube
U Tube
U Tube
Pear-Shap... Flask
Pear-Shap... Flask
Suckion Flask
Canula
Liebig Condenser
Measuring Cylinder
Acid Burette

COMMON ORGANIC LABORATORY APPARATUS

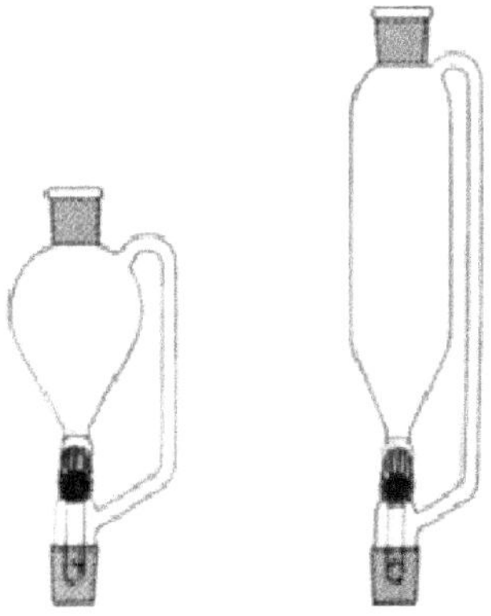

Stillhead

Condenser

Receiver Adaptor

Round bottomed flask

Separating funnels

mark

Volumetric Flask

Filtration funnel

Buchner funnel

Drying Tube

Conical flask.

Beaker

A

Pressure equalized dropping funnels

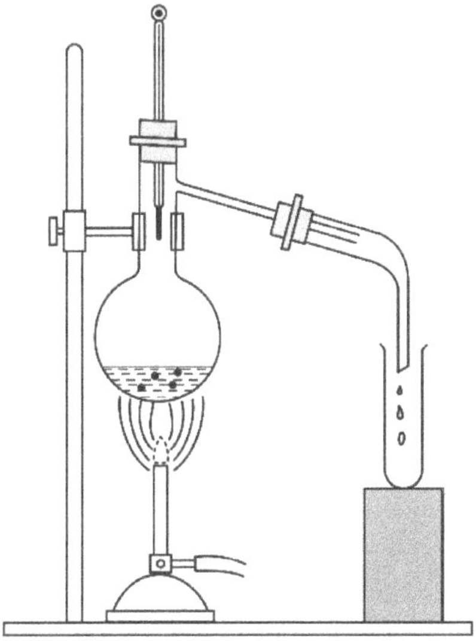

Simple distillation

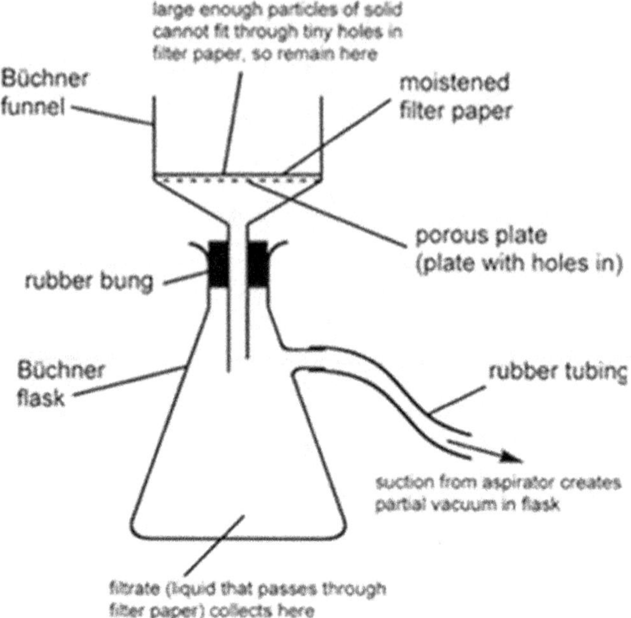

Filteration setup

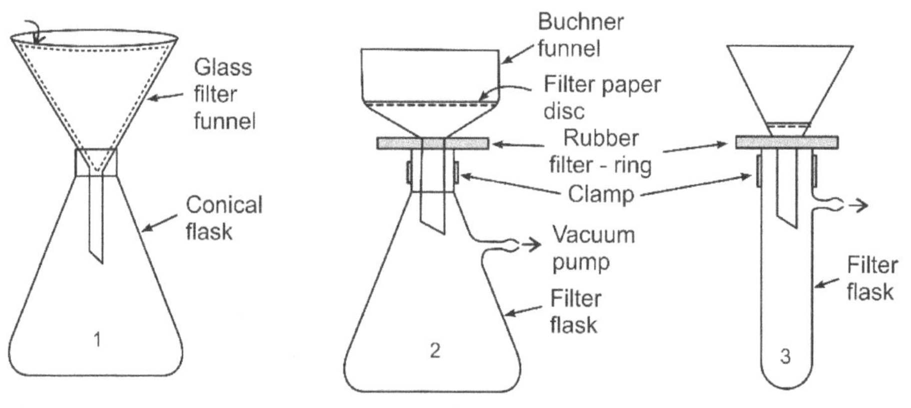

1 – FOR THE FILTRATION OF LIQUIDS

2 and 3 FOR THE COLLECTION OD CRYSTALS BY SUCTION FILTRATION

Fiteration apparatus

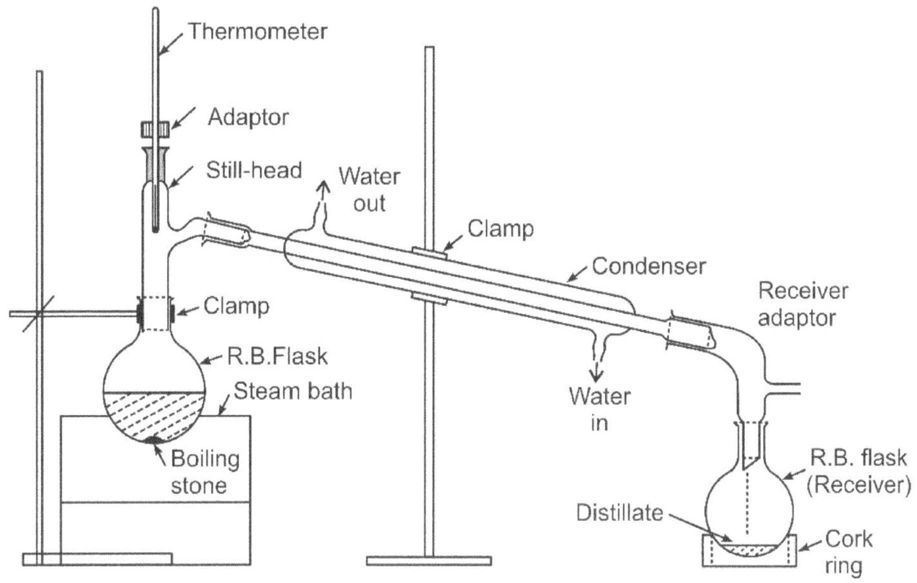

Simple distillation apparatus

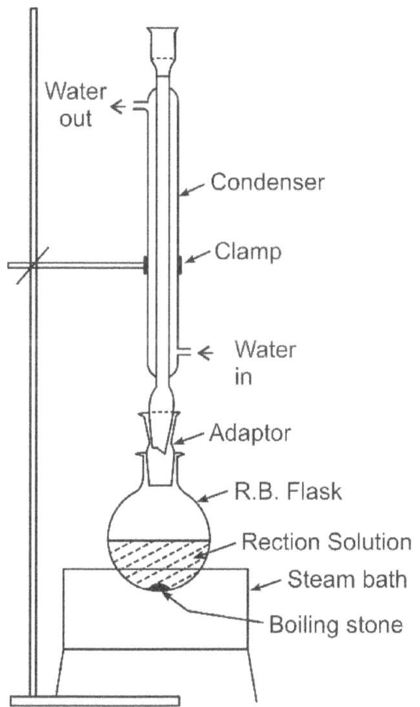

Reflux apparatus

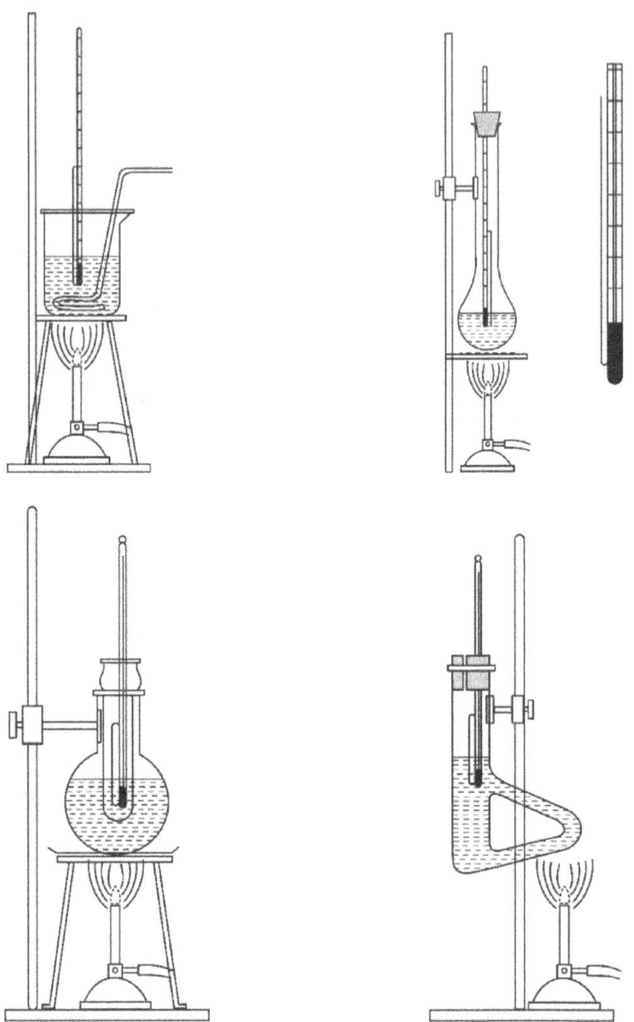

Different types of melting point & boiling point apparatus

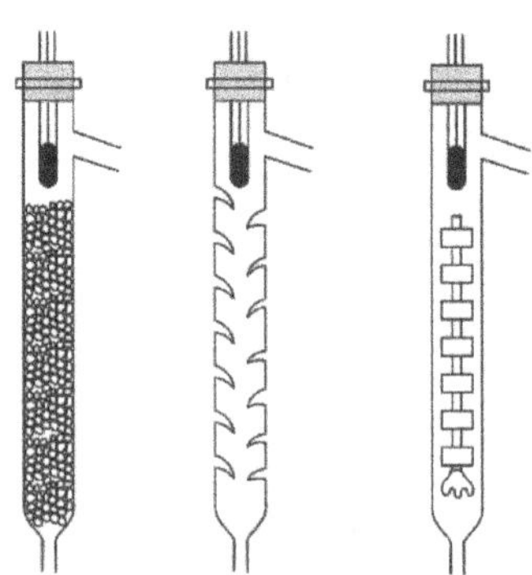

Fractionating columns commonly used

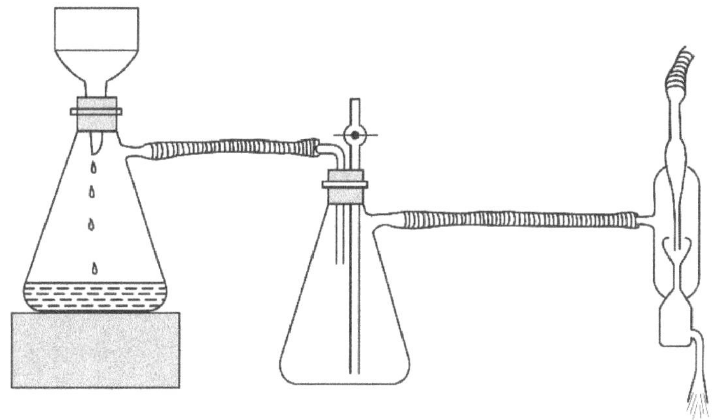

Filteration under water vacuum pump

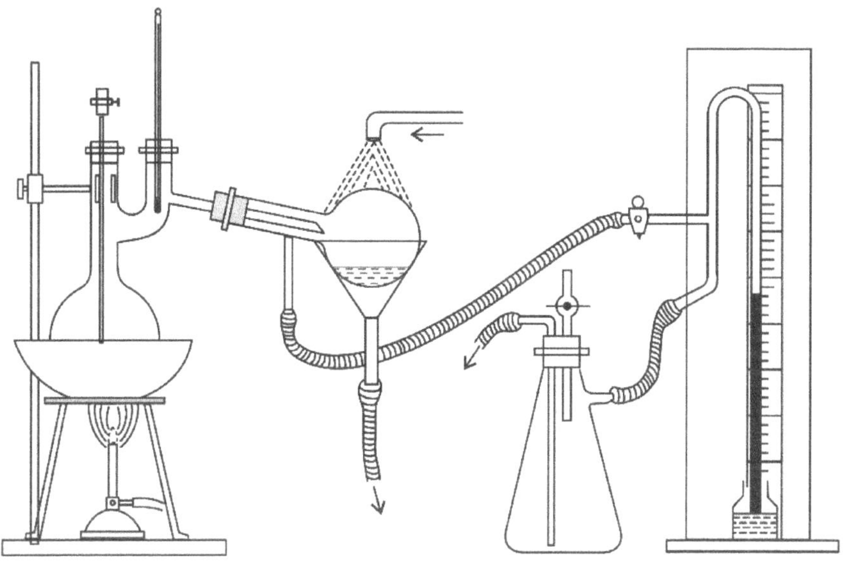

Vacuum Distillation

REFERENCES

Suggestions for Further Reading

1. I L Finar 'Organic Chemistry (2 volumes), by Longmans Group UK Ltd., edition 2000.

2. A Guidebook to Mechanism in Organic Chemistry by Peter Sykes, Longman Group Ltd., (Pearson education), 2004.

3. C K Inglod "Structure and Mechanism of Organic Chemistry', G. Bell and Sons, Ltd., 2nd edition 1969.

4. Instrumental Methods of Drug Analysis by G. V. Sagar, B S P Books, 2017.

5. Vogal's Test Book of Quantitative Chemical Analysis, J. Mendham et. al., Addison Wesley Publishing Co., 6th ed., 2000.

6. Practical Pharmaceutical Chemistry, 4th ed.,, The Athlone Press, London, 1988.

7. The Indian Pharmacopoeia Vol. I & II, Indian Pharmacopoeia Commission, 2007.

8. Brian S. Furnis et al., 'Vogel's Textbook of Practical Organic Chemistry' 5th ed., 2004, Pearson Education Pte. Ltd.

9. 'The Merck Index', Merck & Co ., Inc., Whitehouse Station, NJ., 12 ed., 1996.

10. Jerry March, 'Advanced Organic Chemistry', John Wiley & Sons, New York, 4th ed., 1992.

11. Mann and Saunders 'Practical Organic chemistry' Orient Longman Ltd., New Delhi, 4th ed., 1986.

12. Laboratory 'Exercises Organic Chemistry' by James M. Sugihara, Burgess Publishing Company 4th ed., 1969.

13. Remington: 'The Science and Practice of Pharmacy", Vol. II, Mack Publishing Company, Easton, Pennsylvania, 20th ed., 2000.

14. Qualitative Organic Analysis by Kemp, McGrow-Hill publishing company Ltd. 1970.

15. Dictionary of Organic Compounds by F. G. Mann, Eyre and Spottiswoode Publishers Limited, London, 1966 (five volumes plus annual supplements).

Selected References

1. Introduction to Spectroscopy by Donald L. Pavia et al., Cengage Learning India Private Ltd, 5th ed., 2015.

2. Spectrometric Identification of Organic Compounds by R M Silverstein and G C Bassler, Wiley International Edition, 1967.

3. Organic Structures from Spectra by L D Field, H L Li, A M Magill and Patrick, John Wiley & Sons International, 2012.

4. Foundation of Spectroscopy by Duckett and Gilbert, Oxford University Press, 2000.

5. Organic Structures from Spectra6th Ed,L. D. Field, H. L. Li, A. M. Magill, Wiley, 2020

6. Spectroscopic of Organic Compounds by P S Kalsi, New Age International Publishers, 2002.

7. Organic Spectroscopy by William Kemp, Published by Palgrave, 3rd Edition. 1991.

8. Spectroscopic Methods in Organic Chemistry, D. H. Williams and Ian Fleming, McGraw-Hill Book Company (UK) Limited.

9. NMR in Chemistry, A Multinuclear Introduction by William Kemp, MacMillan Education Limited, 1986.

10. J. B. Stothers , Carbon-13 NMR Spectroscopy, Academic Press, New York, 1972.

11. Elementary Organic Spectroscopy, Y R Sharma, S Chand & Company Limited, 2005.

12. Introduction To Mass Spectrometry, H C Hill,*Publisher*, *Heyden* and *Son. 1969*

13. Fred W. McLafferty, Interpretation of Mass Spectra, W.A. Benjamin, Inc., New York, 1966.

14. Spectroscopy,   D.   H.   Whiffen,   Published   by   Longmans Green And *Co, 1968*.

15. McLafferty and F. Turecek, Interpretation of Mass Spectra, 4th ed., University Science Books, Mill Valley, C A, 1993.

16. D. H. Williams and R. D. Bowen, Mass Spectrometry, Principles and ApplicatioOns, 2nd ed., McGraw-Hill Book Co., London, 1981.

17. The Sadtler Handbook of Ultraviolet Spectra, published by Sadtler ResearchLaboratories, Philadelphia, 1979

18. Stern, E S. and Timmons C. J., Introduction to Electronic Absorption Spectroscopy,   Edward Arnold Publ. Ltd, London 1970

19. William Kemp NMR in Chemistry : a Multinuclear Introduction, Macmillan Education London, UK, 1986.

20. Protons Spectra: NMR Spectra Catalog, Volumes 1 and 2 Varian Associates, Palo Alto, California, 700 Spectra.

21. Sadtler Handbook of Proton NMR Spectra, Philadelphia : Sadtler,1978, 3000 Spectra.

22. Aldrich Library of NMR Spectra, Aldrich Chemical Co., Milwaukee, 2nd, 2 Volumes, 8500 Spectra.

23. George B and McIntyre P Infrared Spectroscopy, Wiley, Chichester 1987

24. Pouchert, Charles J., Aldrich Library of Infra Red Spectra, Aldrich Chemical Co. Inc., Milwaukee, 3rd edn., 1984, over 12,000 dispersive spectra.

25. Laboratory Manual, Dept. of Pure and Applied Chemistry, University of Strathclyde, Glasgow, Scotland (UK), 1985.

26. G. R. Barrow, Introduction to Molecular Spectroscopy, McGraw-Hill, New York, 1962. L. J. Bellamy, The Infrared Spectra of Complex Molecules, Vol. Chapman and Hall, 1975.

27. Aldrich Library of Infrared Spectra, Aldrich Chemical Co., Milwaukee, 1975.

28. L.M. Jackman and S. Sternhell, Applications of N M R Spectroscopy in Organic Chemistry, Pergamon Press, London, 1969.

29. F. W. Wehrli and T. Wirthlin, Interpretation of Carbon-13 N M R spectra, Heyden, London, 1978.

30. J. H. Beynon, R. A. Saunders and A. E. Williams, The Maas spectra of Organic Molecules, Elsevier, London, 1968.

31. J. K. M. Sanders and B.K. Hunter, Modern NMR Spectroscopy, 2nd Ed., Oxford University Press, UK, 1993.

Index